Core Higher
GEOGRAPHY

Yla Garvie

Kenneth Maclean and
Norman Thomson

Hodder & Stoughton

A MEMBER OF THE HODDER HEADLINE GROUP

The illustrations were drawn by Peters & Zabransky Ltd

The publishers would like to thank the following individuals, institutions and companies for permission to reproduce photographs in this book. Every effort has been made to trace ownership of copyright. The publishers will be happy to make arrangements with any copyright holder whom it has not been possible to contact.

© AFP/CORBIS 220, 223; Alan Devlin 104 (left); © Annie Griffiths Belt/CORBIS 236, 237, 238, 240; Associated Press 19, 30, 195, 199 (both), 206; © Bettmann/CORBIS 215; Bruce Connolly 209, 218; Cambridge Newspapers Ltd 292 (top); Caroline Field/Life File 27 (right); © Colin Garratt, Milepost/CORBIS 186 (left); © David Simpson 275 (top); Fife County Council 286 (bottom); First Direct Bank 298; Fotografia Inc./CORBIS 242; GeoScience Features Picture Library 82 (left), 83, 89 (bottom), 97, 108 (bottom), 149 (both), 161 (both), 162 (both); © GSR/Dundee City Council 295; Heather Angel 150; © Hulton-Deutsch Collection/CORBIS 91 (left); Holt Studios 243 (bottom), 244, 253; Hutchinson Picture Library 243 (top); IBM 263; © IGN, Belgium 278 (top); Kennan Ward/CORBIS 27 (left); Life File 203; © Lorne Gill/SNH 89 (top), 102 (top), 108 (top), 144; Michele Lambremont 278 (bottom); Nissan 285 (top right and bottom); Nokia 290 (bottom), 292 (bottom); North News and Picture Agency 285 (top right), 286 (top); NRSC Ltd/Science Photo Library 23; © Patricia and Angus Macdonald/SNH 57, 105 (top), 110 (left), 112, 261, 280; Paul A. Souters/CORBIS 82 (right), 185 (bottom); Perthshire Advertiser 50; © Peter Turnley/CORBIS 275 (bottom); © Reuters Newmedia Inc./CORBIS 215 (top); © Robert van der Hilst/CORBIS 232; © Roger Ressmeyer/CORBIS 186 (right); Scott Camazine/Science Photo Library 14; Scottish Enterprise, Lanarkshire 281; © Skyscan Photolibrary/Pitkin Unichrome Ltd 311; © Tim Wright/CORBIS 185 (top); Tiziana and Gianni Baldizzone/CORBIS 49; Tom Bean/CORBIS 235; Tourist Development Corporation of Malaysia 249; Wolfgang Kaehler/CORBIS 46

All other photographs were supplied by the authors

The publishers would like to thank the following for permission to reproduce material in this book:

Phillipe Descola for extracts from The Spears of Twilight, published by HarperCollins Publishers Ltd.

Maps reproduced from Ordnance Survey mapping with the permission of the Controller of Her Majesty's Stationery Office, © Crown Copyright

Map on page 98 © Harvey 2000

Orders: please contact Bookpoint Ltd, 78 Milton Park, Abingdon, Oxon OX14 4TD. Telephone: (44) 01235 827720, Fax: (44) 01235 400454. Lines are open from 9.00 – 6.00, Monday to Saturday, with a 24 hour message answering service. Email address: orders@bookpoint.co.uk

British Library Cataloguing in Publication Data
A catalogue record for this title is available from The British Library

ISBN 0 340 75839 2

Published by Hodder & Stoughton Educational Scotland
First published 2000
Impression number   10 9 8 7 6 5 4 3
Year   2006 2005 2004 2003 2002 2001

Copyright © 2000 Kenneth Maclean and Norman Thomson

Cover photos from Photodisc
Typeset by Fakenham Photosetting Limited
Printed in Italy for Hodder & Stoughton Educational, a division of Hodder Headline Plc, 338 Euston Road, London NW1 3BH.

FOREWORD

This book has been written to cover the requirements of the eight core topics that must be taught for Core-Unit 1 of the Higher Still syllabus in geography as prescribed by the Scottish Qualifications Authority. It is further designed to complement the Hodder & Stoughton text *Higher Geography Applications*, which covers the Higher Applications Optional Units.

As with any textbook, the authors have covered the mandatory topics but, inevitably with their own slant and emphasis as appropriate. This reflects their own teaching experience, the resources available and the changing world, with which no syllabus, no matter how new, can ever keep up-to-date. Consequently, the textbook has the following broad characteristics:

◆ Each of the four core areas in both physical and human geography is introduced by a lead-in chapter. Initially, the interplay of the core topics of the Atmosphere, Hydrosphere, Lithosphere and Biosphere is reviewed through an outline of the physical character of the Manor Valley near Peebles. In contrast, a macro-scale, global approach introduces the chapters on Population, Rural, Industrial and Urban Geography.

◆ In places there have been a few additions to the prescribed syllabus. For example, in the chapter on the Atmosphere, attention is paid to the El Niño/La Niña phenomenon with all its human impact. Secondly, Biosphere topics have been expanded to cover, for example, trophic levels, heather moorlands and the role of people in transforming ecosystems. It is hoped that this may assist teachers wishing to examine some of the issues prescribed in the new subject area of Managing Environmental Resources. Thirdly, themes such as the Parallel Roads of Glen Roy, periglaciation and the Storegga Slide are included within sections labelled 'To take you further'. It is recognised that some teachers go beyond the prescriptions of the syllabus, if only to encourage senior pupils to read for themselves and not just be thirled to the mechanistic demands of Higher Still.

◆ The emphasis as far as possible has been on real case studies as much as on models and systems diagrams. In this way it is more difficult to segregate 'human' from 'physical' geography. For example in the Biosphere chapter, Tentsmuir is taken as a case study of a psammosere, in which the role of management plans in shaping its ecology is discussed. A study of derelict mining land not only demonstrates the principles of plant succession, but also the value of such land, despite the ravages of industrialisation, to the local community. Secondly, all the prescribed physical landscapes—glaciated uplands, chalk scarp and clay vale, and upland limestone, have their own social and economic opportunities briefly discussed. Rural case studies have been chosen to provide what the authors consider to be the best possible examples of extensive arable farming, namely shifting cultivation and intensive peasant farming (the latter from Malaysia rather than a Punjab with terraced rice-fields apparently cultivated by Chinese pioneer farmers!).

◆ Where appropriate, Scottish examples are used, for instance the Tay and Earn basins in relation to glaciated scenery, soils and vegetation, and hydrology. Three of the five topographic maps are Scottish, including a 1:40 000 extract. Scottish material illustrating the model of demographic transition and migration is discussed in the chapter on Population, several of the industrial case studies are drawn from Central Scotland, while the chapter on Urban Geography uses Edinburgh as an extended case study.

◆ There is a recognition that senior pupils may have to work on their own, particularly (heaven forbid!) if multi-level teaching is necessary. Questions, therefore, are designed to cover the key sections of each of the chapters as comprehensively as possible. It is a myth that Credit Level at Standard Grade effectively prepares candidates for the more factual type of questions they encounter at Higher Grade. Thus the questions attempt to cover key facts essential for both internal and external assessment purposes, and encourage the use of appropriate diagrams as an integral part of an answer. Questions on the prescribed geographical methods and techniques are integrated with the other questions.

◆ Key ideas are noted at the end of each chapter, as are suggestions for further reading. Sample websites are cited, although their ephemeral nature and their propensity to multiply like rabbits, pose an interesting challenge to both teacher and pupil alike. Maximum use has been made of maps, diagrams and photographs, as well as extracts from newspapers and books. Such varied source materials are the very stuff of geography.

In conclusion, it is recognised that working through the prescribed core material is very demanding for senior pupils, but it is hoped that this textbook successfully covers the topics as prescribed, caters for a widening range of ability among candidates, and yet still conveys something of the richness of geography.

Kenneth Maclean and Norman Thomson
September 2000

The authors would like to thank the following individuals and organisations for the help given during the writing of this book:

For advice: Bill Chapman, formerly Principal Teacher of Geography, Perth Grammar School; Paul Ewing, Principal Teacher of Geography, Arbroath High School; Robin Murray, formerly Principal Teacher of Chemistry, Queen Anne High School, Dunfermline; Margaret Murray, formerly Lecturer in Biology, Craigie College of Education, Ayr; Jim Hutchison, Mary Lewis, Jean Pringle and Jackie Yuill—all of the Geography Department, Perth Academy, and the Perth Academy Higher Geography candidates during session 1999-2000 for all their frank but constructive comments after 'trialling' sections of the text.

For information and photographs: John Rogers, Principal Teacher of Computer Education, Perth Academy; Tom Cunningham and Gordon Wardrope of the Warden Service, Tentsmuir Point National Nature Reserve; the hydrology staff of SEPA, East Region, Perth; the staff of Scottish Natural Heritage, Battleby; the late Andrew Watson, Dalkeith, Midlothian; Jenny Honey, The Alba Centre; Iona McCullagh, Sun Micro Systems; Alan Stewart, Abal Studios; the Finnish Consulate, Edinburgh; Frank Sullivan, L.E.E.L; the staff of Scottish Enterprise, Lanarkshire Development Authority, and Economic Development, Fife Council; Sandy Robertson, University of Strathclyde, and the late Andy Hocking,formerly Senior Lecturer, Moray House Institute of Education.

Contents

The Physical
ENVIRONMENT

The Manor Valley:

The Manor Water is a right-bank tributary of the River Tweed (Figure 1.1). Its sources are between 750 and 800 metres above sea level on the plateau-like summits of the Tweedsmuir Hills in the Southern Uplands.

The Manor flows almost due south to north for some 15 km to its confluence with the eastward-flowing River Tweed. The valley floor drops in height from about 400 metres in the south to less than 200 metres in the north. The floor is flat, except for one isolated, steep-sided hill at Posso about half-way down the valley, and widens from less than 500 m in the south to over 2 km. The many tributary valleys are mostly narrow and V-shaped in cross-section (Figure 1.3). The valley sides are uniformly steep, usually convex in profile i.e. steepest near the valley floor, and gentlest near the summits (Figure 1.2 and Figure 1.4).

The Manor Water is fed by 20 smaller streams, the biggest being the Glenrath Burn. The smaller tributaries, with short straight courses, flow rapidly

down to join the slower flowing Manor Water, which eventually meanders across a wide flood plain in the north. The valley narrows and deepens as the Manor changes course before joining the Tweed (Figure 1.2).

The underlying rocks are intensely folded old sedimentary rocks (Silurian shales) which were swathed in boulder clay during the Ice Age, while the Manor and its tributaries have deposited silt as their gradients have slackened at the foot of slopes and on the valley floor. Ice sheets have moulded the smooth, rounded summits along the watershed.

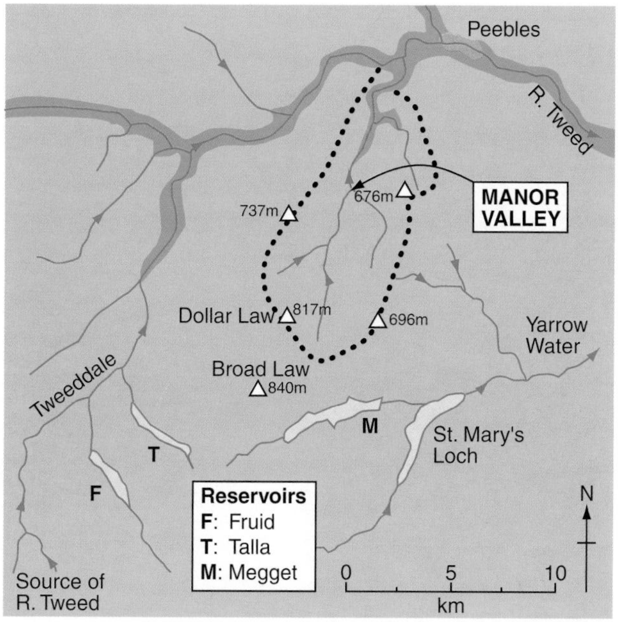

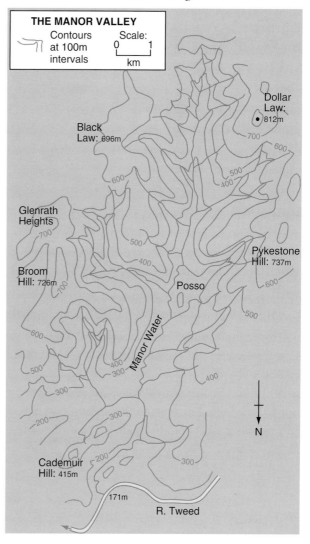

Fig 1.1 Location Map – The Manor Valley

Fig 1.2 The Manor Valley

CROSS-SECTIONS OF THE MANOR VALLEY

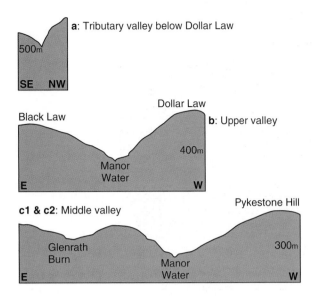

a: Tributary valley below Dollar Law

500m

SE   NW

Dollar Law

Black Law

b: Upper valley

400m

Manor
Water

E   W

c1 & c2: Middle valley

Pykestone Hill

Glenrath
Burn

300m

Manor
Water

E   W

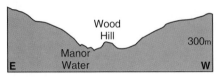

Wood
Hill

300m

Manor
Water

E   W

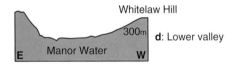

Whitelaw Hill

300m   **d**: Lower valley

E   Manor Water   W

Fig 1.3 Cross-sections of the Manor Valley. The scale is 1.2cm = 1km.

The river basin is located in an area of water surplus with high annual precipitation (from 1000 mm to 2000 mm) in the catchment area. Snow covers the summits for more than 40 days per year.

Wet, rather acid, soils (podzols) have therefore formed in the valley. They are deepest on the valley floor, thinning out on the steeper slopes. The original vegetation of mixed woodland has long gone as a result of a deterioration in climate and clearance for cultivation and grazing. Today, the lower valley floor is cultivated, but most of the valley floor, sides and summits are covered in rough grazing of coarse hill grasses and heather. Conifers have been planted as shelter belts and in large blocks on the lower slopes (Figure 1.4).

The passage above and Figures 1.1 to 1.6 describe the physical environment of one small area of Scotland. As with all other parts of the Earth's

surface, the physical environment of the Manor Valley is composed of, or influenced by, four interwoven elements (Figure 1.6) i.e.

1. The Atmosphere

2. The Hydrosphere

3. The Lithosphere

4. The Biosphere

The next four chapters are about the Earth's physical environment. By studying them, you will learn about these four elements, the processes which take place within the elements, and of how they have shaped the environment in which all human activity takes place.

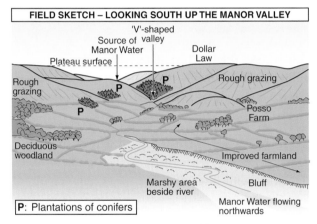

FIELD SKETCH – LOOKING SOUTH UP THE MANOR VALLEY

'V'-shaped
valley

Source of
Manor Water

Dollar
Law

Plateau surface

Rough
grazing

P

P

Rough grazing

P

Posso
Farm

Deciduous
woodland

Improved farmland

Marshy area
beside river

Bluff

Manor Water flowing
northwards

P: Plantations of conifers

Fig 1.4 Annotated field sketch of the Manor Valley

Fig 1.5 Looking north over the Manor Valley from Wood Hill

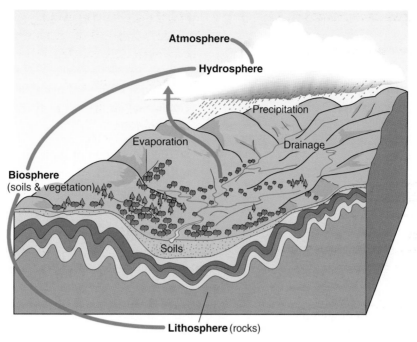

Fig 1.6 The Physical Environment

THE ATMOSPHERE

After working through this chapter, you should be aware that:

◆ the atmosphere has a distinctive composition and structure

◆ the atmosphere controls the amount of solar energy received

◆ the energy received varies with latitude, resulting in areas of energy surplus and areas of energy deficit

◆ the redistribution of energy is achieved mainly through the global patterns of atmospheric circulation, about which ideas have changed

◆ the redistribution of energy is also achieved by oceanic currents

◆ in the tropical areas of Africa, the main climate regions are influenced by the convergence and divergence of tropical maritime and tropical continental air masses, each with distinctive characteristics

◆ the interaction of these air masses and the related seasonal movement of the intertropical convergence zone influences the pattern of rainfall and life in West Africa

◆ various physical and human factors influence global warming, with many different views regarding possible consequences and solutions.

You should also be able to:

◆ interpret climate maps, diagrams and graphs

◆ construct and analyse sample climate graphs.

2.1 Introducing the changing atmosphere

One thing we can be certain of is that the climate of the world is ever changing. Whether we talk of hours, days, seasons, years or long geological periods, change in our atmospheric envelope should be accepted as normal. In the past 30 years, many of the discussions involving climate were concerned with change: in the 1970s there were warnings of potential global cooling and the onset of another ice age; by the 1990s global warming was the main concern. Such interest is inevitable given the many ways in which daily atmospheric change affects us. From death and destruction across several continents in 1998 caused by the El Niño effect to more localised Alpine avalanche disasters in 1999, climatic hazards can be critical in their impact. An ever changing climate, therefore, impinges on many areas of our life.

Traditionally, climatology analysed many years of atmospheric data to provide a global overview of the main climatic regions and their workings. That view has been transformed by:

1 the development of ever more sophisticated satellite observations thanks to a wide range of remote sensing devices,

2 the increasing use of even more powerful computer technology which allows storage of more data and the ability to create more sophisticated forecasts. Such technology has forced us to realise the importance of the total climatic system. Not only does atmospheric change in one part of the Earth affect another, it is also intimately linked with changes in the oceans and on land.

The Atmosphere: Composition and Structure

Originally formed as the Earth cooled, the atmosphere is a blanket of gases which contains solid particles, such as volcanic dust and blown soil, and is attached to the Earth by the force of gravity. Many gases make up

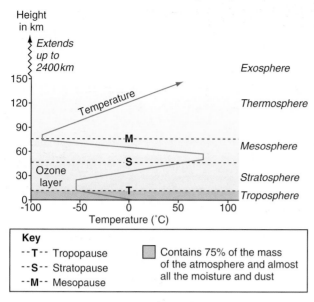

Key
- --**T**-- Tropopause
- --**S**-- Stratopause
- --**M**-- Mesopause

Contains 75% of the mass of the atmosphere and almost all the moisture and dust

Fig 2.1 The structure of the atmosphere

this atmospheric blanket, including **nitrogen** (78.01% by volume), **oxygen** (20.95%), **argon** (0.93%) and small variable amounts of **water vapour** and **carbon dioxide**. Water vapour and carbon dioxide are two of the most important **greenhouse gases**, so-called because of their ability to trap a portion of the Earth's outgoing infra-red radiation, thereby keeping the Earth's surface warmer than it would be otherwise.

Figure 2.1 outlines the vertical structure of the atmosphere and the different names applied to the zones. Of these, the **troposphere** and **stratosphere** are the most significant for our purposes. The troposphere contains most atmospheric moisture and some 75% of all air, water vapour and dust. It is the zone where temperature falls with altitude and where most weather systems develop. Above it, the stratosphere is dry and temperatures begin to rise again. The stratosphere contains large amounts of **ozone gas**. Ozone helps to protect life on Earth by filtering out most of the Sun's harmful ultraviolet rays. Figure 2.1 also shows the **tropopause**, which is the boundary between the troposphere and the stratosphere. The height of the tropopause varies with latitude and season: it can be up to 8 km high at the poles, and up to 16 km at the equator.

The Atmosphere and the Global Heat Budget

The Sun is a huge ball of gas with violent surface explosions. It is our nearest star, 150 million km away, and if it went out, we would have a few problems! The Sun is the primary source of energy for the Earth. This energy is derived from thermonuclear processes which raise surface temperatures up to 6000 °C. In fact, we only intercept a tiny fraction of all its **radiation** (radiant energy) as we progress through our yearly orbit. The solar energy we receive is critical because it drives the atmospheric system and sustains life in the biosphere.

The atmospheric system involves inputs and outputs. In broad terms, incoming solar energy is balanced by outgoing **terrestrial** (Earth based) energy. The balance between input and output is usually referred to as the **heat budget**.

Input begins when **short-wave solar energy** (also called **insolation**) enters the atmosphere and is transferred downwards, interacting with the various components of the atmosphere, until the Earth's surface is reached. Figure 2.2 shows that:

◆ some insolation is (i) **reflected** by clouds and the Earth's surface, and (ii) **scattered** by the gas particles in the air. Taken together, reflection and scattering make up what is called the **albedo** (the reflectiveness of a surface) and this is responsible for a 30% radiation loss.

◆ some insolation is **absorbed**. Figure 2.2 shows that clouds, water vapour and dust absorb another 19% of solar radiation.

◆ only 51% of the input of short-wave solar radiation is absorbed by the Earth's surface.

Balancing this input is an equal **output** of heat (terrestrial radiation) from the Earth. Mainly, this takes the form of **long-wave radiation** or infra-red radiation emitted from the Earth. Long waves are radiated because the Earth's surface is not so hot as the Sun. These long waves radiate back energy, of which 6% goes directly back to space. The remaining 94% is readily absorbed by the water vapour and carbon dioxide in the atmosphere and re-radiated. Eventually all the long-wave energy returns to space but the atmosphere's **greenhouse effect** slows things down. It acts like a huge blanket, limiting the heat energy lost. But for this greenhouse effect,

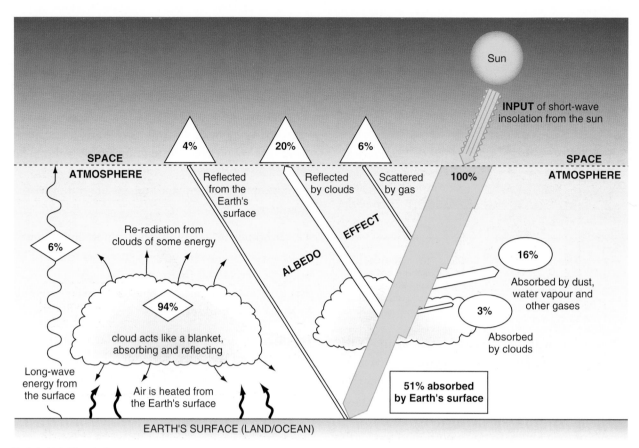

Fig 2.2 Global heat budget

global temperatures would be 33 °C lower, the many atmospheric processes would not function and life could not be sustained.

Energy Receipt and Latitude

As well as a vertical transfer of energy between Earth and space, there is also a horizontal transfer of energy between high and low latitudes. This is the result of latitudinal variations in the global heat budget. Figure 2.3 shows that:

◆ between approximately 35° N and 35° S there is a surplus of energy because incoming insolation exceeds outgoing radiation.

◆ polewards from approximately 35° N and 35° S there is a deficit of energy because outgoing radiation exceeds incoming insolation.

These energy variations are most extreme between the intensely heated tropics and the very cold polar regions. Such marked contrasts are referred to as the

global temperature gradient. The following are among the possible interrelated explanations:

◆ because of the Earth's curvature, the planet's surface slopes further away from the Sun with distance from the equator. Figure 2.4 shows P and Q, beams of equivalent insolation: P striking a small area near the equator (low latitudes) but Q spreading over a more extensive area near the poles (high latitudes).

◆ between the tropics, the rays from the noonday Sun are high in the sky throughout the year, thereby focussing energy. At high latitudes, however, energy is spread over a larger area because the rays strike the ground at a much lower angle.

◆ a related factor is that, with distance from the equator, energy is further dissipated because it travels further through the atmosphere. This results in more absorption and scattering of insolation in the higher latitudes than in the tropics.

◆ albedo is also involved. In polar regions, ice and

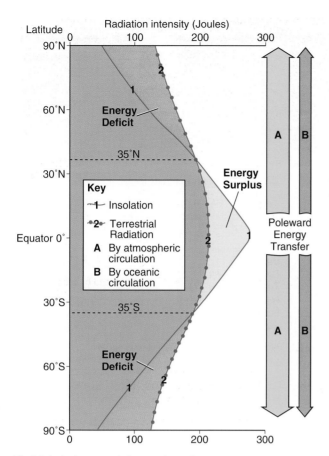

Fig 2.3 Latitude, energy balance and transfer

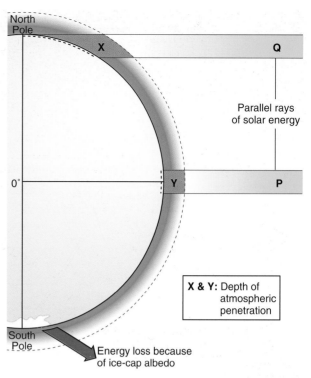

Fig 2.4 Latitude and insolation receipt (at the equinoxes)

snow cover reflect solar radiation back into space. In the equatorial regions, however, the albedo effect is much lower because of areas of dense vegetation.

Energy Transfer

Theoretically, such an imbalance in energy receipt could result in the lower latitudes becoming warmer and the higher latitudes becoming ever colder. In reality, however, energy is transferred from areas of surplus to areas of deficit in two linked ways:

◆ by atmospheric circulation. Basically, all winds, storms and cyclones are the moving parts of a giant atmospheric heat transfer system or 'engine'. In fact, the greater the temperature gradient, as in winter, the faster the 'engine' works to transfer energy. This explains the stronger winter winds and storms in countries such as Britain.

◆ by oceanic currents. The great ocean currents such as the North Atlantic Drift also help to transfer

heat from the tropics towards the poles. Overall, atmospheric circulation accounts for 80% of heat transfer, oceanic circulation the remaining 20% (see Figure 2.3).

To take you further

As well as long-wave radiation, one of the other ways in which energy is transferred to the atmosphere involves **latent heat**. Water molecules require energy to leave the Earth's surface as a gas. The energy that is stored during this change of state from liquid to gas is called latent heat. When these vapour molecules condense at high levels in the troposphere, the latent heat is released. This process warms the air temperature as part of energy transfer. For example, a great deal of latent heat is released in the rising limb of the Hadley Cells as the warm moist air over the oceans rises in enormous cumulus towers (see page 20).

2.2 Energy Transfer and Atmospheric Circulation

Having stated that there is a need for energy transfer and that atmospheric circulation plays a critical role,

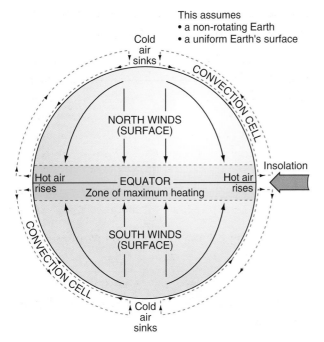

Fig 2.5 Single cell model of atmospheric circulation

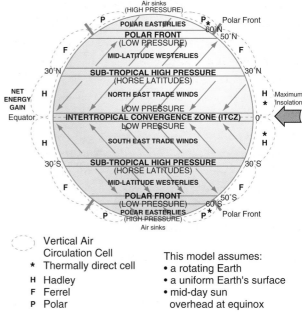

Fig 2.6 The three cell model of atmospheric circulation

exactly how does this happen? Over many years, theories become more complex, observation techniques become more sophisticated and ever more data accumulates.

A single-cell model was proposed in 1735 by George Hadley and is shown in Figure 2.5. It was based on **convection currents** which are upward movements of warm air, which is less dense and therefore lighter. He suggested that in each hemisphere there is a large convection cell powered by the heating of the equatorial regions. Warm air rises vertically into the atmosphere, spreads north and south of the equator, eventually sinks over the poles and returns as north and south surface winds to the equator. By means of these convection cells, surplus energy is transferred polewards. Fundamental to his argument is a non-rotating Earth. Later observations demonstrated that these cells exist but only extend between latitude 0° to 30° north and south of the equator.

A three-cell model was formulated by William Ferrel in 1856 (see Figure 2.6) and modified by Rossby in 1941. This model allows for a **rotating Earth** (from west to east) and the resulting **coriolis force** (Figure 2.7). Surface winds are deflected to the right in the northern hemisphere, and to the left in the southern hemisphere because of this coriolis

force. Consequently, the winds blow from a north-easterly or a south-westerly direction in the northern hemisphere (see Figure 2.7).

Ferrel's model suggested that the three interlocking atmospheric cells ensure energy redistribution in the following ways:

1   the **Hadley convection cells** between 30° N and 30° S are powered by equatorial heating which

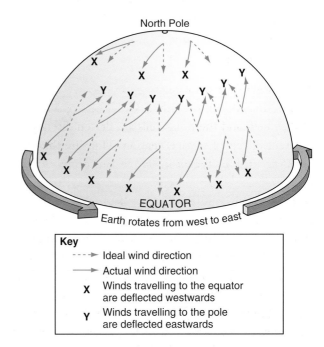

Fig 2.7 The Coriolis effect in the northern hemisphere

causes warmed air to rise to the tropopause. As the warm air rises, low pressure results at the Earth's surface. The rising air carries substantial amounts of moisture which cools, resulting in deep 'towers' of cumulo-nimbus clouds and heavy precipitation. The air spreads polewards and sinks to the subtropics at about 30° north and south of the equator. At these subtropical latitudes (sometimes called the horse latitudes), the sinking air creates high pressure areas resulting in normally clear skies and minimal precipitation. These help maintain the great hot deserts such as the Sahara, Kalahari, Atacama and Australian at these latitudes.

The descending air diverges, splitting up in two ways:

◆ some is drawn by the low pressure back towards the equator (deflected to the right by the coriolis force) as the north-east and south-east trade winds and meet at the **ITCZ** i.e. the **inter-tropical convergence zone** (see page 20).

◆ the remainder moves polewards as variable warm, moist mid-latitude south-westerlies in the northern hemisphere and north-westerlies in the southern hemisphere (see Figure 2.6).

2 the much smaller **polar cells** are each formed by a dome of cold, dense, descending air which moves outwards as the variable polar easterlies to lower, warmer latitudes. Figure 2.6 shows that the polar air is moving towards an areas of low pressure where it meets the warmer subtropical air.

3 the **Ferrel cells**, according to the three-cell model, lie between the other two. The tropical Hadley cells and the polar cells are driven directly by different effects of surface heating. They are therefore called **thermally direct cells** (air ascends over a warm area and descends over a cold area). Ferrel cells, however, are described as **thermally indirect** because they are powered by the other two. This allows them to transfer warm air from the Hadley cells to the high latitudes and transfer cold air back to low latitudes for warming. Figure 2.6 shows that the warm, moist, surface air, transferred by the Ferrel cells as the westerlies (which may actually blow from anywhere between SW and NW), meets the cold air from the polar cells. Their clash creates the mid-latitude, rain bearing, polar front depressions which affect so much of British weather patterns.

Current Thinking: The three-cell model has had a long life and serves as a useful initial tool to discuss atmospheric circulation and energy transfer. Today, however, with the advantage of a great deal more knowledge of the upper troposphere, limitations are acknowledged in the model. Observations based on radio-sonde balloons and high-flying aircraft, show that there is not a mid-latitude Ferrel 'cell' as such. In place of the Ferrel cell, it is now argued that there are:

1 alternating patterns of high and low pressure which travel at relatively low levels.

2 a series of associated high level, horizontal wavelike motions called **Rossby waves** (see Figure 2.8 and below).

To summarise, over the years, single-, then three-cell models attempted to explain atmospheric energy transfer. Today, some authorities suggest that it is better to think of a three-part model of a Hadley cell, Rossby waves and polar cell. Even that picture, however useful, is too simplified to represent a very complex and efficient atmospheric engine capable of transferring such vast amounts of energy.

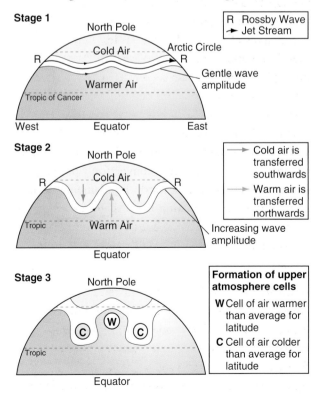

Fig 2.8 Rossby waves in the northern hemisphere

Rossby Waves and Jet Streams

Rossby waves were first detected by the meteorologist Carl Gustav-Rossby in 1940 as part of a programme of upper troposphere research. It was demonstrated that these are very large scale, high velocity belts of winds operating in the upper troposphere. With a distinctive wave-like motion of peak and trough, these Rossby waves snake their way across the globe. At their core are long, relatively narrow cylinders of very fast-flowing air called **jet streams**. First detected by high-flying aircraft during World War Two, jet streams have speeds between 200 km per hour in summer and 450 km per hour in winter. Visually, they can be recognised by the high bands of cirrus clouds racing across the sky.

Figure 2.8 shows varying meandering paths that can be followed by Rossby waves in the northern hemisphere. Some meander loops can be very pronounced and play an important role in energy transfer: by looping southwards, cold air is transferred into lower latitudes; by looping northwards, warm air is introduced to higher latitudes. Sometimes, just like a river forming an oxbow lake, the jet stream crosses the meander neck. This results in stranded pools of either warm air in high latitudes or cold air in low latitudes. Often these pools are slow to shift and create abnormal weather conditions for those latitudes.

Jet streams form because of the marked temperature gradients between (i) cold polar air and subtropical air and (ii) equatorial air and subtropical air (see Figure 2.9). Figure 2.9 also shows that the two main jet streams are located at the interface between these areas of contrasting temperatures. The polar jet stream is the faster of the two because of more pronounced temperature differences at these latitudes. It plays a critical role by influencing the patterns of mid-latitude anticyclones and depressions in the lower atmosphere.

Rossby waves and associated jet streams, therefore, are very important. Their strategic location in the atmosphere allows them to transfer warm tropical air polewards and cold polar air to lower latitudes. (Jet streams are also very useful when planning to fly around the world in a balloon, as in the first circumnavigation of the globe in March 1999).

Global Wind Circulation: Theory and Reality

Whatever the actual workings of energy transfer, broad global patterns of winds and pressure can be distinguished. Figure 2.10 is a simplified model diagram which shows:

◆ the trade winds converging towards the equator from approximately 30° N and S.

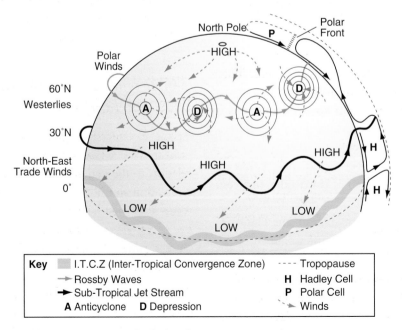

Fig 2.9 Generalised atmospheric circulation in the northern hemisphere

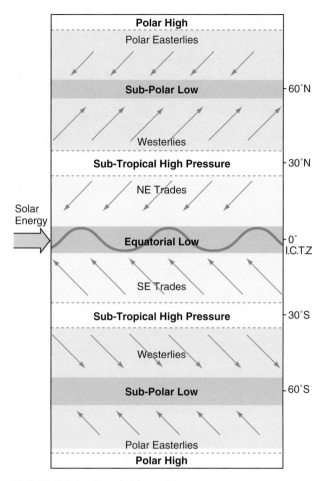

Fig 2.10 Global surface winds: a model

◆ the westerlies moving polewards from approximately 30°–60° N and S.

◆ the polar easterlies blowing to 60° N and S from the Arctic and Antarctic regions.

Related to this latitudinal pattern of global (or planetary) winds are contrasting pressure belts. These are: the polar highs, the subpolar lows, the subtropical highs and the ITCZ (intertropical convergence zone).

This model diagram is useful but, inevitably, a gross simplification. Compare the model with the maps in Figure 2.11 showing the contrasting patterns of global winds and pressure between January and July. The differences between model and maps can be explained mainly by the two following factors:

1 The Earth's tilt and consequent seasonal contrasts: Figure 2.12 is a useful reminder of the Earth's rotation around the Sun (365¼ days) and that the Earth is always tilted at the same angle as it

orbits. In June, in the northern hemisphere, the mid-day sun is directly overhead at the Tropic of Cancer. This is because the Earth's axis is directly tilted towards the Sun. In December, because the axis is tilted away, the mid-day sun is directly overhead at the Tropic of Capricorn. Between the summer and winter solstice, the Sun apparently 'moves' to be overhead at the equator on 21 March (spring equinox) and 23 September (autumn equinox).

The apparent movement of the overhead sun, therefore, is important in two related ways:

◆ it controls a belt of maximum surface heating that moves with the overhead sun and is called the **thermal equator**.

◆ it is the basis of global seasonal contrasts.

2 The distribution of land and sea also makes a profound impact on the pattern of winds and pressure. Land and sea heat up and cool down at quite different rates. In summer, the land heats up at a faster rate than the sea and in winter it cools more quickly. Consequently, because the sea retains more heat, it has a moderating effect on coastal areas, especially in temperate regions. Further inland, particularly deep in the interior of continental land masses, such as central Siberia, huge cells of high pressure form in winter because of the rapid temperature drop. In summer, on the other hand, temperatures rise rapidly, the warmed air rises and large areas of low pressure form compared with the higher pressure over the sea.

Global Wind Circulation: Reality and Seasonal Variations

Trying to understand global wind belts and associated pressure zones, therefore, is best considered in relation to the general pattern of atmospheric circulation and how it has been influenced by seasonal changes and the distribution of land and sea. Look at Figures 2.10 and 2.11, and refer to the climate pages of a good atlas. The following broad patterns can be seen:

◆ In the model, the trade wind belt is located between 30° N and S of the equator. In this belt, winds blow from the subtropical zone of high

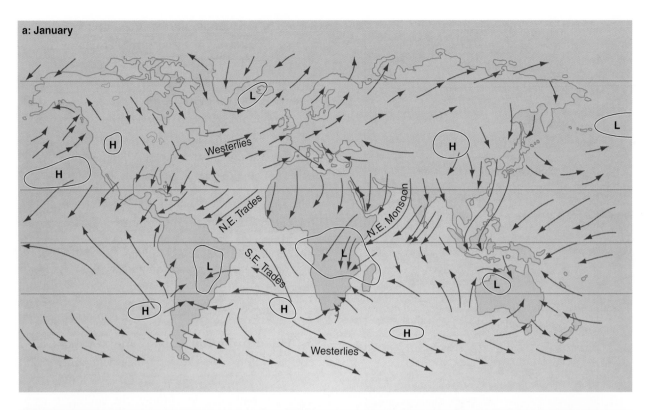

Fig 2.11 Global pressure and winds

pressure and converge on the equator. In reality, they focus on the **thermal equator** and the associated zone of generally calm, hot, rising air and low pressure called the intertropical convergence zone. (At sea, this calm area is called the doldrums).

As the thermal equator migrates seasonally north and south of the equator, it moves the pressure belts and trade wind systems as well. In January, because of greater heating over the interior of Brazil and central and southern Africa, the north-east trades migrate

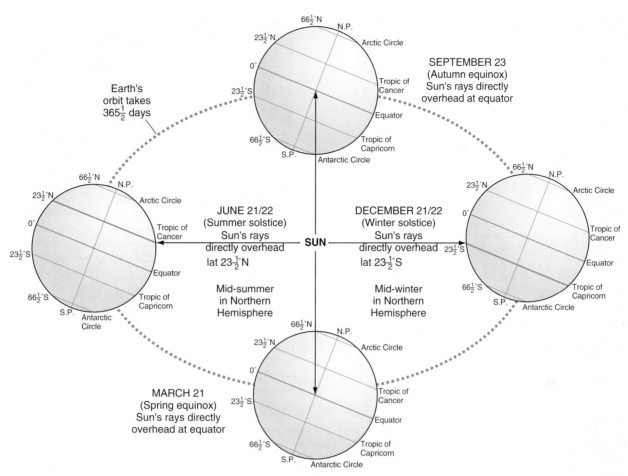

Fig 2.12 Earth's rotation around the Sun

further south over land than sea. The opposite effect can be seen in July as the south-east trades are drawn over West Africa (as south-westerlies) as far north as 20° N.

◆ The mid-latitude 'westerlies' spread polewards from the subtropical high pressure cells. Unlike the trades, they are more variable, particularly in the northern hemisphere, because of an alternating series of depressions and anticyclones, both of which move in a north-easterly direction. In the southern hemisphere, the westerlies and associated depressions are generally stronger (the 'Brave West Winds' or 'Roaring Forties') and more persistent because there are fewer land masses to impede their progress.

◆ The polar easterlies which appear on the model are not nearly so constant as it suggests. Some authorities question the existence of an all-year-round polar cell in the Arctic but cold dense air does move equatorwards, especially in winter. Perhaps the polar easterlies should be thought of as cold easterly winds on the poleward side of the mid-latitude depressions.

◆ Finally, the model breaks down in Asia, east of 60° E. Instead of the north–south trends of oceans (Pacific/Atlantic) and continents (Europe/Africa/North and South America), there is the Indian Ocean in the south and the large land mass of Asia in the north. In summer, for example, a large low pressure cell develops over Southern Asia instead of the expected high pressure (confined to the oceans, e.g. the Azores high) of the model. The intensity of this low pressure system and a northward movement of the subtropical jet stream encourages the inflow of the south-west monsoon. 80% of South Asia's rainfall arrives in this period. By contrast, in winter, an extensive and intense high pressure cell with associated dry conditions develops over the interior of Asia. This should be compared on the maps with the low pressure cells over the north Atlantic (the Icelandic low) and the north Pacific (the Aleutian low).

Therefore, global winds and pressure cells with all their seasonal variety form the mechanism which transfers 80% of energy from areas of surplus to areas of deficit.

Energy Transfer and Oceanic Circulation

The Importance of Ocean Currents

The swirling oceanic currents of this planet also play an important part in redistributing energy, making sure that the low latitudes do not overheat and that the high latitudes do not become too cold. The warm waters of the Gulf Stream thrust north-eastwards, pass the coast of the USA and continue towards the British Isles, Iceland and Norway as the North Atlantic Drift (Figure 2.13). The sea temperature in these areas is warmer than would be expected. Also shown is the cold Canary Current which transfers water cooled at high latitudes back to equatorial waters. Its effect can be seen by the cooler water off the north-west coast of Africa.

The Gulf Stream and the Canary Current are just two of the slow moving, river-like movements of water called ocean currents. Figure 2.14 shows the world's main surface ocean currents, classified as warm and cold. These warm and cold currents assist global energy transfer. Warm currents transport warm

Fig 2.13 A computer model of global sea surface temperature. Temperatures are colour coded from red (warmest) through yellow and green to blue (coolest). It shows the effect of the NAD on the British Isles

water polewards, whereas cold currents take cooler water into lower latitudes. Oceanic currents influence climate in various ways, as suggested in the table below.

A cautionary point should be made about the terms 'warm' and 'cold'. As anyone going for a dip in the waters of the North Atlantic Drift washing the shores of north-west Scotland knows, the word 'warm' is only relative. These adjectives indicate that the water temperature in the current is warmer or cooler than expected for that latitude. Measurements show that water temperatures in the 'cold' Canary Current are warmer than those in the 'warm' North Atlantic Drift off the Norwegian coast. Such relative differences, however, ensure an ice-free coast as far north as the Russian port of Murmansk, while the coasts of west Greenland and Labrador are ice-bound.

The Climatic Effect of Oceanic Currents	
Warm Currents	**Cold Currents**
◆ Warm currents (e.g. the North Atlantic Drift) warm prevailing onshore winds. This ensures that, for its latitude, north-west Europe has a mild winter climate, and harbours in Norway and Iceland are ice free.	◆ Cold currents (e.g. the Canary, the Peruvian) cool onshore winds. This lowers temperatures and shortens growing seasons in the coastal regions. The likelihood of frozen harbours and icebergs is increased by, for example, the Labrador Current.
◆ They help to increase precipitation in adjoining coastal regions, e.g. the SE Trades passing over the warm Brazilian Current.	◆ They help maintain the aridity of coastal deserts by cooling and drying onshore winds, e.g. coastal winds crossing the Peruvian Current to the Atacama Desert. Often, cooling causes coastal fog.
◆ They raise the temperature of the waters to which they are flowing.	◆ They lower the temperature of the waters to which they are flowing.

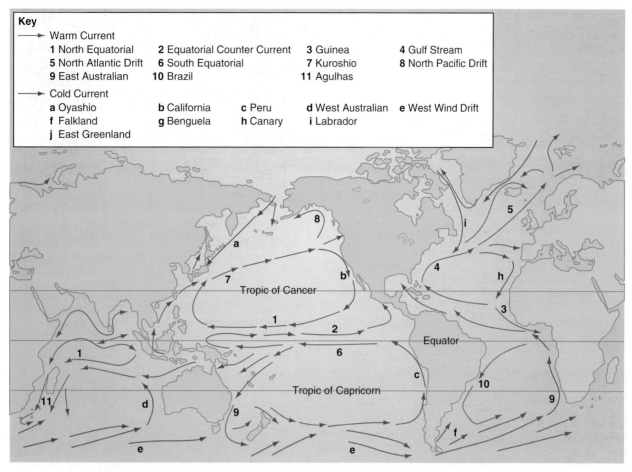

Key

→ Warm Current

1 North Equatorial	**2** Equatorial Counter Current	**3** Guinea	**4** Gulf Stream
5 North Atlantic Drift	**6** South Equatorial	**7** Kuroshio	**8** North Pacific Drift
9 East Australian	**10** Brazil	**11** Agulhas	

→ Cold Current

a Oyashio	**b** California	**c** Peru	**d** West Australian	**e** West Wind Drift
f Falkland	**g** Benguela	**h** Canary	**i** Labrador	
j East Greenland			

Fig 2.14 World ocean currents

Ocean Currents: Patterns and Causes

A close look at Figure 2.14 shows the global pattern of surface ocean currents. These currents are deflected by the continental land masses. Consequently, the major oceanic basins have huge, roughly circular shaped loops of water called **gyres** with an average speed of 5 km/h. In contrast, the water at the centre of a gyre, for example, the Sargasso Sea, is relatively slow moving. Figure 2.15 shows that:

◆ In the northern hemisphere, gyres have a clockwise circulation pattern.

◆ In the southern hemisphere, gyres have an anticlockwise circulation pattern.

◆ The main exception is the Antarctic Circumpolar current (also called the West Wind Drift). This is the only current which flows unimpeded around the globe from west to east.

Various related factors are responsible for these patterns:

◆ The surface ocean currents directly respond to the prevailing global pattern of winds. The overlying trade winds and the westerlies create a frictional drag as they blow over the water surface. For example:

1 the westerlies, in the Atlantic, help to direct the Gulf Stream/North Atlantic Drift north-eastwards towards NW Europe; in the Pacific, the counterpart is the Kuro Siwo.

2 the NE/SE trade winds blowing equatorwards help to guide the westerly flowing North and South Equatorial currents.

3 The West Wind Drift, as the name implies, is mainly a response to the prevailing 'Roaring Forties'.

◆ The Earth's rotation and related coriolis force encourage the surface currents, like winds, to move towards the right in the northern hemisphere,

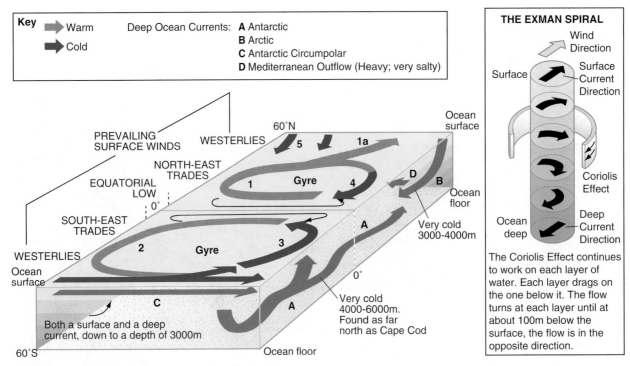

Fig 2.15 Model of ocean currents (based on the Atlantic Ocean)

causing the clockwise gyres in the North Atlantic and North Pacific oceans. In the southern hemisphere, surface currents move towards the left, with resulting anticlockwise gyres in the South Atlantic and South Pacific oceans.

◆ The shape of the land masses can block and direct the movement of the currents. For example, the North Equatorial Current is deflected off the coast of NE Brazil. It then builds up in the Gulf of Mexico before moving north-east as the Gulf Stream.

◆ Temperature differences are a final factor, particularly as they influence the density of water. Initially, the cold waters of the high latitudes flow equatorwards as surface currents, for example, the Labrador, the East Greenland and the Oyashio Currents. Because such water is denser, it sinks to the ocean floor to form very slow-moving, powerful, deep-sea currents. Over possibly hundreds of years, they return towards the low latitudes. There they displace upwards the less dense, surface waters by upwelling.

In these ways, the global pattern of oceanic currents contribute to, and respond to, global climatic energy transfer. Figure 2.16 shows the ocean surface temperatures, derived from satellite observations. The map confirms: (i) the large heat storage capacity of

the oceans, and (ii) their ability to function as a great heat 'engine'.

El Niño/La Niña and the Southern Oscillation

Two features which show the close relationship between the atmosphere and ocean are:

El Niño – the periodic warming of the waters of the tropical Pacific Ocean, and La Niña – the periodic cooling of similar waters of the tropical Pacific Ocean.

Of these two climatic phenomena, El Niño is the better known. Named by Peruvian fishermen, it means 'the Christ Child'. This is because the normally cold offshore waters of the Peruvian Current are replaced by a warm current – El Niño – from roughly Christmas till March. Written evidence of past El Niños dates from the sixteenth century and records confirm:

1 the disappearance of normally plentiful shoals of anchovies

2 heavy rainfall in a normally very dry desert

3 associated flooding.

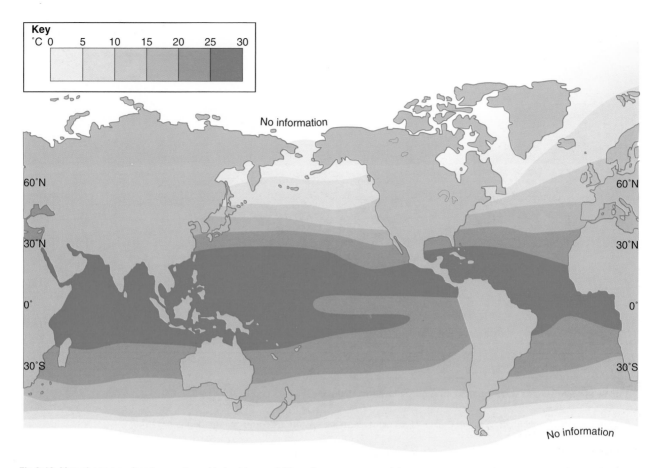

Key
°C 0   5   10   15   20   25   30

No information

60°N   60°N
30°N   30°N
0°   0°
30°S   30°S

No information

Fig 2.16 Map of ocean surface temperatures (derived from satellite radiance measurements)

Since the 1970s, however, with improved data sources, it is now realised that El Niño and La Niña do not confine their impact to the Pacific coast of Peru. They are, in fact, part of a truly global climatic system which can cause a great deal of chaos and suffering.

El Niño: Cause and Global Impact

Look at Figures 2.17 and 2.18. What are the causes of this phenomenon which occurs every three to seven years? Essentially, there is a weakening in the 'normal' pressure gradient in the Pacific. This process is called the Southern Oscillation. Lower than average atmospheric pressure develops over Tahiti and higher than average air pressure builds up over northern Australia. The trade winds weaken and no longer push warm surface water and associated heavy rains westwards towards Australia and New Guinea. Instead, warm surface water spreads eastwards and takes over from the cold Peruvian current. The global environmental results vary but since the 1970s have included:

1   heavy rainfall, flooding and the collapse of the fisheries along the coastal areas of Peru, as happened in 1997/98. The warm water (up to 8 °C warmer) blocked the upwelling of cooler, nutrient rich waters and drove fish away.

2   at the same time, the high pressure system over Australia contributed to extensive drought in Australia and South East Asia. Consequent bush fires in Australia were matched by fierce forest fires in Malaysia, Borneo and Sumatra. These were mainly started by 'slash and burn' forest clearing but were exacerbated by the drought induced by El Niño.

3   a weaker monsoon in South Asia in 1997 with consequent drought and impact upon crops in that region.

4   drought in the Sahel zone of Africa. In many respects this region is still suffering from the effects of prolonged drought which occurred in 1972/73, 1983/84 and 1987. Along with land mismanagement, these droughts contributed to severe famine. As will be seen, rainfall occurs almost entirely between July

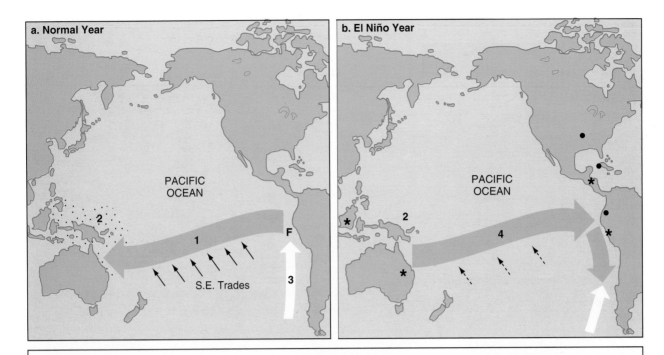

Key

1 Flow of warm surface water from South America towards Australia

2 Water piles up, making the Western Pacific 45cm higher than the eastern side and several degrees warmer

3 Cold sub-surface water, rich in nitrates, phosphates on which plankton thrive, replaces surface water
   F Valuable coastal fishing grounds

4 Every 2-7 years, the S.E. Trades weaken and warm water surges eastwards to South America. Upwelling of cold water prevented so fish stocks are decimated.

   Other possible consequences: ● Floods ✻ Drought

Fig 2.17 Causes of El Niño

and September and is related to the movement of the intertropical convergence zone. There is evidence that a strong El Niño pattern in the Pacific means less converging air over the Sahel and, consequently, less upward air movement and less rainfall.

5 in compensation, however, during El Niño years, hurricanes in the Caribbean and Atlantic are reduced in strength and frequency.

La Niña: Cause and Global Impact

La Niña ('The Girl') is triggered by quite opposite conditions. This time, abnormally high pressure conditions develop over Tahiti and lower than normal atmospheric conditions form over northern Australia. Consequently, warm surface water is

pushed further westwards than normal. The global impact has included the following:

1 more than normal amounts of deep, cold water and nutrients upwell, resulting in an increase in fish stocks and even drier weather conditions in Peru and Ecuador.

2 on the other side of the Pacific, there are wetter than normal conditions reflected in heavier monsoon 'bursts' in South Asia causing flooding (as in Bangladesh in 1987) and tropical storms in northern Australia and in Mozambique in February 2000.

3 rainfall in the Sahel zone, so critical for this part of the world, is usually heavier. Above average rainfall brings economic and social benefits.

4 hurricanes, however, are more likely to strike the

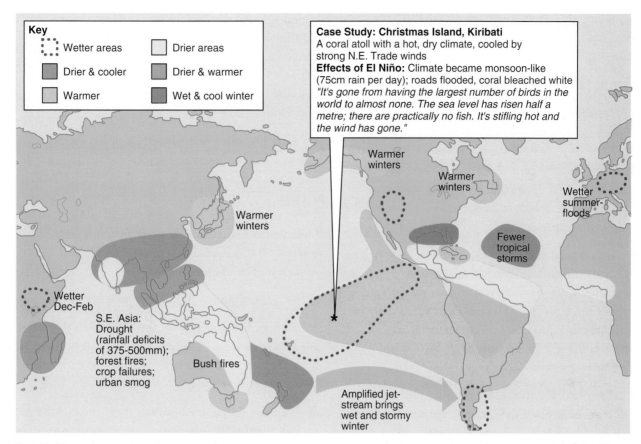

Key
- ⬚ Wetter areas
- ☐ Drier areas
- ▨ Drier & cooler
- ▨ Drier & warmer
- ☐ Warmer
- ▨ Wet & cool winter

Case Study: Christmas Island, Kiribati
A coral atoll with a hot, dry climate, cooled by strong N.E. Trade winds
Effects of El Niño: Climate became monsoon-like (75cm rain per day); roads flooded, coral bleached white
"It's gone from having the largest number of birds in the world to almost none. The sea level has risen half a metre; there are practically no fish. It's stifling hot and the wind has gone."

Warmer winters

Warmer winters

Warmer winters

Wetter summer-floods

Fewer tropical storms

Wetter Dec-Feb

S.E. Asia: Drought (rainfall deficits of 375-500mm); forest fires; crop failures; urban smog

Bush fires

Amplified jet-stream brings wet and stormy winter

Fig 2.18 Climate changes 1997 – the effects of El Niño

southern USA and the Caribbean. Autumn 1998 was a deadly hurricane season for this region, particularly Hurricane Mitch which devastated Honduras, Nicaragua and El Salvador. Severe winds, prolonged massive rainfall and extensive landslides killed thousands and left over a million homeless and facing the threat of disease. Effectively, the economies of these countries were set back many years (Figure 2.19).

Fig 2.19 Some of the devastation left by Hurricane Mitch in the capital of Honduras, Tegucigalpa

More remains to be learned about El Niño and La Niña and their exact causes. What is not in doubt is their impact. Some areas such as South East Asia can suffer a climatalogical double blow with one following the other – severe drought one year, followed by a year of flooding. Both phenomena are a salutary reminder of the enormous amounts of energy transferred by the interactions of atmosphere and ocean.

2.4 Tropical Africa: Climate

The earlier discussion of the atmosphere's structure and the need for energy redistribution was important not just for its own sake but to help us appreciate the critical role of climate in the modern world. Climate matters, and this can be illustrated from tropical Africa where people have a well developed understanding of their often difficult climatic environments. These are areas where people have to make decisions about food and water shortage, land degradation, poor grazing land and rural-urban migration. Such decisions are influenced, directly and indirectly, by climate, particularly the seasonality and unreliability of the rainfall.

Tropical Africa: Climatic Regions

The importance of rainfall in tropical Africa can be illustrated from three of the broad climatic regions shown in Figure 2.20. It must be stressed, however, that this is a very generalised map. There are many ways of classifying climate. Essentially, climatic boundaries are based on combinations of temperature and/or precipitation.

1 **Equatorial Climate:** broadly, this zone occurs up to 10° north and south of the equator. It has rain throughout the year, often with two maxima (see graph for Lagos, page 34), and can exceed 2000 mm, especially in upland areas. Temperatures are consistently very high—around 26 °C – with, therefore, a very limited range.

2 **Tropical Continental:** broadly, this region forms an arc around the equatorial zone. Much of Kenya, although straddling the equator, falls into this category because of the altitude of its interior plateau. Temperatures are equally high, but with a more pronounced range than the equatorial type. The key contrasts are: (i) the marked seasonal variation in

rainfall – there are distinct wet and dry seasons, and (ii) the total rainfall is less. (Sometimes this type of climate is referred to as 'Savanna'. Strictly speaking, the word is best reserved as an overall term to describe the 'natural' vegetation of grasses and deciduous trees.)

3 **Hot Desert:** essentially, areas of less than 250 mm of rain per year, they lie north and south of the tropical continental regions. Temperatures can average 35 °C in the hot season, dropping to below 15 °C in the cool season. The diurnal range, however, is particularly marked, from a possible midday 50 °C dropping to freezing at night (a result of the absence of cloud cover).

A few general points should be made about the map:

◆ Although the map gives the impression that there are abrupt changes between climate regions, there are none. Rather, they gradually merge into each other (as do vegetation zones).

◆ The key difference between these three climate regions is **rainfall** rather than temperature. Africa, as a whole, is a hot continent.

◆ It is for this reason that it is better to forget terms such as 'winter' and 'summer'. To the people of tropical Africa, **wet season** and **dry season** are much more meaningful in their everyday life.

Hadley Cells, Air Masses and the ITCZ in Africa

Critical to our understanding of the varying rainfall totals and its seasonal distribution in tropical Africa is the seasonal movement of the ITCZ (intertropical convergence zone). It is useful first to remind ourselves about the Hadley cells (see pages).

Figure 2.21 reminds us that air flows equatorwards at low levels in each of the Hadley cells in the form of trade winds. The trade winds, therefore, feed air from the subtropical high pressure systems and this converges on the thermal equator to form the ITCZ. The ITCZ can be defined as the belt of low pressure produced by the combination of equatorial heating and the convergence of trade winds, and migrates in response to the changing position of the thermal equator.

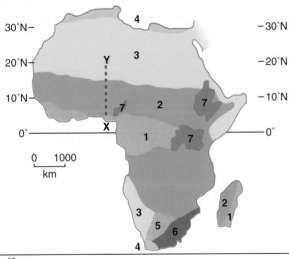

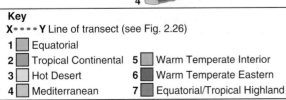

Key
X ●●●● Y Line of transect (see Fig. 2.26)
1 ▢ Equatorial
2 ▢ Tropical Continental 5 ▢ Warm Temperate Interior
3 ▢ Hot Desert 6 ▢ Warm Temperate Eastern
4 ▢ Mediterranean 7 ▢ Equatorial/Tropical Highland

Fig 2.20 Africa: climate regions (generalised)

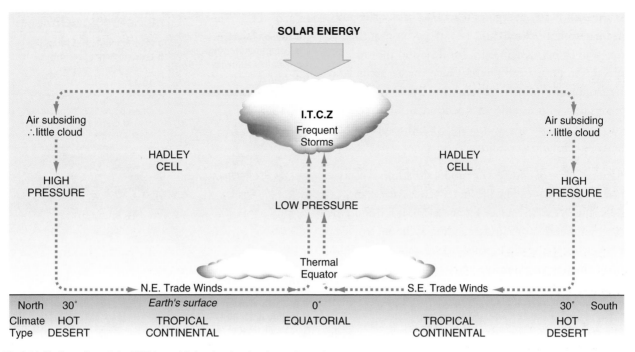

Fig 2.21 Hadley cells and the ITCZ (a model showing the situation at the equinoxes)

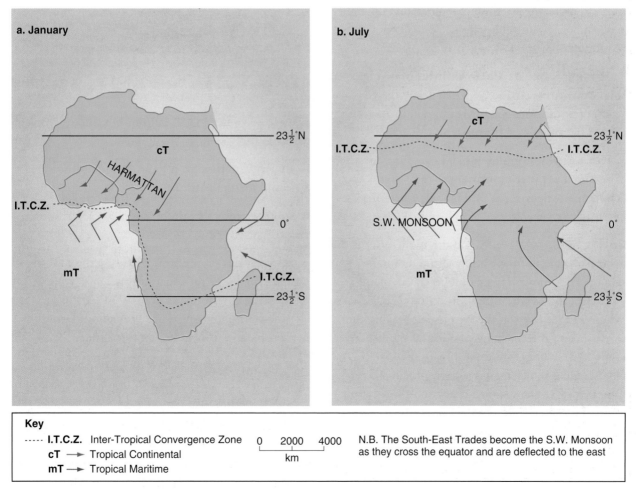

Fig 2.22 Features of tropical air masses

As the airflows converge at the ITCZ, they rise and create a zone of clouds and rainfall because of the injection of massive amounts of heat and moisture into the tropical atmosphere. Once the air ascends to the tropopause, it diverges and flows polewards, descending over a wide area centred around 30° N and S. As it descends, it is warmed and results in dry, generally cloudless conditions.

Figure 2.21 is, of course, the model situation. Reality is more complex as can be illustrated from Africa, and especially West Africa (see Figure 2.23). It is important to remember that the air descending at the subtropics will be affected by the Earth's surface underneath. The name given to an extensive body of air influenced by the Earth's surface is an **air mass**. The area where an air mass develops is known as its **source area**.

The Earth's surface is, of course, mainly covered with oceans. Air masses which develop over oceans generally contain more water vapour than those forming over land. As well as maritime/continental contrasts, there are also distinct temperature difference between polar and tropical areas. Consequently, there are four types of air mass which affect Africa:

1 least important are **Polar Continental (cP)** and **Polar maritime (mP)**. Generally, only the coastal fringes of South Africa and Mediterranean Africa are affected by mP and cP air masses.

2 by far the most important and extensive air masses are **Tropical Continental (cT)** and **Tropical maritime (mT)**. Figure 2.22 summarises the detailed characteristics of mT and cT air masses.

Another reason why reality and the model vary is the seasonal change in the position of the thermal equator. As it moves in response to the changing position of the overhead Sun, so do the Hadley cells and the ITCZ. Look at the two maps showing the air masses and the changing position of the ITCZ (Figure 2.22). These are critical maps in understanding the changing pattern of rainfall over tropical Africa. The seasonal change in the location of the ITCZ governs the supply of rainfall.

In July, that is, during the northern hemisphere's summer, the ITCZ has reached its most northerly extent (about 20° N) because the Sahara receives its maximum insolation at this time. The associated low

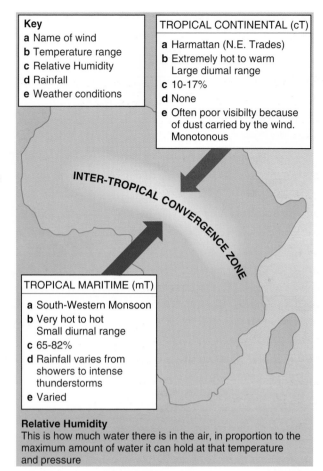

Key
a Name of wind
b Temperature range
c Relative Humidity
d Rainfall
e Weather conditions

TROPICAL CONTINENTAL (cT)
a Harmattan (N.E. Trades)
b Extremely hot to warm
   Large diumal range
c 10-17%
d None
e Often poor visibilty because of dust carried by the wind. Monotonous

INTER-TROPICAL CONVERGENCE ZONE

TROPICAL MARITIME (mT)
a South-Western Monsoon
b Very hot to hot
   Small diurnal range
c 65-82%
d Rainfall varies from showers to intense thunderstorms
e Varied

Relative Humidity
This is how much water there is in the air, in proportion to the maximum amount of water it can hold at that temperature and pressure

Fig 2.23 Africa: tropical air masses and the ITCZ

air pressure of the ITCZ pulls in hot, moist, tropical maritime (mT) winds. These include:

◆ south-westerlies from their Atlantic source area to West and Central Africa.

◆ south-easterlies from their Indian Ocean source area to East Africa

In this way, the moisture provided by the mT air (within which convectional storms can develop freely) brings the wet season to West Africa. Although welcome, excessive rainfall can bring its own problems, as suggested in this extract.

But the year had gone mad. Rain fell as it had never fallen before. For days and nights together it poured down in violent torrents, and washed away the yam heaps. Trees were uprooted and deep gorges appeared everywhere. Then the rain became less violent. But it went on from day to day without a pause. The spell of sunshine which always came in the middle of the wet season did not appear. The yams put on luxuriant leaves, but every farmer knew that without sunshine the tubers would not grow. That year the harvest was sad, like a funeral.

Source: *Things Fall Apart* C. Achebe

By January, in response to the changing position of the thermal equator, the ITCZ has migrated southwards to the Tropic of Capricorn during the southern hemisphere's summer. Part of it, however, is still found about 8° N, along the coastal belt of West Africa. This is because of the larger extent of land in northern Africa (compared to the south) whose temperatures, even in January, are able to exert an influence. With the exception of East Africa and the coastal belt of West Africa, most of Africa north of the equator experiences its dry season at this time. Figure 2.22 shows that West Africa is influenced by a north-easterly flow of hot, dry cT air with its source region in the central Sahara. Rainfall is exceptionally rare and the hot, dry north-east wind, which brings with it large amounts of dust, is locally known as **the Harmattan**. This extract gives some idea of the nature of the Harmattan:

The dry parched breath of the harmattan coming from the

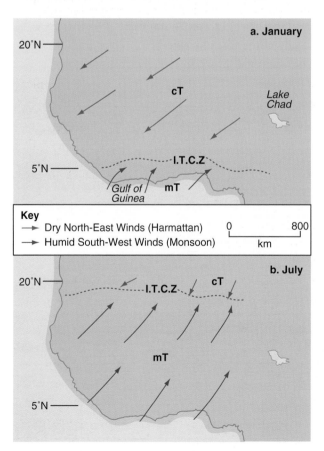

Fig 2.24 West Africa: tropical air masses and the ITCZ

north-east licks up every trace of moisture. . . . All the young plants are dried up; the long grass withers and becomes converted to hay, and is then fired . . . to ensure a fresh crop in the following rains. . . . The earth cracks, and doors and furniture split and open out, warping into grotesque shapes. The effect upon the system is unpleasant; the skin chaps, and the face has the feeling of being drawn up. The lips are parched and the skin peels off them, while an intolerable thirst can scarcely be allayed.

Source: *Dahomey As It Is* J.A. Skertchley.

Rainfall, the ITCZ and West Africa

Rainfall is such a critical part of the African climate that its character and importance need to be considered in more detail in relation to West Africa. Key aspects of rainfall – its annual total, its seasonality, its variability and its reliability – are all related to the seasonal migration of the ITCZ.

Figure 2.24 shows the changing location of the ITCZ in relation to the dominant air masses. During the northern hemisphere summer, the ITCZ reaches its northern limit and all of West Africa is affected by warm, humid mT air (the south-west winds). Because of its oceanic source area, this air mass has large amounts of potential precipitation. By January, the ITCZ is located near the coast of West Africa and dry cT air (the Harmattan) predominates over most of the region. Only the coastal area is still affected by a narrow wedge of mT air (Figure 2.25).

Fig 2.25 .This Meteosat image brings out the yellow expanse of the Sahara and the Rain Forest belt, partly obscured by the ITCZ. What evidence is there that the data for the image was collected in January?

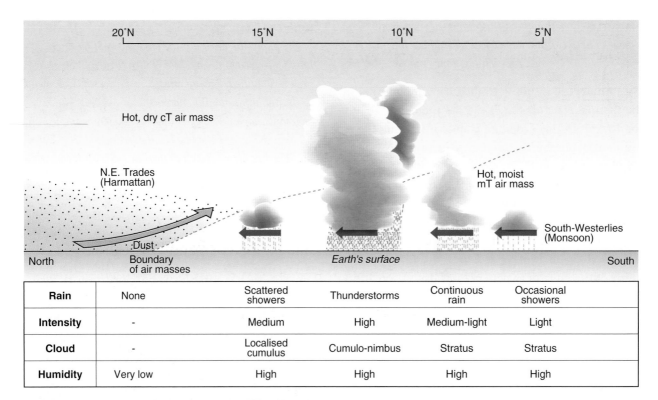

Rain	None	Scattered showers	Thunderstorms	Continuous rain	Occasional showers
Intensity	-	Medium	High	Medium-light	Light
Cloud	-	Localised cumulus	Cumulo-nimbus	Stratus	Stratus
Humidity	Very low	High	High	High	High

Fig 2.26 Transect showing weather bands within the ITCZ in West Africa

Figure 2.26 is a very generalised cross-section through the ITCZ. It shows the situation in July when the mT air predominates over most of West Africa. It also emphasises the extent of the ITCZ (look at the latitude scale) and that it is not a solid 'belt' of rainfall. Instead, depending on the moisture content and the stability of the winds, there are various degrees of rainfall intensity. The first rains usually come from localised cumulus clouds. This type of rainfall is not so reliable or intense as the

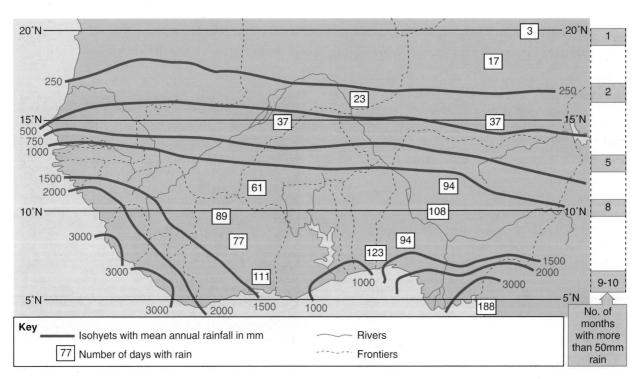

Fig 2.27 Rainfall in West Africa – isohyets

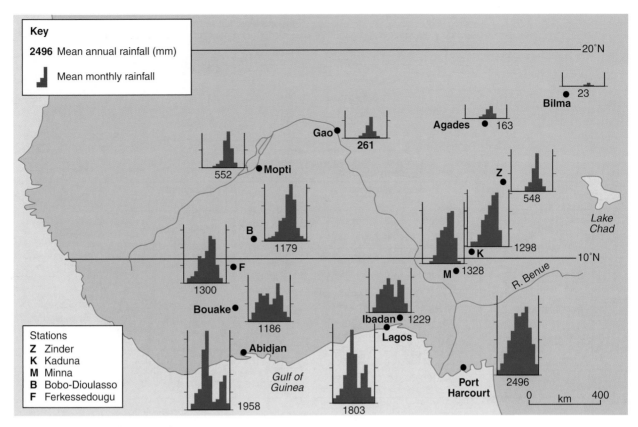

Fig 2.28 Rainfall in West Africa – graphs

thunderstorms associated with the cumulo–nimbus cloud clusters. As the ITCZ reaches its northern limit, the southern areas receive more continuous rain.

If you now look at Figure 2.27 showing isohyets and Figure 2.28 showing sample rainfall graphs in West Africa, the following generalisations can be made about rainfall as you move north from the Gulf of Guinea:

1 the further north, the smaller is the total mean annual rainfall. In the south, for example, at the Niger delta, the mean is over 2500 mm; well inland at Bilma, it is less than 25 mm.

2 the further north, the more the rainfall is concentrated in the summer months. August is usually the peak month. In the south, by contrast, the wet season extends for most of the year. Lagos, for example, has no dry months on average but it has two maximum rainfall peaks which are related to the migration of the thermal equator and the ITCZ.

3 finally, the further north, the more **variable** is the rainfall. This means that rainfall can deviate from the monthly and annual mean total by a considerable amount. The Sahel zone, on the southern Sahara fringe, has particularly been affected by variability.

For most of the 1950s, this area benefited from above average rainfall. Since 1969, however, there has been a series of drought years. The graph (Figure 2.29) show the percentage departure of rainfall in this region from 'normal' conditions. The causes of such drought conditions are:

1 the ITCZ, with associated inflow of mT air, has not moved so far north as usual,

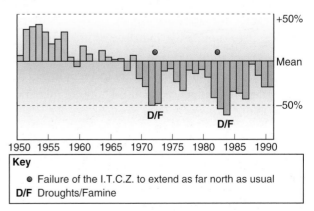

Key

● Failure of the I.T.C.Z. to extend as far north as usual
D/F Droughts/Famine

Fig 2.29 Rainfall fluctuations for stations in the Western Sahel 1950–1991 (expressed as a percentage departure from the long-term mean)

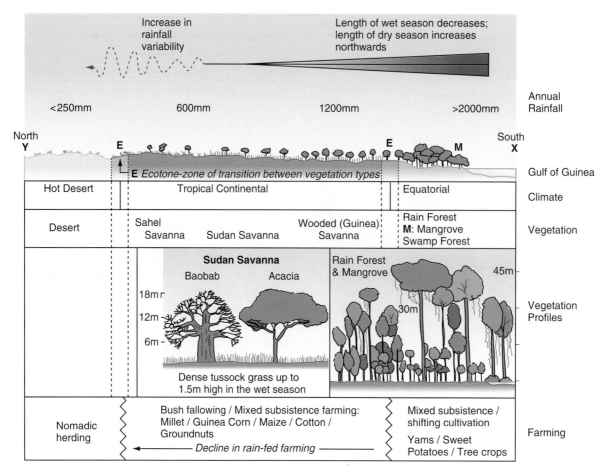

Fig 2.30 Transect diagram showing changes in climate, vegetation and farming in West Africa

2   the El Niño phenomenon.

For this area of already marginal climate, such drought was a key contributary cause of much suffering, particularly among the nomadic pastoralists and farmers, with the death of up to five million cattle and several hundred thousand people in the 1970s alone.

West Africa: A Summary Transect

To conclude: the transect diagram of West Africa (Figure 2.30) summarises the changes in climate regions, vegetation and land use from the Gulf of Guinea in the south to the Sahara in the north. The key underlying influence is rainfall and the migration of the ITCZ.

The transect shows the changing landscapes of the zones which run parallel with the coast. Each zone has its distinctive climate, ranging from equatorial through tropical continental to hot desert. Each has its distinctive vegetation, the rainforests giving way to the several types of savanna, with the

northernmost, the scrub grasses and thorn bushes of the Sahel, merging imperceptibly into the even sparser desert plants. The diagram also shows the broad zonation of agriculture as it merges into the pastoralism which predominates in the Sahel. The limiting factor for rain-fed agriculture is the growing season (a period of rain effective for crop growth followed by a period when water is stored in the soil). Rainfall, then, is the key to much of landscape and life in West Africa (Figure 2.31).

2.5 Global Warming

Headlines discussing the trend shown in Figure 2.32 are increasingly common and the evidence from the graph appears to confirm that global warming is an inescapable fact of life. Perhaps we should not be surprised. After all, this chapter began with the statement that global climate patterns have always been changing.

Fig 2.31 The two photographs bring out the contrasts between the wet and the dry seasons in the savanna

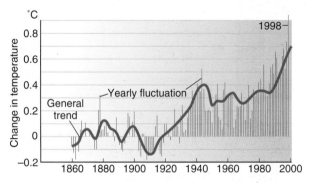

Fig 2.32 Changes in annual mean global surface temperature 1860–2000 (Relative to average temperature in 1900)

Global Warming: Enhancing the Greenhouse Effect

Global warming is the result of an enhanced greenhouse effect. Figure 2.33 reminds us (see page 00) that the greenhouse effect is the warming of the Earth's surface from heat that is trapped in the atmosphere. The atmosphere, therefore, acts rather like a blanket without which the planet would be some 33°–35°C cooler and unable to support life.

In this process, a critical role is played by certain trace gases, for example, water vapour (which accounts for 1% of the air's composition) and carbon dioxide (which accounts for 0.03%). Such trace, or greenhouse, gases trap the long-wave radiated energy from the Earth which causes the atmosphere to heat up. Advocates of global warming argue that changes in the amount and types of greenhouse gases have caused global warming of some 0.7°C since 1900. Furthermore, in the last 40 years, the rate of growth in mean global atmospheric temperature has accelerated. Ten of the 11 warmest years of the twentieth century have occurred since 1980, while computer modelling forecasts further warming of between 1°C–3.5°C by 2100.

Doubts, however, are expressed by certain climatologists. The evidence for global warming is interpreted in different ways and there is debate about the causes, consequences and solutions.

Causes of Global Temperature Change

Various physical and human mechanisms are suggested as driving forces behind global temperature change.

Physical causes can include:

◆ **Changing output of solar radiation**. Peaks of sunspot activity seem to raise global mean temperatures. The warm temperatures of the 1940s, for example, represent a period of high solar activity (and low volcanic activity). Alternatively, the lower temperatures of the 'Little Ice Age' coincided with a period of abnormally low sunspot activity.

◆ **Volcanic eruptions.** After particularly violent eruptions, enormous quantities of volcanic dust particles are ejected high into the atmosphere. Global winds then redistribute the particles, thereby helping to reduce temperatures by shielding the Earth from incoming insolation. When Toba, in Sumatra, erupted over 73 000 years ago, it put out enough material to help initiate an Ice Age. More recently, examples include Tambora (1815), Krakatoa (1883), El Chichon (1982) and Pinatubo (1991). The Tambora eruption, for example, caused 'the year without a summer' in 1816 with June frosts in New England. The 30 million tonnes of volcanic dust thrust into the atmosphere by Mount Pinatubo set back the global warming trends of the early 1990s until the billions of particles fell back to Earth by the end of 1993.

◆ **'Wobble, roll and stretch' theory.** Milutin Milankovitch, a Yugoslavian geophysicist, suggested that:

1 the Earth wobbles in space like a spinning top. As a result there is a variation in the time of year when the Earth is nearest the Sun.

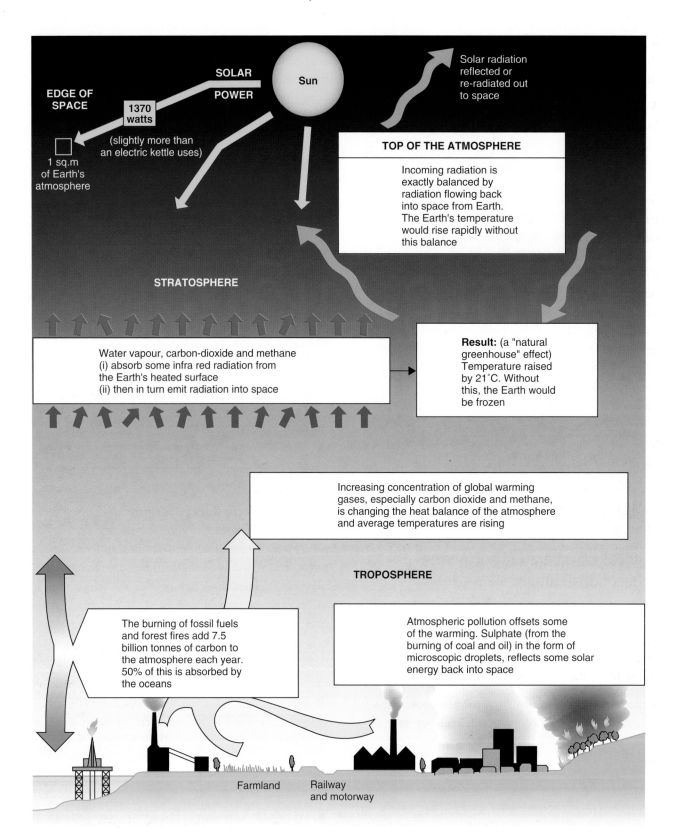

Fig 2.33 Global warming

2 every 41 000 years, there is a change in the tilt of the Earth's axis. A greater tilt or 'roll' means more sunlight for polar regions.

3 over a 97 000-year cycle, the Earth's orbit 'stretches', affecting the amount of energy received.

These cycles of change, therefore, affect the amount of insolation reaching the Earth.

◆ **Changing oceanic circulation and land masses.** In the short term, changes in oceanic–atmospheric heat interchange cause the El Niño/La Niña effect. Variations in the operation of the North Atlantic Drift (NAD) probably helped generate polar conditions in Scotland during past glacial advances. Normally, heat is released as the warm saline waters of the NAD warm up the air streams passing eastwards to Scotland. There is evidence, however, that in the past the NAD died out somewhere off the coast of Spain with consequent cooling by up to 5 °C on average. Over a much longer time span, plate movements can shift land masses into new latitudes and push up mountain ranges, thereby altering climates.

Such physical causes of global temperature change have long been in operation. The changing intensity of the greenhouse effect is nothing new. In recent years, however, human activities are increasingly seen as important factors in global warming by contributing greenhouse gases to enhance the greenhouse effect. Figure 2.34 shows the main greenhouse gases and their relative contribution to global warming. Water vapour is the dominant greenhouse gas in the atmosphere but little of it is the result of human activities.

Human factors are really the result of growing global population and economic developments in industry, transport and agriculture, and usually include:

◆ **Increased consumption of fossil fuels.** Since the Industrial and Transport Revolutions began some 250 years ago, we have been burning carbon-based fuels (coal, oil and natural gas) at an increasing rate. Thermal power stations, factories, the ever increasing numbers of road vehicles and domestic heating have all contributed to global CO_2 levels. To the predominant emissions of the so-called developed

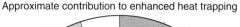

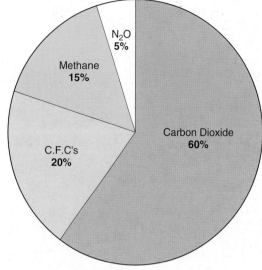

Source: *Inter-governmental Panel on Climate Change*

Fig 2.34 Greenhouse gases

countries, notably the USA (with the highest per capita emissions in the world), are increasingly added those of rapidly industrialising countries, notably China. This reflects that country's abundance of coal deposits, its rapid economic growth and less efficient power stations. Figures vary, but according to one source (NASA), the release of CO_2 from combustion is responsible for 55% of greenhouse gas build up due to human activity. Once in the atmosphere, CO_2 is chemically stable and can remain there for a long time.

◆ **Deforestation.** The clearance and burning of the tropical rainforests augments the level of atmospheric CO_2 by reducing the amount that is recycled during photosynthesis. Logging, burning and clearing tropical rainforest for arable land, pasture and plantations has increased in South East Asia and the Amazon basin. During 1997, major forest and peat swamp fires in Sumatra, Borneo and parts of Amazonia created massive smogs over extensive areas, often hundreds of kilometres away (Figure 2.35).

◆ **Increased output of methane, nitrous oxides and CFCs.** Molecule for molecule, some of these other trace gases are much more harmful than CO_2. The levels of methane naturally emitted as 'marsh gas' from decaying vegetation have risen as a result of: (i) increased rice production from methane-

Fig 2.35 A thick multi-coloured blanket of smog covers the CBD of Kuala Lumpur, January 30 1997

releasing paddy fields; (ii) ever-increasing numbers of ever-flatulent, ever-belching domestic cattle; and (iii) waste disposal sites which leak methane. Nitrous oxide (N_2O) is even more powerful. Its atmospheric concentration is rising because of car exhaust emissions, the increasing use of nitrogen-containing fertilisers and deforestation. Most potent of all are the CFCs (chlorofluorocarbons). 177 000 times more powerful than CO_2, CFCs do not exist naturally.

Found in solvents, refrigerators and foam production, they are thought to contribute some 20% of the greenhouse effect. The CFC world ban following the 1989 Montreal Protocol has been vital in attempting to slow the greenhouse effect. CFCs, unfortunately, have a long lifetime.

Consequences of the Greenhouse Effect

Many climatologists argue that the global rise in temperature over the past 100 years reflects mainly human influences. They maintain that further temperature rises will continue, and various environmental and human effects are forecast. Figure 2.36 shows some possible impacts on Scotland while the broader global picture is outlined below.

◆ Air temperatures will rise between 1 °C and 3.5 °C but with varying regional impact. Maximum warming will be most likely to occur over land in the high northern latitudes, e.g. Scotland, while some countries, e.g. China, may possibly become cooler.

◆ Equally varied will be the changes in precipitation and frequency of storms. Globally, precipitation is expected to increase overall, particularly during northern winters in the higher

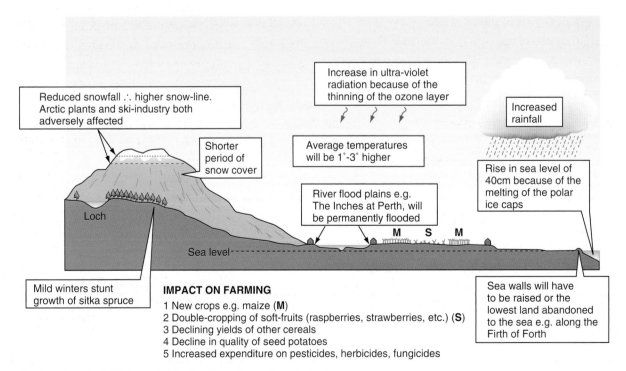

Fig 2.36 The impact of global warming: Scotland 2030 AD – a changed climate

latitudes, e.g. Scotland. Interior continental areas, e.g. the Great Plains of the USA, by contrast, will have less precipitation and, therefore, lower grain production. Some climatologists predict that global warming will increase the frequency and intensity of tropical hurricanes and tornadoes, such as the 48 'twisters' which ripped through parts of Oklahoma and Kansas in May 1999, leaving a trail of death and destruction in their wake.

◆ Global sea levels will rise. Partly, this is the result of the expansion of the sea as it gets warmer. More dramatic is the potential melting of ice resting on land, that is, glaciers and ice caps, especially in Greenland and West Antarctica. During the twentieth century, the global sea level has risen by 10–25 cm. 'Worst case scenarios' suggest sea levels rising by 3.6 m by 2100. If so, extensive areas of low-lying, densely-populated areas such as the Mekong, the Nile and the Ganges-Brahmaputra deltas will be flooded, fresh water will be contaminated, shorelines eroded and people will be forced to move. Low-lying coral islands, such as the tourist-dependent Maldives and Florida Keys are at risk, as are some of the world's major coastal cities. Venice, Bangkok, New Orleans and Tokyo are already affected by subsidence because of dropping groundwater levels. Rising sea levels will make matters worse.

◆ Tropical diseases such as malaria, dengue fever and yellow fever are already spreading beyond their traditional locations as warmer areas expand.

◆ Overall, there will be gains and losses but the poorer, developing states often located in the lower latitudes will bear the brunt of the negative results of global warming compared to the more prosperous temperate states.

Global Warming: Solutions

In 1992, at the Rio Earth Summit, 162 countries agreed to stabilise greenhouse gas concentrations in the atmosphere. Later conferences at Kyoto (1997) and The Hague (2000) further aimed to counter climatic change by urging the adoption of policies such as those shown in Figure 2.37. However, issues of policy, ethics and science are involved.

◆ Policies of cutting emissions of industrial gases, particularly CO_2, from power stations and cars have been agreed in principle by wealthier countries, hoping that the developing nations will follow. For example, the USA (responsible for 25% of global emissions from 4% of the world's population) has agreed, in theory, to cut back CO_2 and five other gases to 7% below their 1990 level by 2010. But, given the dependence of American society on the car, air conditioning and central heating, how easy is this to implement politically in the USA? Remember that the USA increased its energy consumption by 15% in the last decade of the twentieth century! One suggestion is 'carbon offsets'. By building the latest model of a 'clean' power station in India, for instance, with its rapidly developing economy, the

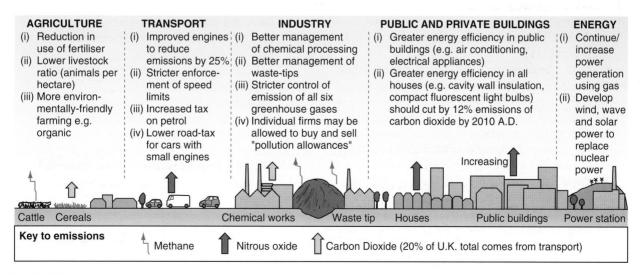

AGRICULTURE	TRANSPORT	INDUSTRY	PUBLIC AND PRIVATE BUILDINGS	ENERGY
(i) Reduction in use of fertiliser (ii) Lower livestock ratio (animals per hectare) (iii) More environmentally-friendly farming e.g. organic	(i) Improved engines to reduce emissions by 25% (ii) Stricter enforcement of speed limits (iii) Increased tax on petrol (iv) Lower road-tax for cars with small engines	(i) Better management of chemical processing (ii) Better management of waste-tips (iii) Stricter control of emission of all six greenhouse gases (iv) Individual firms may be allowed to buy and sell "pollution allowances"	(i) Greater energy efficiency in public buildings (e.g. air conditioning, electrical appliances) (ii) Greater energy efficiency in all houses (e.g. cavity wall insulation, compact fluorescent light bulbs) should cut by 12% emissions of carbon dioxide by 2010 A.D.	(i) Continue/ increase power generation using gas (ii) Develop wind, wave and solar power to replace nuclear power

Cattle   Cereals   Chemical works   Waste tip   Houses   Public buildings   Power station

Key to emissions			
⌐ Methane	⬆ Nitrous oxide	⇧ Carbon Dioxide (20% of U.K. total comes from transport)	

Fig 2.37 Global warming: tackling the problem

Americans get an emissions credit and the world benefits by reducing CO_2.

◆ Such an approach may help with the ethics of blame for emissions, past, present and future. Since the Industrial Revolution, the developed economies have not really paid for their past sins of uncontrolled emissions. Yet voices in developed countries urge others such as India and China (potentially the world's biggest producer of CO_2 by 2015) to limit the use of their own fossil fuels and slow down their rate of economic development. Is that fair? Equally, having deforested vast swathes of native forests in North America and Western Europe, is it fair that we deny debtor countries such as Brazil or Indonesia the opportunity to clear their own forests as part of their programme of economic development?

◆ Finally, there is dispute among scientists. Certainly they believe in the greenhouse effect but debate centres on the scientific data and their interpretation. How reliable are the data? For one thing, many of the land-based measurements are urban-based and may reflect an accelerating 'heat island' effect on temperature as settlements grow. In addition, satellite and balloon measurements, contrary to computer modelling predictions, show a slight cooling trend in the lower troposphere. Some scientists argue that greenhouse heating will actually lead to more cloud cover which would shield the Earth's surface and cool the planet. Likewise there are those who argue for short-term warming and a long-term 'icehouse effect'.

ASSIGNMENTS

The Atmosphere and Energy Transfer
Read pages 4 to 7

1 What is the **atmosphere**, and of which **main gases** is it composed?
2 a) With the help of Figure 2.1, **describe** the main features of the **troposphere** and the **stratosphere**, bringing out their differences.
   b) What is the **tropopause**?
3 a) **Draw** a large copy of Figure 2.2.
   b) **Explain** the terms **insolation** and **albedo**.
   c) **Fully explain** why only 51% of **insolation input** reaches the Earth's surface.
   d) **Describe and explain** the presence of the surface of the Earth's main energy **outputs**.
   e) What is the **'greenhouse effect'**, and why is it important?
   f) Write a **brief note** on the **Global Heat Budget**.
4 Study Figure 2.3. This shows that there are latitudinal variations in the **global heat** budget. Just like the world economy (certain areas rich, others poor), some areas have surplus energy, others have a deficit.
   a) Mentioning latitude, **describe** the areas of energy surplus and areas of energy deficit.
   b) **Copy** Figure 2.4. Give **FOUR** reasons explaining why there is a net gain of energy in the low latitudes and net energy loss in the high latitudes.
   c) Why does surplus energy need to be transferred between low and high latitudes? What accounts for 80% of energy transfer, and what is responsible for the remaining 20%?

Energy Transfer and the Atmosphere
Read pages 7 to 14

Ideas about the atmosphere and energy transfer have changed over the years.

Study Figures 2.5–2.8.
1 Who proposed the **single cell model** of energy transfer? Briefly **outline** its workings and its shortcomings (see Figure 2.5).
2 Hadley did not allow for a rotating earth. With the aid of Figure 2.7, **briefly explain** the effect of the **coriolis force** on surface winds.
3 a) Make a large copy of Figure 2.6.
   b) **Describe and explain** the workings of the **three-cell model** originally proposed by Ferrel. Your answer should refer to the differences between **thermally direct** and **thermally indirect** cells.
   c) **Explain** why the Ferrel three-cell model of atmospheric circulation is criticised. Briefly indicate an alternative three part model.
3 Study Figures 2.8 and 2.9.
   a) What are: **Rossby waves**, and **Jet streams**?
   b) **Explain** how they assist in the atmospheric transfer of energy.
   c) Why are jet streams useful when planning to fly around the world in a balloon?
4 Study Figure 2.10. This shows a very simplified model of global surface winds.
   a) **Describe** the three **main global wind groups** in

relation to broad latitudinal zones and belts of high and low pressure.

b) Where, and what, are **zones of convergence**, and **zones of divergence**?

5 Study Figures 2.11 and 2.12.

a) Suggest two reasons why the model of global winds does not correspond to the patterns shown in Figure 2.11.

b) Look again at Figure 2.10. Does this diagram show the global pattern of winds: **(i)** on June 21/22? **(ii)** on Dec 21/22? **(iii)** on March 21/September 23?

Explain your answer.

c) Why does the text state that the sun 'apparently "moves"'?

d) Explain, as fully as possible, the term **thermal equator** and the letters **ITCZ**. What causes the seasonal migration of the thermal equator and the ITCZ?

Energy Transfer and Ocean Currents
Read pages 14 to 19

1 From page 11, **explain** why land and sea heat up and cool at different rates.

2 a) What is an ocean current?

b) Why should care be taken over the meanings of **warm** and **cold** currents?

c) Briefly **explain** how warm and cold currents assist in the transfer of global energy.

3 Describe, with reference to named currents, the climatic effect of ocean currents.

4 a) On an outline map of Europe, **plot** the climate stations shown in Table 1. For each location **show** the average minimum and maximum temperatures. Using a bold arrow as a symbol, show the North Atlantic Drift.

b) Explain the differences in temperature from west to east.

5 Study Figures 2.14 and 2.15.

a) What is a **gyre**? What circulation patterns are followed by gyres in the northern hemisphere and the southern hemisphere?

b) With reference to the Atlantic Ocean, **outline** the main factors responsible for the circulation patterns of the gyres.

Tropical Africa: Climate
Read pages 19 to 26

1 a) On a blank map of Africa **show**:
  (i) lines of latitude 0°, 10° N, and 20° N
  (ii) the equatorial, tropical continental and hot desert climate regions

b) Briefly **outline** the broad climatic features of each of these climate regions.

2 a) Revise the following key terms: **Hadley Cell**, **thermal equator** and **ITCZ**.

b) Look at Figures 2.10 and 2.21. Fully **describe and explain** the workings of the Hadley Cells, showing the links between the cells, the trade winds, the thermal equator and the ITCZ.

3 Figure 2.21 shows the theoretical model. In reality, the air descending at the subtropics is influenced by the surface beneath. Does the air descend over land or sea?

a) Explain the terms **source area** and **air mass**.

b) Which are the two main air masses that influence the tropical areas of Africa?

c) Study Figure 2.23. **Draw up a table** to summarise the differences between **Tropical Continental** (cT) and **Tropical maritime** (mT) air masses.

d) Explain the term **relative humidity**.

4 Study Figure 2.22. Note the changing locations of the ITCZ in Africa north of the equator. Figure 2.24 gives a more detailed picture of these changing locations over West Africa.

a) Look at the situation in July. Which wind system, mT or cT, dominates West Africa?

b) Describe and explain the location of the ITCZ in July. Which season characterises West Africa at this time?

c) Look at the situation in January. Which wind system, mT or cT, dominates West Africa?

d) Describe and explain the location of the ITCZ in January. Which season characterises West Africa at this time?

e) Using the extracts, **pick out** phrases to describe the character of the wet and dry seasons.

5 a) On an outline map of West Africa, **plot** the location of Lagos, Sokoto and Timbuktu.

b) Using the data provided in Table 2, **draw** climate

Table 1

Station	Min temp	Max temp	Station	Min temp	Max temp
Valentia	7 °C	15 °C	Cambridge	3 °C	17 °C
Hamburg	0 °C	17 °C	Berlin	−1 °C	18 °C
Warsaw	−4 °C	18 °C	Kiev	−6 °C	20 °C

Table 2

	Lagos 6° 27N			Sokoto 13° 04N			Timbuktu 16° 52N		
	A	B	C	A	B	C	A	B	C
Jan	27	28	65	20	0	13	22	0	22
Feb	29	41	69	22	0	13	24	0	19
Mar	29	96	72	31	0	11	28	2	18
Apr	28	143	72	34	10	17	32	0	15
May	28	274	76	33	51	31	35	5	18
June	26	460	80	30	89	41	36	23	31
July	25	282	80	28	147	55	32	79	45
Aug	25	69	76	26	236	64	30	81	57
Sep	25	140	77	27	145	59	32	38	45
Oct	26	208	76	28	13	37	31	3	23
Nov	27	69	72	27	0	18	28	0	17
Dec	28	25	68	25	0	15	22	0	19
Year	27	1835	74	28	691	31	29	231	27

A: Average monthly temperature °C
B: Mean monthly rainfall mm
C: % Relative humidity at mid-day

graphs to show the mean monthly temperatures and mean monthly rainfall.

c) Referring to the graphs, **describe and explain** the changes in rainfall northwards mentioning total and seasonal distribution.

d) On a common frame, draw three line graphs to show the northwards change in relative humidity. Comment on the results.

6 Study Figure 2.29. It is a reminder that rainfall variability increases northwards.

a) **Explain** the variations in annual rainfall in the Sahel from 1950 to the early 1990s.

b) Identifying actual years, what problems faced pastoralists and farmers in the Sahel from 1950 to the early 1990s?

Global warming
Read pages 26 to 32

1 Study Figures 2.32–2.34.

a) **Revision.** What is meant by the greenhouse effect?

b) **Explain** the link between global warming and the greenhouse effect?

c) In rank order of importance, which are the main greenhouse gases? What are the main sources for the main greenhouse gases?

2 a) Refer to Figure 2.32. **Describe** the changes in global temperature from 1860 to the late 1990s.

b) Using a summary spider diagram, **outline** the main physical factors believed responsible for global warming?

c) Using a summary spider diagram, **outline** the main human factors believed responsible for global warming.

d) Human activity has encouraged an increased output of methane, nitrous oxides and CFCs. Why are these more worrying than the build-up of CO_2?

3 Study Figures 2.36 and 2.37.

a) With regard to either Scotland or the global scene, **discuss** the possible effects of global warming.

b) **Outline** possible methods that might reduce emissions of greenhouse gases. What are some of the difficulties involved in adopting such strategies?

Extra Assignments

1 'The three-cell model of atmospheric circulation is long past its sell-by date'. Is this a fair comment? Give reasons for your answer.

2 What are the causes of El Niño and La Niña? Select either El Niño or La Niña and describe its environmental impact. Does poverty make their impact worse?

3 With the help of Figure 2.30, **describe and explain** the changes in climate, vegetation and farming from the Gulf of Guinea to the Sahel.

4 Read page 112 about the North Atlantic Conveyor System. **Explain** the link between global warming and the possible relatively rapid onset of another Ice Age.

5 No one doubts that global warming is a fact of life. The dispute is about the causes: which are the more important factors—physical or human?

Key Terms and Concepts

- air mass (p. 22)
- albedo (p. 5)
- atmosphere (p. 4)
- coriolis force (p. 8)
- El Niño (p. 16)
- Ferrel cell (p. 9)
- greenhouse effect (p. 5)
- gyres (p. 15)
- Hadley cell (p. 8)
- heat budget (p. 5)
- intertropical convergence zone (p. 20)
- jet streams (p. 10)
- La Niña (p. 16)
- long-wave radiation (p. 5)
- North Atlantic Drift (p. 14)
- ocean current (p. 14)
- polar cell (p. 9)
- Rossby waves (p. 10)
- short-wave radiation/insolation (p. 5)
- stratosphere (p. 5)
- source area (p. 22)
- thermal equator (p. 11)
- thermally direct/indirect (p. 9)
- tropopause (p. 5)
- troposphere (p. 5)
- tropical continental (p. 22)
- tropical maritime (p. 22)
- variability of rainfall (p. 25)
- 'Wobble, Roll and Stretch' theory (p. 27)

Suggested Readings

Atkinson, BW and Gadd, A (1986) *A Modern Guide to Forecasting Weather* Mitchell Beazely.

Broadley, E and Cunningham, R (1991) *Core Themes in Geography: Physical* Oliver & Boyd.

Henderson-Sellers, A and Robinson, P (1986) *Contemporary Climatology* Longman.

Kemp, D (1990) *Global Environmental Issues: A Climatological Approach* Routledge.

Moore, P, Chaloner, B and Stott, P (1996) *Global Environmental Change* Blackwell Science.

Waugh, D (1995) *Geography: An Integrated Approach* (2nd edition) Nelson.

White, ID, Mottershead, DN and Harrison, SJ (1984) *Environmental Systems* George Allen & Unwin.

Wright, D (1983) *Meteorology* Basil Blackwell.

Internet Sources

Climate change
www.ucsusa.org/
NASA: www.nasa.gov
Professor D Rosenberg's home page:
www.middlebury.edu/~rosenber/

World weather links
www.geocities.com/SiliconValley/3452/weather1.html
World Climate: www.worldclimate.com (good for weather satellite images e.g. of the ITCZ)

General Web sites with a wide range of recommended sites.
Encyclopaedia Britannica's Eblast: www.eblast.com
BBC Education Web site: www.bbc.co.uk/education

After working through this chapter, you should be aware:

◆ that water is a basic and critical resource obtained by utilising the hydrological cycle

◆ of the key characteristics of the hydrological cycle, and how it functions

◆ of the movement of water within drainage basins

◆ that some drainage basins, e.g. the Tay basin, are more prone to flooding than others, with important consequences for river management

◆ that the Savanna regions of Africa experience marked seasonal variations in river flow with profound effects upon human activity

◆ of the effect of flowing water shaping the landforms of the upper, middle and lower sections of river basins by the processes of erosion, transportation and deposition

You should also be able to:

◆ construct and analyse flood hydrographs

◆ distinguish and explain the characteristic landform and river features from an OS map

The hydrosphere, basically the Earth's surface waters, is important for two main reasons. First, it has been suggested that, if states went to war over oil in the twentieth century, water might be a bone of contention in the twenty-first (Figure 3.1). For water is an unevenly distributed yet critical resource. Too often it is taken for granted, particularly by countries with more than their fair share of the hydrological cycle (see below). Second, even peaceful streams and rivers have the power to shape much of the Earth's surface through their ability to erode, transport and deposit. The energy potential of those same streams and rivers, however, can be transformed during floods with at times dramatic consequences for life and landscape. Water is both life-giving and life-threatening.

Water and Life: An Extract

The use of rivers and their valleys is never straightforward, in spite of experiences dating back to earlier river-based civilizations such as Egypt and Mesopotamia. Effective river management can provide a water supply for domestic and industrial purposes, for agriculture, for power, for waste disposal and for recreation. **Hydrology**, the scientific study of water at the Earth's surface and its links with the atmosphere, is a key aspect of geography that impinges on many aspects of life and landscape.

Danger Zone: States where water withdrawal will be more than 40% of total water available

Source: *GEO 2000. U.N. Global Environment Outlook*

Fig 3.1 Water wars of the future? 2025 AD

Some idea of the significance of water can be illustrated from the following extracts. These describe features of climate, hydrology and farming in southern Spain, near Granada, as seen by Chris Stewart, former drummer with the group *Genesis*, who now lives and farms in that water-deficient region.

During our first year at El Valero the weather had been more or less predictable. The summers were hot and the winters were mild. Although a feeling of nervous

anticipation would set in when we contemplated the onset of the fierce summer heat, we were surprised when it actually happened by how well we adapted to it ... Winter weather was comfortable, cool and sunny, though with not quite enough rain to keep the flora of the hills in good fettle. Even during our own short time here we had noticed that the winters had seemed to become just slightly drier.

The river ran along easily and inoffensively through winter and summer alike, swelling briefly as the June heat melted the mountain snow, then returning to its lazy summer level. The rain and the river muddled along in their own way, apparently reluctant to give us any trouble, until the summer when we had our first taste of serious drought.

Almost no snow had fallen that winter on the mountains, and the spring rains fell feebly and dried up with a spate of hot winds coming up from the Sahara. By June the river was no more than a few brackish puddles among the boulders, and then in July, for the first time in living memory, the trickle of water in the Cadiar river stopped altogether.

Dead fish lay rotting in the dry pools and the paths of the valley were ankle deep in hot dust. The grass in the fields at El Valero withered to brown and crackled beneath our feet, and the leaves of the trees shrivelled and curled.

It's the Greenhouse Effect, said some ... the hole in the ozone layer ... El Niño ... an unfortunate alignment of planets. The old men shook their heads and predicted dark times to come. The drought affected the whole of Andalucia and most of Spain. Rivers and springs dried up all over the province; wells were down to the salty sludge at the bottom; whole forests of trees, even the hardy Alleppo pines, withered and died. Orgiva was limited to an hour of water a day, and there were bush fires breaking out right across Spain. Ana and I felt somehow let down by the river.

Then in mid-September it rained. A few heavy drops fell, sporadically at first, each one making a little crater in the dust. Little by little the drops coalesced into a steady drizzle ... It rained lightly for three days, enough to settle the dust and build up the flow of the river, and then it stopped.

September moved into October with no more rain, though something kept the river going. And then in November

the downfall began, not with a deluge, just a nice steady downpour that kept on coming day and night, day and night. By the morning of the second day there was a terrifying flow of dark water racing down from the gorge. Effortlessly it swept the bridge out of the way, pulverising the stone piers and sweeping the beams far down the river. And with each passing hour it rose still more, bringing with it boulders the size of small buildings thundering like cannon as they moved through the awful tumult.

The days of rain became weeks and our roof started to leak, the solar power died, and all the firewood was so soaked it was useless. The river thundered on, filling the valley with a sense of foreboding. As the earth became saturated with water, the hills began to crumble into the valleys. We would hear a roar and watch as hundreds of tonnes of sodden earth avalanched down the mountainside, bringing trees and bushes along with it ... I had never imagined such awesome erosion; the mountains were literally being swept down to the sea.

Source: *Driving over Lemons* by Chris Stewart.

3.1 The Global Hydrological Cycle

Perhaps the most important of all nature's cycles is the **hydrological (or water) cycle**. It explains the global distribution and movement of water and, unlike the geological cycle, it can be completed much more quickly. Essentially, the model involves a continual circulation of water (as a liquid, solid or vapour) between the oceans, the atmosphere, vegetation and the land (Figure 3.2).

Solar energy powers this cycle. By heating the oceans, lakes, rivers and the Earth's surface, water is **evaporated** into the atmosphere. Most of this water vapour comes from the oceans, especially from the tropical waters. Added to this is water vapour **transpired**, that is, lost through the leaves of plants which have drawn water from the soil.

All this moisture-laden air is transferred by the global winds. Unable to retain all the water vapour, **condensation** results in **precipitation** as rain or snow, according to altitude.

Much of this precipitation falls as rain over the oceans and represents a quick completion of the

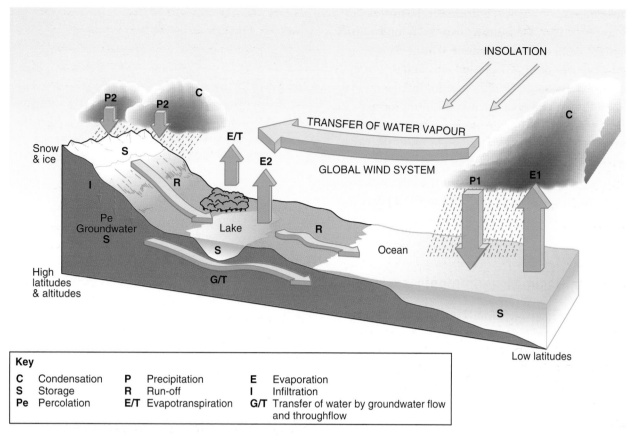

Fig 3.2 The global hydrological cycle

cycle for some water molecules. Over high latitudes and mountains, however, the input of snow into glaciers and icecaps may mean thousands of years of **storage** for water molecules until they are finally released during warmer weather.

In a variety of ways, water makes its way overland back to the oceans. First, it **infiltrates** the soil and, if the soil is on a slope, the water slowly moves downslope as **throughflow**. If the underlying rock is permeable, the water **percolates** downwards, filling it until it is saturated. Such saturated rock is called the **groundwater zone** and acts as a store. Like a sponge, the saturated rock slowly releases the water, allowing it to flow back to the sea in springs and rivers.

Second, **surface water** or **runoff**, though faster, is really quite insignificant in the grand scale of things. Major rivers, such as the Amazon, Chang Jiang or Mississippi, however important locally, are just minor trickles within the never-ending motion of the hydrological cycle.

There have been changes throughout the Earth's history in the distribution of water in the system. During the ice ages, for example, sea level was lower because more water was locked in ice caps. The hydrological cycle is a good example of a **closed system**: the total amount of water is the same, with virtually no water added to or lost from the cycle. Water just moves from one storage type to another.

3.2 The Drainage Basin

Defining the Drainage Basin

In examining the hydrological cycle, we have seen that rivers are the critical link between atmosphere and oceans. Rivers and their tributary streams form networks, not unlike a giant natural guttering system capable of transporting water and eroded material (**load**), sometimes at high speed and in vast volumes, from land to sea. The area drained by such a guttering system is called a **drainage basin** (or catchment). Figure 3.3 shows that the boundary of a drainage basin is the **watershed**. This is the perimeter rim of higher ground separating neighbouring basins. Within a basin, there is an ever-diminishing number of catchments, from those of the smallest streams draining into larger streams draining into ever larger tributary rivers right up to the main or **trunk river**. Basin size varies markedly: the Amazon is the world's largest basin, while, at the other end of the scale, is the basin of the Manor Water, somewhat miniscule in comparison (see Chapter 1).

Drainage Basin Measurements

The size of a drainage basin is not so important as the amount or volume of water that flows through it. Think of the question, Which is Britain's biggest river? The Thames is one answer but the Tay, with Britain's third largest drainage basin, produces far more water than the Thames. This is a reflection of its higher **drainage density** which measures how frequently streams and rivers occur on the surface of the land. Given the importance of flow measurement for river management and flood control, many quantitative measures are used to allow comparisons

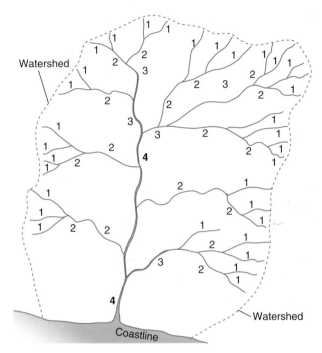

Fig 3.4 Stream ordering: an imaginary 4th order drainage basin

between basins. Two elementary but useful measurements are:

1 Stream ordering (after Strahler) analyses the actual network of streams. Figure 3.4 shows a large number of unbranched tributaries designated as first-order streams. Where two first-orders meet they form a second-order and so on until the main or trunk river is reached. Usually, the drainage basin as a whole is ranked according to the number of the highest order stream that flows in it. Major world rivers such as the Mississippi attain tenth-order status, a high figure for Britain is a fourth-order.

2

$$\text{Drainage density} = \frac{\text{total length of all streams within the drainage basin (km)}}{\text{area of the basin (km}^2\text{)}}$$

This is probably the most significant measurement that can be made. As with all such comparative techniques, care must be taken that definitions and measurements are consistent between basins. For example, when making such a calculation, maps of the same scale and, preferably, of the same series must be used.

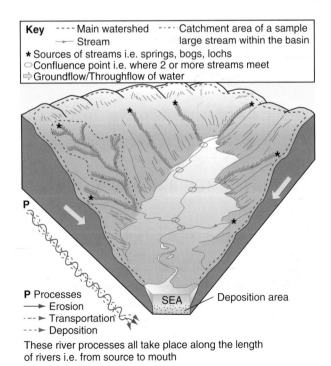

Key
- - - - Main watershed - - - - Catchment area of a sample
→ Stream   large stream within the basin
★ Sources of streams i.e. springs, bogs, lochs
○ Confluence point i.e. where 2 or more streams meet
⇨ Groundflow/Throughflow of water

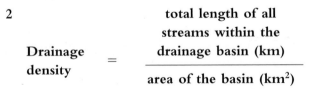

P Processes
→ Erosion
- - ▶ Transportation
- - - ▶ Deposition

These river processes all take place along the length of rivers i.e. from source to mouth

Fig 3.3 Main features of a drainage basin

In Britain, drainage density figures range from 2–4 km per km². The higher density is found where precipitation is heavy and constant, slopes are steep, rocks are impermeable (i.e. 'waterproof') and vegetation is sparse. Higher drainage density also means that there is a greater likelihood of flooding taking place.

The Drainage Basin as an Open System

Drainage basins are part of the global hydrological cycle. As already mentioned, the cycle is a **closed system** dealing with a fixed amount of water, rather like a household central heating system. On the other hand, each drainage basin is an **open system**, with independent inputs and outputs, as in a household domestic water supply – an **input** of clean water from a reservoir is **stored** in a tank, **transferred** internally by pipes, and then exits as **output** through the drains.

Applying these ideas to a drainage basin, therefore, the following elements can be identified:

inputs of water from the precipitation that falls within the catchment; water **storage** in, for example, lochs, in the soil and in saturated rocks;

transfer of water by several processes, such as surface runoff, throughflow and groundwater flow, before the final release by various **outputs** that include rivers flowing into the sea and evapotranspiration.

Each of these components is interlinked, the output from one serving as input to another. Figure 3.5 highlights the main processes discussed below.

1   Precipitation is the main input and hydrologists are interested not just in the total amount but in the type (snow, rainfall, sleet, etc.), the intensity (from gentle drizzle to intense convectional downpour), the duration and the frequency.

2   Evapotranspiration includes (i) the direct **evaporation** of moisture from the soil and from different water surfaces including pools, ponds, rivers and lochs, and (ii) **transpiration**, by which plant leaves lose water vapour to the extent that roots may draw up all the water available to them.

3   Interception is the ability of vegetation to affect the amount of precipitation reaching the ground. Dense foliage with large leaves is capable of temporarily holding the rain, allowing some to

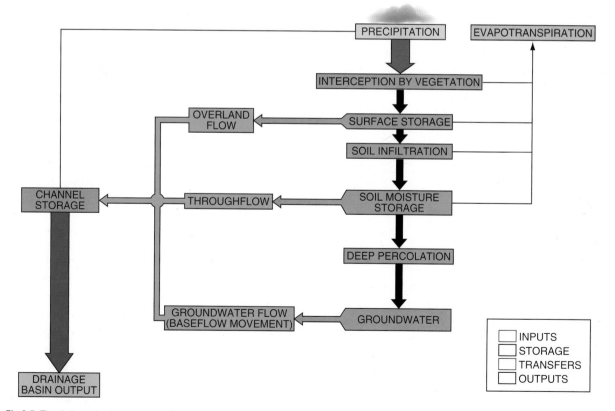

Fig 3.5 The drainage basin: a systems diagram

evaporate, then drip directly to the ground as **throughfall** or run indirectly via branches and trunk as **stemflow**. The effectiveness of interception is best appreciated once vegetation is cleared. Just think of the links between deforestation of tropical forests and consequent soil erosion. In Britain, winter leaf fall markedly reduces the ability of deciduous trees to intercept precipitation. Interestingly, even in summer, coniferous trees are more effective at interception because drops of water find it easier to cling to needles.

4   **Infiltration** is the movement of precipitation into the soil. From there, precipitation may percolate down to the water table or below. How much water actually moves into and is stored by the soil and rocks depends on their porosity and permeability. Porous soil (e.g. sand) and rock (e.g. chalk) permit greater infiltration compared to relatively impermeable soil (e.g. clay) and rock (e.g. gneiss) (Figure 3.6). The infiltration rate is also influenced by how much water is already in the soil's pore spaces, and whether or not the soil has been compacted by machinery or livestock.

The water that has infiltrated soil and rock is eventually released into rivers, often much later. This subsurface transfer involves:

◆   **throughflow**, the downslope movement of water through the lower soil or subsoil towards streams and rivers. Throughflow plays an important role in the redistribution of precipitation into river systems.

◆   **groundwater**, which is stored beneath the water table in the pores, joints and bedding planes of permeable rocks resting above an impermeable layer. It is stored at depth as a result of **percolation**, which is the downward movement of water. Storage is long term and any groundwater or **baseflow movement** below the water table is very slow, often lagging behind precipitation by weeks, months and even years! Groundwater plays a critical long-term supply role by sustaining valley springs and river flow during long dry periods.

5   **Runoff** is the sum of all the rainwater that flows over the surface of the basin. It takes two forms: **streamflow** is the water flowing through permanent

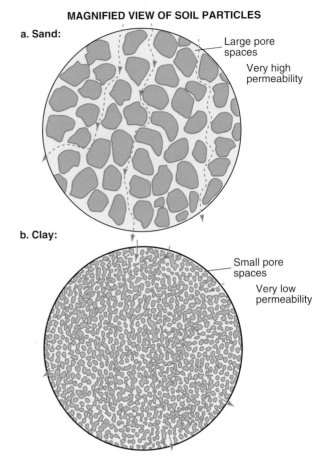

MAGNIFIED VIEW OF SOIL PARTICLES

a. Sand:

Large pore spaces

Very high permeability

b. Clay:

Small pore spaces

Very low permeability

Fig 3.6 Comparing the permeability of clay and sand

stream and river channels; **overland flow** (sometimes called **surface runoff**) can take the form of either (i) sheetwash, that is sheets of water flowing across the ground, particularly if its surface cannot be infiltrated or (ii) water following lots of tiny channels called rills, only a few centimetres deep. Either way, overland flow very quickly transfers water to the stream and river channels whose rapidly rising waters are raised a little more by direct rainfall known as **channel precipitation**. The water is finally discharged to sea, mainly by the trunk river (aided by groundwater seepage) and is lost to the local basin system to be incorporated into the global hydrological cycle.

Figure 3.7 summarises the different routes taken by water within a drainage basin. It is a complex system, made even more so by human activities such as urban development, deforestation and the ever increasing use of water for domestic, industrial, power and agricultural purposes.

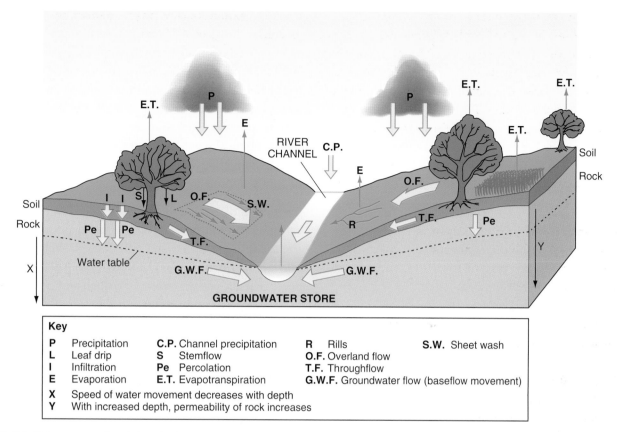

Fig 3.7 Water movement in a drainage basin

3.3 Flows and Flood (Storm) Hydrographs

Rivers behave in different ways depending on the pattern of precipitation and the size and character of the drainage basin. Nothing is equal in nature and this is reflected in differences even in adjacent drainage basins and their discharge patterns. Compare two basins: in one of them, the rivers are very placid because their basin has gentle slopes, permeable rocks, deep soil and is extensively forested; in the other, the rivers are 'flashy' (water levels rise and fall very rapidly) reflecting steep slopes, impermeable rocks and limited cultivation – all factors which discourage infiltration of any heavy rainfall, therefore promoting rapid runoff and a potential flood hazard.

Such differences in **discharge** (the velocity and speed of rivers) needs to be carefully measured as part of river basin management. At various gauging stations, discharge data are collected and used to construct short-period (storm) **hydrographs**. Not

only do these show how rivers respond to rainstorms but they are also a fingerprint of the physical and human geography of differing drainage basins. Figure 3.8 shows a model hydrograph and you should pick out the following features:

◆ while there is a link between rainfall and river flow, river levels do not immediately increase after rain. The gap between maximum rainfall and the river reaching its maximum level (**peak discharge**) is called **lag time**. Although there is some direct channel precipitation, most takes time to feed through as surface runoff and throughflow. The speed can range from minutes to days depending on the size and character of the basin.

◆ the **baseflow** is a fairly constant feature responding only slowly to rainfall change and has a very slow time lag.

◆ the steepness of the **rising limb** reflects how rapidly the water reaches the river channels. Particularly rapid overland flow results in a very steep rising limb.

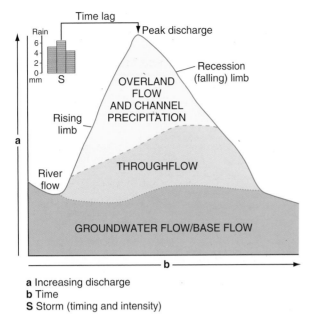

a **Increasing discharge**
b **Time**
S **Storm (timing and intensity)**

Fig 3.8 A model storm hydrograph

trunk river more rapidly. Shape, though not so reliable an indicator of flood potential, suggests that an elongated basin will produce a lower peak flow and longer lag time than a circular one of the same size. Essentially, water from the far end of an elongated basin takes longer to reach the gauging station compared to a rounder basin with easier access. Slope also acts as a control: other things being equal, channel flow is faster down a steep slope than in gently sloping areas.

2   Types of rock also affects the speed of runoff and groundwater flow depending on its permeability. **Permeable** rocks let water pass through them and may be: (i) **porous** (e.g. chalk), with innumerable pores that allow water to sink through and be stored. (ii) **pervious** (e.g. carboniferous limestone), allowing water to flow down joints and along bedding planes.

◆ the **recession** or **falling limb**, however, is less steep. The shape of this limb reflects the amount of water in the basin and the decreasing rate of discharge.

◆ once the rising water reaches the top of its channel, it attains what is called **bankful discharge level**. This is critical because flooding occurs if the channel water rises above this level.

◆ after the storm runoff has ended, baseflow again becomes the main source of water in the channel.

Factors Influencing Storm Hydrograph Shape

The size and shape of a hydrograph can be related to two broad groups of interrelated factors: those to do with the character of the drainage basin and those, of a more changeable nature, associated with climatic and tidal conditions (Figure 3.9).

Drainage Basin Controls

1   Area, shape and slope influence hydrograph shape. Large basins receive more precipitation than small basins, resulting in larger runoff. Larger size, however, means a longer lag time compared to smaller basins where precipitation will reach the

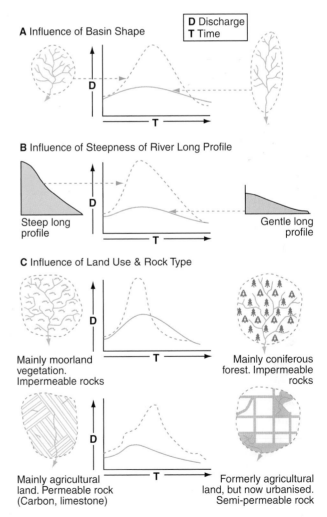

Fig 3.9 Model hydrograph shapes and drainage basin features

Permeable rocks mean rapid infiltration and little surface drainage. **Impermeable** rocks (e.g. gneiss, granite), on the other hand, are effectively 'watertight'. They do not allow water to infiltrate and encourage rapid surface runoff and a high stream density.

(NB: at considerable depths, overlying weight crushes normally permeable rocks and reduces their permeability)

3 Soil influences the rate of infiltration and the speed of throughflow because of its thickness and porosity. Infiltration is greater on thick soils and less on very thin, skeletal soils. Clay, with its smaller pore spaces, is impermeable, supporting surface runoff and potential flooding, whereas sandy soils are permeable and discourage flooding.

4 Land use also affects river discharge rates and reflects:

◆ the extent of **urbanisation**. Concrete and tarmac form impermeable surfaces and heavy rainfall and rapid runoff can quickly overwhelm the capacity of guttering and drains to cope.

◆ the degree of **afforestation**. Areas of mature forest are able to reduce the impact of excessive precipitation, thanks to interception by the canopy, the delaying action of roots and transpiration. Deforested areas and basins with extensive moorland or grassland cover are not so effective at delaying rapid runoff. This explains why afforestation schemes play an important role as part of flood prevention measures.

◆ **Abandonment of arable land**, normally able to help in interception, for urban growth and recreational land use has increased the potential flood hazard.

5 Drainage density is lower on permeable rocks and soils, higher on impermeable rocks and soils. A higher density network, sometimes enlarged by the installation of field drains, allows rapid runoff. Lower density means fewer streams to channel water flow.

Climatic and Tidal Controls

1 Precipitation and temperature are critical controls, often interrelated. A thick cover of snow effectively acts as an extensive store but any rapid thaw will quickly release meltwater. The length and intensity of rainstorms affect how much water forms runoff or throughflow. Intense, short-lived conventional rainfall during a hot summer, with resulting hard baked ground, can produce rapid surface runoff and sudden flash floods. Less intensive rain, over a prolonged period of cooler weather, makes infiltration so much easier and is reflected in a gentler rising limb.

2 Tidal conditions can also be influential. High spring tides, extending quite far inland (see River Tay case study), can pond back water in the flood plain, blocking its normal exit out to sea.

CASE STUDY 1
The River Niger and the Savanna Regions
The Niger Basin and contrasting regimes

Fig 3.10 Aerial view of a stretch of the River Niger and the surrounding savanna landscape

The Niger is the world's seventh longest river (Figure 3.10). With its main tributary, the Benue, it forms an extensive drainage basin that stretches across West Africa's savanna and tropical rain forest zones (Figure 3.11). These distinctive bioclimatic zones were discussed earlier (page 20). We saw that, as a result of the seasonal migration of the ITCZ, life and landscape in the savanna regions are influenced by:

◆ the changes from the wet season (dominated by tropical maritime air) to the dry season (dominated by tropical continental air).

◆ increasingly variable rainfall northwards towards the Sahel savanna.

In the Niger basin, therefore, river discharge responds to these seasonal and variable inputs of the hydrological cycle. Consequently, rivers such as the Niger, the Benue, and their many smaller tributaries display marked variations in flow between the seasons. They swell in size during the wet

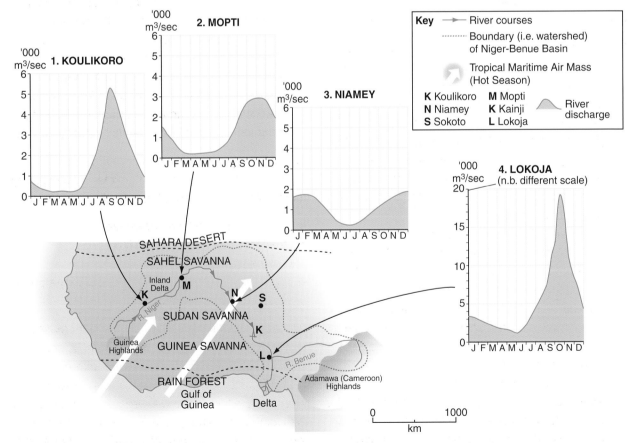

Fig 3.11 Contrasts in mean monthly discharges along the Niger

season and they shrink, often dramatically, during the dry season, with many consequences for the inhabitants of the adjacent areas.

Fluctuations in discharge are shown by a river's **regime**. Regimes may represent a particular year or, using data collected over a long period, an average year. While the flood hydrograph shows short-term conditions, regimes display seasonal variations. The actual extent of these variations depends on the size of the catchment area, the amount of precipitation within it, temperatures, evapotranspiration rates, and how much water is derived from direct runoff and groundwater discharge. Groundwater discharge is especially important during the intense dry season in the savanna of West Africa because it is the only source of river water. Such water comes from storage in (i) various rocks and (ii) the alluvial deposits of flood plains. Groundwater may be several decades old but it plays a critical role in reducing the fluctuations between seasons and years.

Along its various sections, the Niger demonstrates interesting but complicated changes in discharge during its 4080 km journey from source to sea.

1   Figure 3.11 shows that its source rises in the Guinea Highlands, only some 300 km from the ocean. These highland areas of hard crystalline rock and steep gradients experience, on average, an annual precipitation of over 1500 mm, mainly falling from April to June with resulting rapid runoff. By May, river levels are rapidly rising but it takes many months for flood water to reach the lower stretches, and then only in a reduced form.

2   The river's long middle section consists of a large, 2000 km long loop as the river flows first north east then south east through Mali and Niger. Here the valley is broad, its gradient is low, there are few tributaries – most, such as the Bani comes from the south (why?) – but it has a distinctive inland delta with a complex maze of distributaries, shallow lakes, marshes, swamps and man-made channels. Effectively, the inland delta acts as a store, delaying the flood water for almost four months and reducing the amount of discharge. This is because much of the flood

water infiltrates the extensive alluvial plain. Also, evaporation and evapotranspiration rates are high: the river flows sluggishly, allowing water more time to evaporate; the surface area of the water (distributaries, lakes, marshes, etc.) widens and is open to the sun. On top of this natural loss, water is extracted for irrigation purposes (by the Sansanding barrage, for example).

3   After the inland delta, the Niger forms one channel and the gradient steepens. In spite of the delay and reduction in volume, Niger floodwater eventually enters Lake Kainji by late January. This is Nigeria's largest artificial lake, ponded back by the Kainji Dam as part of a major multi-purpose scheme for hydropower, flood control and expansion of irrigated agriculture, improved navigation and fishing. Retaining the flood surplus allows water to be released in a controlled fashion during the dry season.

4   By Lokoja, the river has broadened as it receives more water from the Benue and its tributaries. With its source in the mountains of Cameroon, the Benue doubles the flow of the Niger and ensures seasonal flooding.

5   Finally, some 2000 km south east of its source, the river enters its main delta, with many distributaries, mangrove swamps and coastal creeks. Here the climate is equatorial with high temperatures and heavy rainfall virtually all year.

Figure 3.11 shows sample river regimes for Koulikoro, Mopti, Niamey and Lokoja. By comparing them, it can be seen that from Koulikoro to Niamey:

◆   the peak flow reduces downriver
◆   the length of the peak flow increases downstream
◆   peak flow changes from a wet season feature to a part of the dry season feature.

At Lakoja, the discharge has markedly increased and there is a distinct peak once again. These characteristics can be related to the different stretches of the river. The regime at Koulikoro reflects the effect of the wet season rapid runoff from the Guinea Highlands; at Mopti and Niamey, the regimes relate to the retarding and reducing effects of the inland delta as well as the low precipitation inputs; the regime at Lakoja, however, demonstrates the impact of the wet season in the highlands of Cameroon. In all these regimes, the effect of the time lag between the onset of the rains and the increase in discharge is apparent, and also the importance of groundwater flow during the dry season.

Climatic Variability and Water Utilisation

Throughout the savanna regions of West Africa, seasonal variations in high and low water, and the impact of abnormally prolonged drought on water supply, affect the work patterns and lifestyle of a variety of people – farmers, fishermen and pastoralists.

Farmers throughout the savanna adopt different strategies to deal with water shortages.

◆ In the Sudan savanna, with average annual rainfall ranging from 650 mm–1000 mm, rain-fed agriculture is supplemented by dry season irrigation of flood plains. Some of the features of such 'Fadama' cultivation can be seen in Figure 3.12 which also shows the importance of wells in exploiting water resources. The diagram further shows that intercropping is carried out. Interplanting up to four crops between each other

is a long established farming practice in the Savanna. Not only does it increase output and reduce the chance of disease, intercropping allows farmers to reduce risks resulting from variability of rainfall. A quick growing variety of pearl millet can be resown if the rains fail.

◆ In the drier Sahel savanna, however, the presence of extensive, seasonally large rivers, such as the Niger, means that rain-fed cultivation is not so important as the farming of flooded areas. In the inland delta, **décrue** (French for 'after the flood') agriculture is very sophisticated. Again, local cultivators use not just different crops such as rice, pearl millet, sorghum and guinea corn but many different varieties of each. In this way they skilfully respond to variations in flood height and soil moisture conditions. Figure 3.13 shows the relationship between rainfall, water regime and the various crops. In this complex area of channels, ponds, marshes and lakes, farmers use the rising floods (August onwards) and the moist land exposed as levels recede (February onwards).

Rivers and lakes in the savanna are also important sources of fish. Figure 3.13 shows how fishing activities respond to the Niger regime. Low water and the beginning of the flood period are quiet times for the fishermen. Once levels rise though, the fish spread out over the nutrient-rich, flooded areas, then

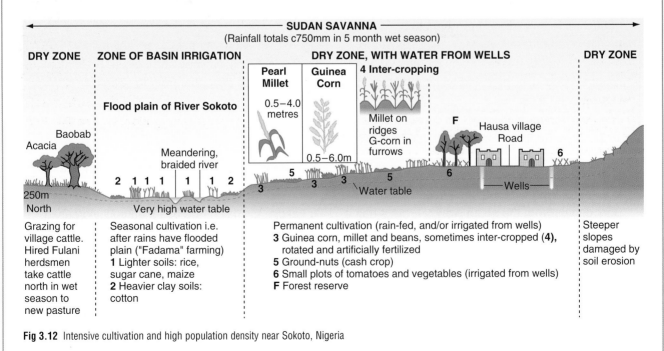

Fig 3.12 Intensive cultivation and high population density near Sokoto, Nigeria

return to the river channels when water levels drop. This is the busiest period. Severe drought, however, can drastically reduce the fish catch because of the loss of breeding and feeding grounds after lower flood levels. Pastoralists and farmers, also affected by drought, can make matters worse by their own fishing activities which further deplete stocks.

Pastoralists, such as the Fulani, move their herds of cattle, sheep and goats into the inland delta and flood plains during the dry months as the waters recede, exposing nutritionally rich 'borgou' grasses and herbs. These herds help manure the farmers' fields, and milk is exchanged for fish and cereals. When the flood waters start to rise again and tsetse flies become active, the Fulani return with their stock along well trodden transhumance routes to the traditional wet season grazing grounds (Figure 3.14).

In these ways, therefore, rainfall variability influences water table levels, seasonal flow of rivers, and the levels of lakes throughout the savanna regions of West Africa. The long established skills with which farming communities successfully manage the risks of rainfall variability are a critical part of water utilisation in the savanna. Such skills include using a wide range of crop species – up to 7 types of pearl millet and 18 types of sorghum. It is all too easy to underestimate indigenous knowledge and skills in utilising seasonally scarce and variable water resources of the African savanna.

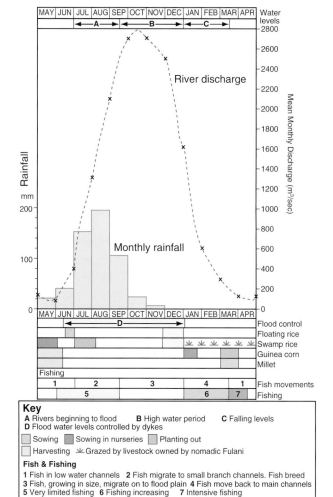

Fig 3.13 Mopti: water conditions, farming and fishing

Fig 3.14 Fulani pastoralists in the inland delta region of the Niger River

CASE STUDY 2:
Flooding in the Tay and Earn Basins

Fig 3.15 Rescue services at work in the North Muirton District of Perth during the January 1993 flood

Account from the year 1210: 'So violent was the torrent that the whole town was undermined, the houses levelled, and many persons of both sexes lost their lives. The Royal Palace did not escape. The King's (William the Lion) youngest son, John, with his nurse were carried down the river and drowned, with about fourteen of the King's domesticks'.

Source: H. Boece

Even in normal times, the River Tay has a stronger flow than the Severn and Thames put together and carries enormous amounts of water to the sea. It is a river which, along with its significant tributaries (the Almond, Braan, Tummel, Isla, Lyon and Garry) and neighbouring river, the Earn, has a long history of flooding, most recently experienced in 1993. Not for the first time, the river was uncontrollable. Rural communities were cut off, farmland was extensively damaged at a cost of over £12 million, and hundreds had to evacuate their flooded homes in low-lying parts of Perth, particularly the North Muirton District (Figure 3.15).

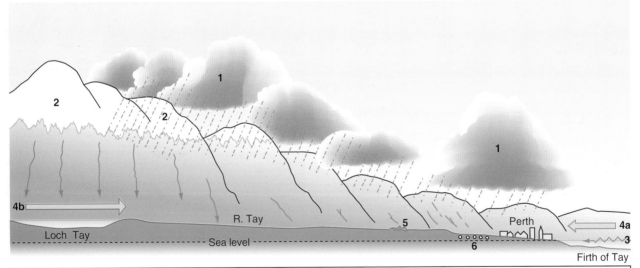

Causes		
1	**Heavy Rain**	Throughout the Tay Basin over a prolonged period
2	**Melting Snow**	A rapid snow melt, especially if the fall has been heavy, can result in the equivalent of several days' or even weeks' rainfall draining into the river system
3	**Tides**	If high spring tides coincide with the river running at its peak flow, the tides act like a dam holding back the river current
4	**Wind** a:	Easterly winds at Perth can help to raise the river level
	b:	A strong South-West wind blowing down the length of Loch Tay can force water to the lowest end at Kenmore. This extra water can cause the river to rise by several metres
5	**Ice Floes**	As a frozen river begins to thaw, ice floes can dam up the water, and when they in turn melt, a surge of water occurs
6	**New Drainage**	When old field drainage systems are improved, water is discharged into rivers much more quickly

Fig 3.16 The causes of flooding on the River Tay

General Causes of Flooding in the Tay Basin

Perth is a settlement whose inhabitants have long had to live with flood hazards. Such extreme flood conditions are perfectly natural and to be expected. Historical records (see the extract for 1210) describe their impact and help explain the various causes of flooding. These are summarised in Figure 3.16. A few general points are worth stressing:

◆ Historically, severest floods were the result of rapid snowmelt (caused by sudden rises in temperatures) often combined with heavy rainfall. Such snowmelt floods were frequent during the later stages of the Little Ice Age (about 1500–1800) with its long cold winters.

◆ Tributaries are important. Because of the sheer size of the Tay basin and the length of time that it takes water from the upper parts to travel to the lower basin, the water in the lower tributaries often rises first. The 1847 Tay flood, for example, was the result of very heavy regional rainfall in the catchments of the Isla and the Ericht.

◆ To date, the worst Perth flood was in February 1814 (7 m at Smeaton's Bridge in Perth). It was caused by ice floes from the upper Tay and its tributaries blocking the arches of the bridge, effectively damming the river.

Causes of Flooding in 1993

Figure 3.17 shows the low lying areas throughout the Tay and Earn basins that were flooded in 1993. In spite of a 24 hour early warning system, installed after floods in 1990, flooding of the extent and impact of 1993 was not expected. In the inevitable post mortems, various causes were reviewed.

The critical factor was the extreme meteorological conditions over such extensive drainage basins. The Tay basin alone covers 6475 km^2 and stretches from its source on Ben Lui some 160 km to where it enters the North Sea. Conditions were exceptionally wet in the first 18 days of January throughout the Tay and Earn catchments. Of particular importance was a very rapid thaw of unconsolidated snow where temperatures quickly rose to unseasonally high

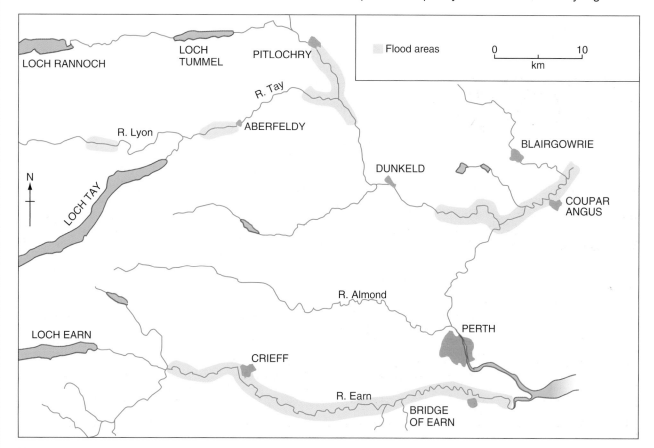

Fig 3.17 Map of areas affected by the 1993 Tay floods

levels. Additionally, much of the lower ground was saturated and runoff was pronounced. Heavy rain, rapid snow melt and a large increase in temperature combined to give weather conditions that could not have been predicted.

The importance of other possible contributory factors was also discussed, such as the flood-intensifying effects of human modifications to the drainage basin. The sample newspaper extracts give an idea of differing views about the role of farming practices and afforestation.

The effectiveness of dams was also debated. Although Scottish Hydro-Electric were accused of not doing enough to retain water behind the dams of the Breadalbane and Tummel-Gary schemes, none of the main reservoirs ever spilled water during the flood period, indeed they helped to slow the discharge.

The 1993 Flood: Timing and Effect

The timetable of events shows that by 15th and 16th January, serious flooding was affecting farmland, villages and communication links throughout the basin. By 18th January, the water level at Perth was peaking, and thundering past at a flow rate of 2060 cumecs (metres^3/second) compared to a yearly

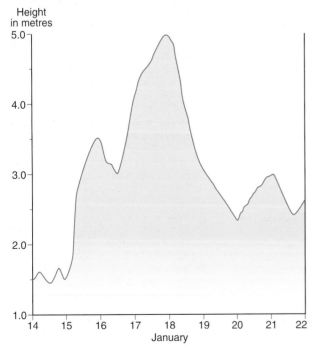

Fig 3.18 River Tay at Perth – changes in water level 14–22 January 1993

Flood prevention in Scotland

From Mr Andy Wightman and others

Sir: Millions of pounds are going to be needed to clear up the mess and pay for the insurance claims after the recent floods in Tayside. Many people must wonder about the cause of this recurrent problem which now affects many river systems in Scotland.

In most countries of the world, watershed managers recognise that the removal of natural forest cover and overgrazing leads to soil erosion, rapid water run-off, siltation and flooding. Scotland is also suffering. Overgrazing by 9 million sheep and 300,000 red deer along with forestry ploughing, draining and clearfelling are critically damaging our upland areas. The consequences are felt by large numbers of people who live and work in the lowlands.

The public is paying twice for this absurd situation. We subsidise most of the land-use practices that cause the flooding and then pick up the bill for the resulting damage. We then spend even more money on expensive engineering solutions in the lowlands in futile efforts to contain the flood waters. Instead of tackling the symptoms of flooding, it is time that the Government examined the origin of the problem and sorted out land-use policy in the uplands. Ecological solutions are needed rather than grandiose engineering schemes.

Overgrazing and the use of intensive afforestation techniques need to be curtailed. Upland vegetation and native woodlands should be restored to many of our upland areas, especially along burn and river courses to increase retention of rain and melt-water. Improved grass and moorland management techniques are also vital in reducing water run-off rates. All our flood-prone catchments are in urgent need of integrated watershed management schemes to bring about the necessary land-use changes.

The Overseas Development Administration is very good at dispatching British experts to developing countries to advise on watershed management. Perhaps the Scottish Office would be well advised to request their assistance in Scotland to help solve our river flooding problems.

Yours sincerely,
ANDY WIGHTMAN, Reforesting Scotland: DAVE MORRIS, Ramblers' Association: MARTIN MATHERS, WWF UK (World Wide Fund for Nature); DAVID MINNS, RSPB: MICHAEL SCOTT, Plantlife; ALASTAIR SOMMERVILLE, Scottish Wildlife Trust

Reforesting Scotland
Edinburgh

Land use not cause of large floods

Sir, – Andy Wightman and his colleagues (Points of View, 23 January) may be right to draw attention to the need for a change in land use in the Scottish uplands. They are wrong, however, to suggest that present land use in the uplands is the root cause of what they perceive to be Scotland's recurrent flood problems.

Fairly frequent flooding is a feature of natural river behaviour worldwide, regardless of catchment land use. In Scotland such floods are caused by heavy rain and/or melting snow. Most of these floods are small.

From time to time, however, extreme weather conditions cause much bigger events. As a result, most Scottish rivers can be expected to experience a peak flood discharge twice the capacity of the natural channel about once every 20 to 30 years on average.

Catastrophic floods may have discharges five or six times channel capacity. When events like this occur in big rivers like the Tay the volume of floodwater amounts to hundreds of millions of cubic metres.

The role of land use in flood generation is a complicated one. A superficial examination of research findings would suggest that the size and frequency of occurrence of floods can indeed by increased significantly by deforestation, forest ploughing and land drainage, and decreased by land use practices that retard the flow of water towards rivers.

More detailed study, however, shows that these changes apply only to the smaller floods. Large floods are simply too big to be influenced by land use, and in Scotland it is these floods that cause the problems.

Land use change in the uplands cannot solve Scotland's flood problems. Land use change in the flood-prone areas, on the other hand, is another matter.

Most of the problems resulting from the recent floods can be attributed to the way in which flood-prone areas have been developed for industrial, agricultural and residential purposes.

If there is a lesson to be learned from recent events it is that there is a price to pay for unwise use of flood-plains.

(Dr) David Ledger
Lecturer in hydrology
University of Edinburgh
Darwen Building
Mayfield Road, Edinburgh
28 January

Dams helped to reduce river's flow

Sir, – I refer to James Rougvie's article today, "Who can tame the might of the Tay?" The answer is that no-one can control or even forecast accurately the maximum flow of water that can be expected to surge down the Tay valley and through Perth.

The flood of 18 January was duly anticipated by the excellent flood warning scheme only recently installed, but not its maximum flow or the fact that breaches would occur in certain flood banks.

I am surprised that you should publish an article on the Tay floods with a meaningless figure. Was the figure of 40,000 million gallons of flood water to which James Rougvie refers passing Perth in an hour, a day or during the whole period of the flood?

It was, of course, per day, roughly equal to the huge and terrifying figure of 2,000 cubic metres per second (cumecs, as units of water flow are conventionally called) as referred to earlier in his article. No-one can say that such a flood or a greater one may not recur next month, next year or next decade.

Thanks to the presence of Scottish Hydro-Electric's storage dams in the hills, the flow passing Perth on 18 January was in fact materially less than it would have been without them. Exactly how much less is impossible to state accurately as it depends on the rate of rainfall and consequent flows in different parts of the catchment, as well as on the water levels in the dams.

Scottish Hydro-Electric controls the water level in its dams using the latest information and forecasts available. It is only thanks to this close control that maximum use can be made of the water available to generate electricity.

Ultimately, it is thanks to this efficient use of water that tariffs in the company's area are kept low. If its role were to be changed to play the part of flood manager, it could indeed hold the level of water in its major storage reservoirs feet or even yards below the economic level. However, this would be very costly and who would pay? In addition, even this measure could not guarantee that recent flood levels would not be exceeded.

Landowners, farmers (and local authorities) naturally use the land they control to maximum advantage, so of course, they build flood banks to prevent flood damage and cannot be prevented from so doing. Certain banks, often in the hollows of ancient river course, were not always well built.

Some were undermined and others, when the water came over the top, simply had their substance scoured away. The flood banks will surely be rebuilt, hopefully made stronger by the use of wire mesh, gabions or even sheet-piling and not forgetting the use of natural vegetation to consolidate them. The further downstream one goes, the higher and better the banks must be: skilful siting and engineering are obviously required.

This does not mean changing the character of the river, or turning it into a canal. For the last century the Tay has been known not only as the largest, but also arguably as the best salmon river in Britain. Long may it so continue.

Alastair Duncan Millar
Reynock, Remony
Aberfeldy, Perthshire

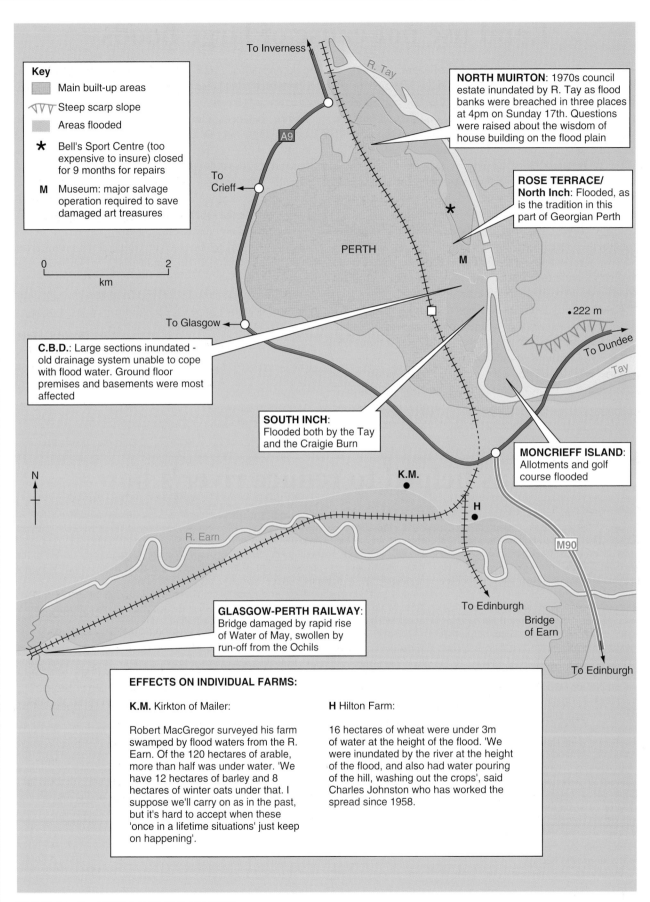

Key

Main built-up areas

Steep scarp slope

Areas flooded

* Bell's Sport Centre (too expensive to insure) closed for 9 months for repairs

M Museum: major salvage operation required to save damaged art treasures

NORTH MUIRTON: 1970s council estate inundated by R. Tay as flood banks were breached in three places at 4pm on Sunday 17th. Questions were raised about the wisdom of house building on the flood plain

ROSE TERRACE/ North Inch: Flooded, as is the tradition in this part of Georgian Perth

PERTH

C.B.D.: Large sections inundated - old drainage system unable to cope with flood water. Ground floor premises and basements were most affected

SOUTH INCH: Flooded both by the Tay and the Craigie Burn

MONCRIEFF ISLAND: Allotments and golf course flooded

GLASGOW-PERTH RAILWAY: Bridge damaged by rapid rise of Water of May, swollen by run-off from the Ochils

To Inverness

To Crieff

To Glasgow

To Dundee

To Edinburgh

Bridge of Earn

To Edinburgh

R. Tay

Tay

R. Earn

A9

M90

K.M.

H

M

•222 m

0   2
km

N

EFFECTS ON INDIVIDUAL FARMS:

K.M. Kirkton of Mailer:

Robert MacGregor surveyed his farm swamped by flood waters from the R. Earn. Of the 120 hectares of arable, more than half was under water. 'We have 12 hectares of barley and 8 hectares of winter oats under that. I suppose we'll carry on as in the past, but it's hard to accept when these 'once in a lifetime situations' just keep on happening'.

H Hilton Farm:

16 hectares of wheat were under 3m of water at the height of the flood. 'We were inundated by the river at the height of the flood, and also had water pouring of the hill, washing out the crops', said Charles Johnston who has worked the spread since 1958.

Fig 3.19 Aspects of flooding in the Perth area (1993)

average of 164 cumecs (Figure 3.18). Perhaps the critical event in Perth was the breaching, in three places, of the floodbanks that protected the North Muirton district (Figure 3.19) some 2 km north of Perth's Central Business District. Built in the 1970s on former farmland, the North Muirton housing estate is sited on a floodplain of the Tay. In spite of floodbanks built after an earlier 1974 flood, some 1200 out of the 1500 houses affected throughout the basin by the floods were located in North Muirton. Chance plays a part in flooding; had the banks held for a few hours more, the estate would have been spared. Arguably, once flood embankments were breached they made the problem worse as the excess water could not drain back into the river channel.

The 1993 flood was costly both in monetary and human terms. Its impact in the immediate Perth area is shown in Figure 3.19 and is a reminder that floods only become a hazard for people when they settle on flood plains.

Solutions Discussed

After a major flood of 1621, the minister of St. John's Kirk urged people to repent of their sins as a means of deliverance from flooding. It seemed to work, as the waters abated. In the post-1993 flood period various proposals were discussed to help solve the flood hazard. These included:

- reducing the height of flood banks on farmland to allow water to spread 'naturally' across stretches of floodplain and act as a safety-valve by slowing its downstream flow. This solution would have financial and land use implications. What would be the appropriate land uses immediately adjacent to the river? Often economic factors are perceived by the farmers as more significant than environmental factors when making land use decisions.

- strengthening and raising flood barriers at vulnerable points such as North Muirton. The question was inevitably posed: why had such extensive building been permitted on a flood plain in the first place?

- increasing the holding capacity of the hydro-schemes, possibly by building dams solely with the purpose of flood control. This would raise questions of cost, especially if the landowners had to be compensated.

- building completely new flood banks along stretches such as the North Inch. This area has a long history of flooding and a good case could be made for building an embankment. If built only on one side, however, what about the knock-on effect on the other? Also, once breached, embankments prevent the natural return flow of water back to the Tay.

- improving drains: a major cause of flood damage in Perth itself was the backing-up drains which were unable to flow into the swollen river. Much of Perth's Central Business District is low-lying so this is always going to be a problem, particularly for buildings with basements.

- straightening of rivers to by-pass pronounced meanders and speed the flow. This can help prevent flooding but is a costly solution and, if carried out on tributaries such as the Isla, would mean a more rapid flow of water into the Tay itself and make matters worse.

The Final Solution for Perth

After much consultation, Perth is to be protected, at a cost of £20 million, by 10 km of almost continuous flood walls, embankments and pumping stations along the west bank of the river, stretching from 4 km west of its confluence with the Almond to Perth Harbour. Due to be finished by 2002, the yet-to-be-completed defences were tested by, and coped with very severe rain (50 mm in 24 hours, with half of that total falling in 75 minutes) in September 1999. Nature is fickle, and the Tay basin with its large size, varied localised catchments areas (see assignment 5, page 69), and powerful discharge will always pose a challenge.

3.4 Rivers: Profiles, Processes and Landforms

Rivers, like glaciers and the sea, are important forces in shaping the landscape. All three agents not only **erode** the land, but also **transport** and **deposit** the eroded material, along with other products of weathering and mass movement (see pages 82–92). In these ways, distinctive landforms are formed.

In this section of the chapter, river features are discussed not only generally but also with reference to the main rivers and tributaries of the Tay and Earn basins. You will see that theory and reality do not always agree but this makes the study of rivers all the more interesting.

Long Profiles and River Courses: Theory and Reality

Figure 3.20A is essentially a sideways look at a river from its source to its mouth, usually the sea. This diagram is an 'ideal' long profile and has:

◆ a typical **concave long profile**, frequently regarded as having 3 sections: a steeper upper course and, with progressively gentler gradients, a middle course and a lower course.

◆ a **base level**: the lowest level to which erosion by fluvial or river action can take place. Usually, base level is the sea.

Figure 3.20B shows the long profiles for the rivers Tay, Earn and Isla showing that theory and reality differ. The smooth graded profile is the ultimate model to which rivers are aiming. That none of these rivers displays a graded profile is for one or more of the following possible reasons:

◆ Local base levels occur. The Earn, for example, within only a few kilometres from its source, flows into Loch Earn, over 87 m at its deepest as a result of glacial erosion. Ice was also responsible for the 161 m maximum depth of Loch Tay.

◆ Waterfalls and rapids also interrupt the smooth profile. Reekie Linn (Figure 3.21) is one of the most impressive waterfalls in the whole Tay basin and formed where the Isla cuts across a band of intrusive igneous rock. Much the same is true of the Tay at Campsie Linn where a great dyke of dolerite cuts across the river's lower course. (Figure 3.22)

◆ Further irregularities in the long profiles of these rivers are the result of **rejuvenation** (being made young again). Basically, rivers start to cut down into their beds again to reach the new base level. (See page 65 for more details about rejuvenation.)

Processes and Energy Along the River

A great deal of energy is expended by a river as its interrelated processes of transportation and erosion shape the landscape. Under normal conditions, most

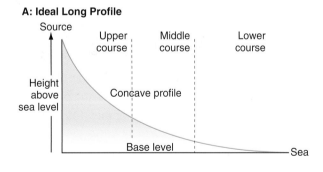

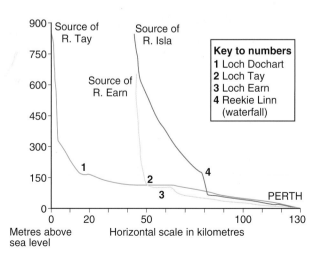

Fig 3.20 Long profiles of rivers: ideal and reality

Fig 3.21 Reekie Linn: This is a double waterfall on the River Isla. It is formed where igneous rock (felsite) intrudes into sandstone and conglomerate.

(estimated at 95%) of a river's energy is required to overcome friction with the beds and banks of the river's channel. Once flooding occurs or after a major tributary has joined there is more surplus energy available to transport material (called **load**), and to erode by various methods.

Methods of Transportation

A river's load is introduced into its channel by the denudation of the surrounding hillsides. Figure 3.23 shows that the load may be transported in four ways.

1 by **traction:** the force of water rolls and drags the larger particles along the bed of the river.

2 by **saltation:** relatively small particles on the river bed are dislodged by similar sized particles into the current of water. After travelling a short distance, the particles again settle on the river bed, and displace further particles. Basically, the bedload bounces along when energy permits.

3 by **suspension:** the finest particles (clay and silt) are carried in suspension, being 'swirled' along by the water. As particles break down, the amount of suspension load increases.

4 by **solution:** the river transports the products of chemical weathering from the hillsides and river bed.

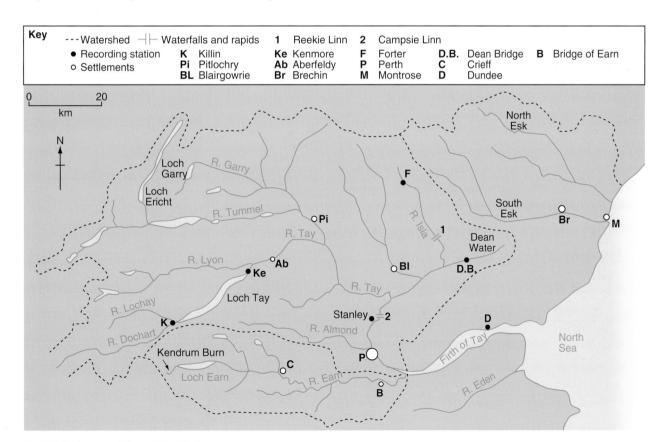

Fig 3.22 Outline map of Tay and Earn Basins

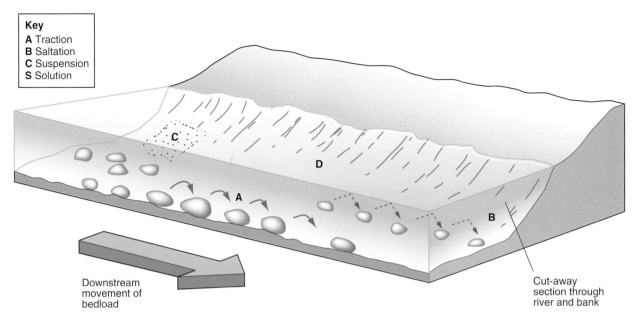

Key
A Traction
B Saltation
C Suspension
S Solution

Downstream movement of bedload

Cut-away section through river and bank

Fig 3.23 Methods of river transportation

This is a dissolved load that is being transported as opposed to a particle load.

As an average, suspended material makes up about 75% of the load of most rivers. Conditions do vary and will depend upon the type of rock through which the channel passes, and the amount of water being discharged. Most particle load is transported during flood conditions when velocity and volume are at their greatest.

Methods of Erosion

As the load is transported, the surplus river energy employs various processes to erode the bed and banks of the river channel. By this means, erosion deepens, widens and lengthens its channel. The methods include:

1 Corrasion/abrasion This is the abrading or wearing away of the river's bedrock and banks by sand, pebbles and stones. This process is particularly potent in a valley's upper course during floods, and assists in the vertical erosion (downcutting) of this section of the valley. One particularly effective form of abrasion is potholing. Boulders, swirled around by turbulent eddies, gradually drill deep, smooth-sided depressions or potholes (Figure 3.24).

2 Hydraulic action The pressure of rushing water mechanically drags away particles of unconsolidated material such as sands and gravels. While this process can happen in the upper course, it is very effective in the lower course of the river where embankments are especially vulnerable.

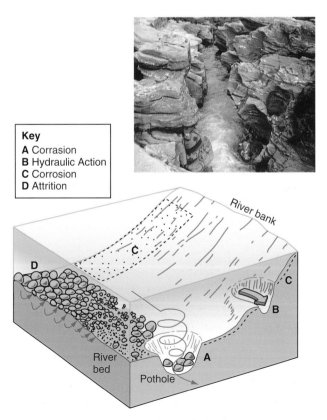

Key
A Corrasion
B Hydraulic Action
C Corrosion
D Attrition

River bank

River bed

Pothole

Fig 3.24 Boulders, swirled around by turbulent eddies gradually drill deep potholes. In the photo they have coalesced to form a narrow gorge.

3 Corrosion can occur anywhere along a river's course and is the chemical erosion of rock surfaces by the flowing water. Not only does carbon dioxide (CO_2) attack limestone but, by weakening its structure, makes it vulnerable to other erosive processes.

4 Attrition involves the various forms of load attacking themselves. By rubbing and impacting against each other, load particles progressively are reduced in size and become increasingly rounded.

Deposition

Deposition by rivers takes place when their velocity decreases and they can no longer transport their loads. Any velocity decrease may occur:

◆ as rivers reach a base level by entering the sea or a loch to form a delta

◆ when rivers overflow their banks

◆ on the inside bend of a river's meander (see page 64)

◆ when there is a reduction in discharge as a result of drought or flood water subsiding

◆ with a sudden decrease in gradient, e.g. immediately below a waterfall.

General Comments on Processes and Energy

Interrelationships: The ability of a river to erode, transport and deposit different sizes of particles along its length depends upon variations in velocity. In 1935, the Swedish geomorphologist Filip Hjulström produced a graph (Figure 3.25) to show the interrelationship between processes, velocity and particle size. It includes:

◆ the **critical erosion velocity curve** marking the boundary between erosion and transport. This shows the minimum velocity needed to pick up particles of varying size lying in the river channel. Initially, it takes a lot of energy to pick up or **entrain** the material but, after that is done, momentum keeps it moving. The curve shows that, as velocity increases, the size of particle able to be carried and eroded also increases. Cobbles and boulders need very high velocities to move

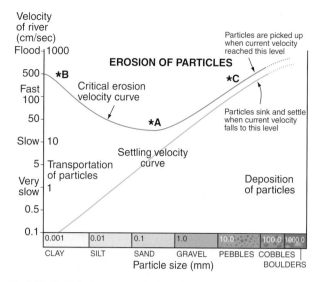

Fig 3.25 Hjulström's curve of velocity and particle size

them. The exceptions, however, are clay and silt particles. Although small, they tend to stick together so that sand-size particles are often picked up first.

◆ the **settling velocity curve** is the threshold between deposition and transportation. When velocity decreases, the suspended particles become too heavy and are deposited on the channel bed. Sand-size particles, therefore, drop down before silt and clay. Clay particles keep moving in suspension at a very low velocity.

Velocity, river course and channel size: As a river flows from source to mouth, it might be expected that the initial rushing upland torrent would become a slow-moving stream as velocity apparently decreases downstream. Accurate measurements, however, demonstrate that velocity not only stays constant downstream but may even increase. This is because:

◆ in the upper course, the small stream channels are filled with large, angular boulders and have unevenly shaped banks. Therefore, water flowing through these channels is affected by friction and, in spite of their steeper gradient, stream velocity is generally reduced (apart from waterfall sites).

◆ by the lower course, although its gradient is much lower, channel size is larger and the river banks are smoother. Consequently, the amount of river flow impeded by friction is proportionately less compared to the upper course. Stream velocity is thus higher than might be expected.

Contrasting River Landforms: Upper Course to Lower Course

Along a river's course there are distinct changes in its channel features, its valley landforms and its long profile from source to mouth.

Upper Course Characteristics

Figures 3.26 and 3.27 show the key characteristics of upper courses, some of which are found around the Kendrum Burn, the main source of the River Earn. (Figure 3.28)

1 Channel features: Typically, the upper stretches of a river channel are rocky, often being covered with boulders of varying shape and size, strewn across an uneven floor. Under 'normal' conditions the discharge is fairly low, fed from throughflow and groundflow. Under flood conditions, river energy is expended on vertical erosion, with hydraulic action and corrosion processes at work. **Potholes** form where pebbles and cobbles, rotated by swirling eddies, grind deep holes in the bedrock. Sometimes potholes coalesce and form a narrow gorge. (Figure 3.26)

2 Valley features: In the upper course, the valley bottom is narrow and the channel often may occupy most or all of the valley bottom. Where conditions allow, deposition by the river may initiate the beginnings of a very narrow flood plain.

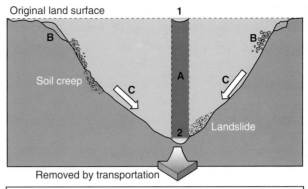

Key
1 Original river-bed   **2** Present day river-bed
A Amount of material eroded by river
B Physical weathering of rock
   (weathered material incorporated into C)
C Mass movement

Fig 3.27 Upper course features and processes influencing V-shaped valleys

Fig 3.28 The Kendrum Burn – the upper course of the River Earn. Notice the size and shape of banks and boulders. Is it a model 'V' shaped cross-section?

The valley sides, typically, are steep and in cross-section are 'V' shaped. Careful consideration of Figure 3.27 shows that such a shape is the result of the interaction of the following processes:

◆ vertical erosion by the river itself, responsible for the gorge-like section in the diagram.

◆ physical weathering (e.g. frost action on the sides of the valley) which provides debris to be moved downslope

◆ mass movement, including soil creep and even occasional landslides, initiated when soil is very saturated.

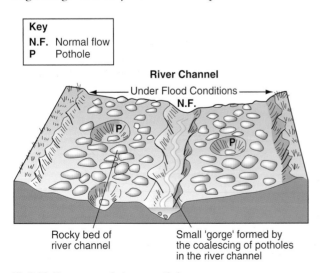

Key
N.F. Normal flow
P Pothole

Fig 3.26 Upper course features – potholes

During periods of heavy rain, material formed by these processes will be rapidly and efficiently removed by the swollen river. Arguably, the river is more important as an agent of transport than erosion in shaping the V-shaped cross-section. It is important to remember that exact shape of the 'V' will also depend upon the resistance of the rock. Typically, the river will have to negotiate bends formed by protruding lower hillside slopes called **interlocking** spurs.

3 Long profile: Generally, the gradient is steep and the profile is uneven, particularly where **waterfalls** and rapids form. Often, they develop where a band of harder, resistant rock crosses a river's channel. On the downstream side, the softer, less resistant rocks are more rapidly eroded. Figure 3.29 shows how the undercutting of the softer rock causes the portions of the hard band to collapse. Thanks to the impact of large volumes of water (especially at times of spate) and the whirling impact of boulders, a plunge pool is excavated at the base of the waterfall. Over the years, as these erosional processes are repeated, the position of the waterfall gradually recedes in an upstream direction. Rapids are characterised by small waterfalls, stretches of broken water and bedrock exposures.

Middle Course Characteristics

These characteristics develop as shown in Figure 3.30.

1 Channel features: Generally speaking, the channel is now wider and deeper, and has smoother banks and floor compared to the upper stretches. Although traction and saltation assist in transportation, gradual attrition of the river's load means that more silt and clay-sized particles are not only carried in suspension but are increasingly deposited.

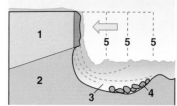

Band of hard rock e.g. lava; Large boulders formed from undercutting of harder rock; Deep plunge pool formed by abrasion and hydraulic action; Soft rock e.g. sandstones and conglomerates; Retreat of waterfall forms a gorge by headward erosion

Fig 3.29 Upper course features – waterfall formation

2 Valley features: Provided the rocks are not particularly hard and resistant, the valley sides of the middle course are not so steep and have been eroded back compared to the upper course. With the general lowering of the land surface and the gentler gradient of the valley sides, the river's erosional energy is now increasingly expended horizontally rather than vertically. **Lateral erosion** by the river's meanders broadens the valley floor into a **narrow flood plain**. **Meanders** (see section below for more detail) gradually shift their courses downstream and steadily blunt the valley spurs, forcing them back into a line of bluffs. (Figure 3.30)

3 Long profile: The gradient, though now much gentler, is still adequate to assist in the downstream migration of meanders. Overall, the profile is smoother but an outcrop of resistant rock (see Figure 3.31 of Campsie Linn) is a reminder that rapids and waterfalls are not just confined to the upper course. As earlier discussed, in spite of the gentle gradient, the river's velocity is not reduced, thanks to its smoother, deeper channel which reduces friction.

Lower Course Characteristics

Typical features are illustrated from the lower Earn valley. (Figure 3.32)

A: Model Upper Course

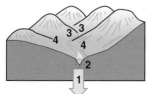

- Interlocking spurs
- Mainly vertical erosion
- 'V' shaped cross-section
- Lateral erosion beginning on the outer bank of the river bend

Fig 3.30 Broadening of a river valley

B: Model Middle Course

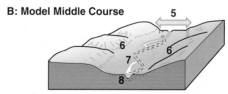

- End of spurs worn away by lateral erosion to form bluffs
- Lateral erosion
- Downstream migration of meanders
- Increased deposition of alluvium to form a small flood plain

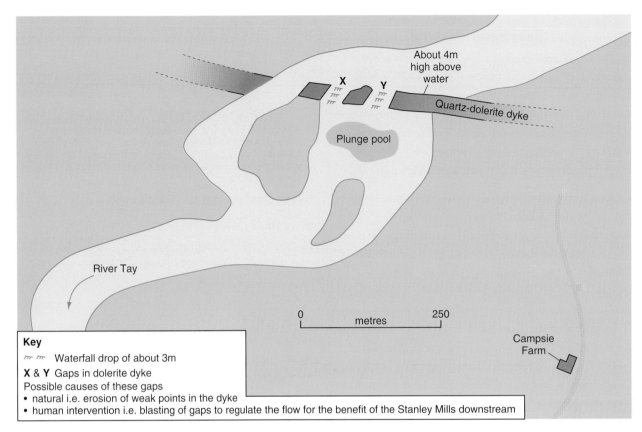

Key

⊤⊤ ⊤⊤ Waterfall drop of about 3m

X & Y Gaps in dolerite dyke

Possible causes of these gaps
- natural i.e. erosion of weak points in the dyke
- human intervention i.e. blasting of gaps to regulate the flow for the benefit of the Stanley Mills downstream

Fig 3.31 Middle course features – Campsie Linn

1 Channel features: The channel is now at its broadest and deepest. Bedload is carried entirely (i) in suspension, consisting of the finer sediments, and (ii) in solution, the 'hidden giant' of transportation processes. Deposition now dominates, particularly during floods when the river's energy is expended in spreading its load over the lower course. Erosion, however, also occurs as is shown by the formation of meanders.

2 Valley features: Thanks to lateral erosion, valley sides may now be several kilometres away, as in the case of the lower Earn. Typically, the broad, low-lying valley floor has the following features (Figure 3.33):

(i) **Floodplain and natural levées:** Over many years, the river has covered the valley floor with enormous quantities of **alluvium** (sedimentary deposits). The alluvium was deposited by migrating meanders (see below) and floodwater. Across the resulting level **floodplain**, variations in relief are very slight so that any rapid increase in discharge almost inevitably results in flooding. When floodwaters spread across the floodplain, friction increases, the velocity of the water slows and is no longer able to transport sediment. The floodwater's speed reduces most quickly at the sides of the channel. Consequently, coarser alluvial sediments – sand and gravel – are deposited at the channel edge, gradually building up into natural ridges or **levées** after successive floods. (Figure 3.34). By preventing water from returning to the main channel flood, more sediment is added to the floodplain. Occasionally, levées act as natural embankments and

Fig 3.32 This shows the eastern section of the River Earn floodplain, as well as the confluence of the Earn and the Tay

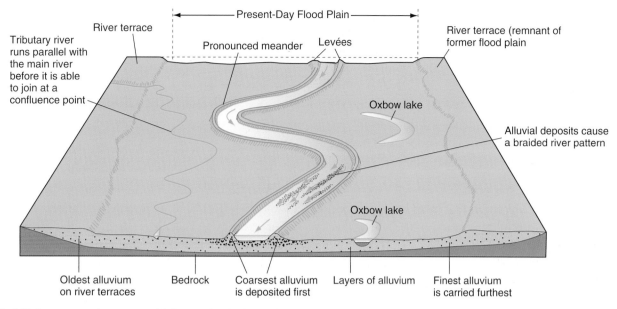

Fig 3.33 Lower course features – model diagram of main characteristics

permit the river to flow at a level greater than the floodplain. Often, they have been strengthened as a part of flood prevention measures. Levées may also divert tributary streams and rivers, forcing them to flow parallel to the trunk river until they gain access.

(ii) **Braided channels:** These are formed by the choking of the main channel by the deposition of considerable amounts of the river's load. The channel, consequently, splits up into several smaller channels which flow around fresh 'islands' of deposited material before rejoining and then further dividing. Such channels are not confined to the flood plain but are often associated with rivers with highly variable discharge. Braided channels do, however, slow down a river's velocity during periods of spate.

(iii) **Meanders:** Rarely do rivers follow straight courses. Rather, river channels follow curves and bends called **meanders** – so named after the 'classic' examples found on the Maiandros (now Menderes) river in Turkey, famed for its twisting course. Although associated with middle and lower courses of a river, they are found also in upper courses, particularly in glacial troughs. Generally, meanders occur more frequently and become progressively larger and more pronounced downstream. Ideas

about their formation vary but can include the following explanations:

◆ in a river bed, even in a straight river channel, a series of alternating irregularities called **pools** and **riffles** develops. The pools are deeper stretches of slow moving water, usually with fine alluvial deposits; the riffles are shallower sections of faster

A. Before Flood:

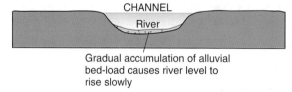

B. During Flood:

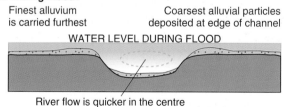

C. After Many Sequences of Flooding:

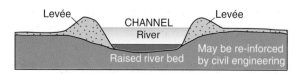

Fig 3.34 Lower course features – levée formation

water, flowing above coarser material. Laboratory and stream measurements suggest that once a pool and riffle pattern forms, channels develop a side-to-side swing. Figure 3.35A shows that alternate pools (A, B and C) have migrated to opposite sides of the channel while riffles (P, Q and R) are found on the straight sections. This tendency for water to follow such a winding or **sinuous** course seems to be part of a naturally occurring process.

◆ once initiated, this process is assisted by erosion and deposition. The meander curves become more pronounced as a result of erosion undercutting the river cliffs in the alluvial deposits of the flood plain. Figure 3.35B shows that this occurs on the outside bend where velocity is strongest and the water deepest. On the other side, deposition takes place in the form of a gentle **slip-off-slope** or **point bar** where the current is weakest.

◆ meander formation is further assisted by the **helicoidal flow** of water. Helicoidal flow, essentially, is a corkscrew like movement of water as the river flows downstream. The cross-section shows the transfer of sediment from the undercut, concave river cliff on the outside of one bend, to the convex point bar on the inside of the next bend.

◆ meanders are **dynamic** (ever changing) features of the alluvial landscape. Speeded up enormously, their movement is not unlike a long rope being flicked. Thanks to helicoidal flow, meanders migrate downstream by (i) laterally eroding any obstacle and (ii) depositing alluvial material upstream of the point bars, thereby assisting in the build up of the floodplain.

◆ continued migration accentuates the meanders and makes more likely the formation of oxbow lakes (see below). The actual rate of meander migration is influenced by the balance between the power of the

A. Sequence of Meander Formation

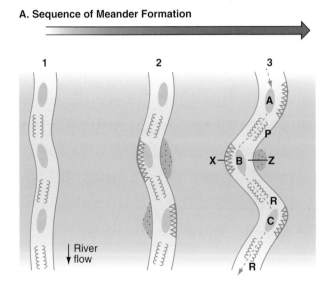

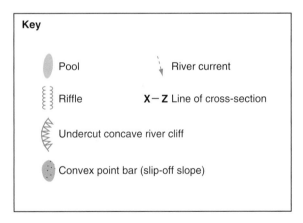

B. Section Through A Meander Channel

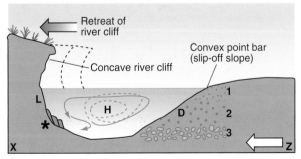

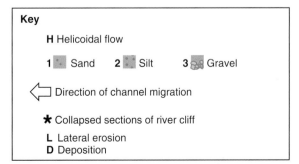

Fig 3.35 Aspects of meander formation

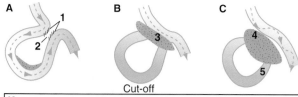

Key
1 Lateral erosion on outside bends of pronounced meander
2 Narrow neck of land gradually becoming narrower
3 Neck is cut through by river during floods
4 Cut-off is sealed by deposition
5 Ox-bow lake starting to silt up

Fig 3.36 The development of an oxbow lake

river on the one hand, and the strength of the bank-forming sediments. Generally, sandy bank material is non-cohesive and falls into the river when undercut. Clay-rich sediment falls away as blocks which help to protect the bank from erosion.

(iv) **Oxbow lakes:** With the migration of meanders downstream, the bends may grow closer together producing pronounced meander loops. Oxbows form during flood conditions when a meander loop is breached as the river takes a shorter, direct route. The abandoned loop is sealed off by the freshly deposited alluvium and then forms a crescent-shaped, water-filled loop. Eventually, it is not unusual for such abandoned bends to silt up. (Figure 3.36)

(v) **Estuaries and deltas:** When a river enters a loch or the sea, its flow characteristics are markedly changed. Velocity is reduced and load is deposited. Where the river's mouth broadens to form an estuary, tidal currents are able to scour out most of the sediments, transferring them to the sea's transportation system. The sediment that is not removed from the estuary forms extensive sand and mud banks, often colonised by salt-tolerant vegetation and exposed at low tide. Deltas are essentially the seaward extension of the floodplain and form when tides are weak. They also grow where streams enter freshwater lochs. Some characteristic features are shown in Figure 3.37.

3 Long profile: As the river reaches base level, the gradient is very gentle in this section. The volume of water is now at its greatest and velocity differs little from that registered in the upper reaches. It is from this interface of land and sea that the impact of rejuvenation feeds back upstream as sea level falls in relation to the land (see below).

Rejuvenation

As mentioned earlier, rivers given a fresh lease of life are referred to as **rejuvenated**. This occurs when the base level (sea level) is lowered, erosional processes are initiated again, and the river regrades its bed. Among the causes of rejuvenation are an uplift of land, such as the slow rise of land after the Ice Ages (**isostatic uplift**) or a fall in sea level (**eustatic change**) as occurred during the Ice Ages. The following are examples of rejuvenated landforms:

◆ **River terraces:** Land, once depressed by a massive overburden of ice, has been steadily rising since the end of the Ice Age. Rivers, such as the Tay and Isla, have adjusted by eroding into their former floodplains. Portions of the former floodplains remain as **river terraces** and flank either side of the present floodplains. (Figure 3.38). Continued isostatic uplift

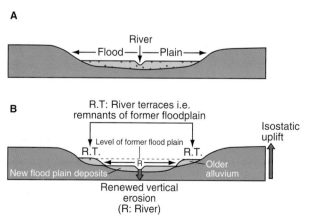

Fig 3.37 Deltaic deposits have formed at the mouth of the River Kishorn, Western Ross

Fig 3.38 Paired river terraces formed by isostatic uplift

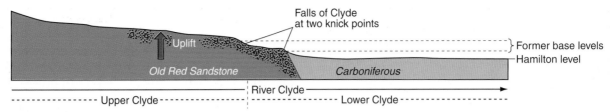

Fig 3.39 River knickpoints: the falls of Clyde

has meant that several pairs of terraces flank such plains. Often, terraces are sought as flood-free sites for settlement and transport links.

◆ **Knickpoints:** As new valleys are eroded as a result of fresh downcutting, the inner section of the former floodplain is removed. Terraces can be traced upstream until a point where the old and the new valley floors converge. This point is called a knickpoint and is often marked by rapids or a waterfall. Figure 3.39 shows such knickpoints at the Clyde Falls (near Lanark) between the Upper Clyde and the Lower Clyde. The Upper Clyde shows the former long profile of the Clyde before it was rejuvenated by over 170 m; the Lower Clyde demonstrates the reactivation of downcutting to a fresh base level. Initially, the knickpoint receded upstream quite rapidly in the relatively soft rocks but the rate of erosion is now considerably slower thanks to a harder outcrop.

◆ **Incised meanders:** Rejuvenation may also affect river meanders. Fresh downcutting can erase the former meanders or, alternatively, they may be incised into the underlying bedrock as the knickpoint retreats upstream along the initial meander line. One of the best exemplars is the River Wear at Durham, while, in Scotland, the Tay has incised a 46 m deep gorge at Stanley. At both Stanley and the Clyde Falls rejuvenation assisted early industrialists of the eighteenth century to utilise the available water-power resources.

To Take You Further

Drainage Patterns and River Capture

This chapter started off with river basins. Within each basin, it is possible to identify different spatial arrangements of the streams and rivers. Such arrangements are called **drainage patterns** and they

respond to the amount of water, the original gradient of the basin and its rock structure. Various descriptive terms are commonly used. (Figure 3.40)

When a drainage system develops on a single type of rock, two patterns are:

◆ **Parallel:** often found on newly uplifted land or on the steep sides of a glacial trough, the initially parallel tributaries eventually meet at an acute angle.

◆ **Dendritic:** (from the Greek 'Dendron' meaning 'a tree') is characterised by branches rather like the limbs of a tree and is probably the most common pattern. Slope, again, is the main determining factor.

When, however, there are different underlying belts of rock in parallel outcrops of varying hardness, then a trellised pattern is more likely to result.

◆ **Trellised:** often develops when a main trunk river flows directly to the sea across parallel bands of hard and soft rock, dipping in the same direction. This main river is called a **consequent** as a result (consequence) of the initial slope of the land. Figure 3.40 shows how the tributaries of the trunk river flow parallel to the edge of the hard bands of rock, gradually eroding headwards (backwards) along the softer rock, and meet the consequent river at right angles. Such a tributary is called a **subsequent**. At a later stage, this drainage pattern is enhanced by the addition of short, spring-fed, **obsequent** streams feeding at right angles to the subsequent. (See pages 123–125 on escarpment scenery.)

◆ **Radial:** often develops on a single large peak such as a large dormant volcano with the streams and rivers radiating outwards, not unlike the spokes of a wheel. On a larger scale, such a pattern characterises the granite domes of Dartmoor and northern Arran, where streams flow outwards in all directions. In the Lake District, a radial pattern can also be discerned where

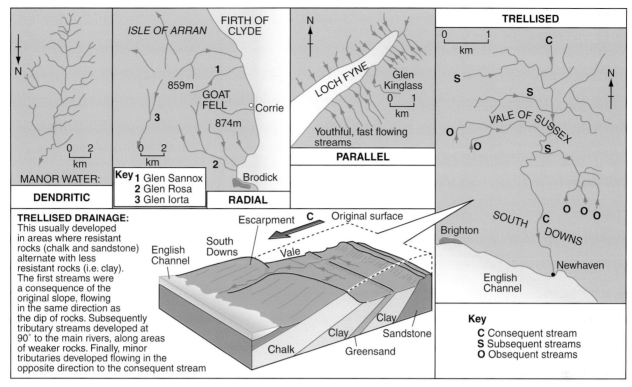

Fig 3.40 Sample drainage patterns

the drainage pattern initially developed on a former uplifted dome, covered with homogeneous (similar) rock. This covering layer has been removed, but the streams and rivers continue to downcut into the various underlying, older rocks and broadly maintain the early pattern. In other words, the initial pattern is now **superimposed** on the present day surface.

It should be noted that using such descriptive terms has been criticised by those who maintain that quantitative measures are more appropriate to the discussion of drainage basins. One of the main difficulties is that one geographer's 'dendritic' may be 'parallel' or 'trellised' to another.

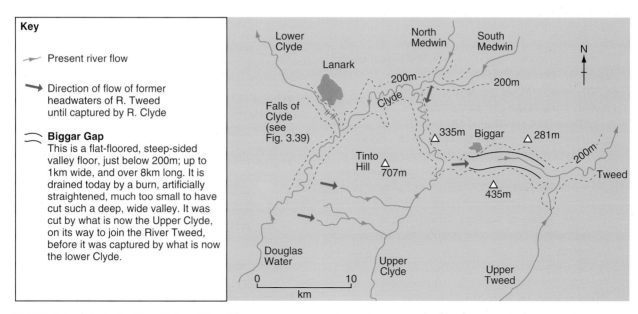

Fig 3.41 River capture – the Rivers Clyde and Tweed

Debate is not unknown when it comes to the term '**river capture**'. Essentially, river capture involves the re-arranging of river courses when rivers attempt to reach a new base level and, in the process, capture the headwaters of their neighbours. Some would suggest that glaciation instead is often the key factor in disturbing river courses in appropriate areas. Figure 3.41 shows the hypothetical links between the pre-glacial Clyde and Tweed. It is believed that, formerly, headwaters of the Clyde, such as the Douglas Water, the North Medwin and the Upper Clyde itself, were all tributary headwaters of the River Tweed. As a result of the relatively easy headward erosion of the Lower Clyde exploiting the less resistant rocks underlying its course, it was able to enlarge its own basin by 'beheading' the one-time Tweed tributaries. This can be seen in the largely stream-free Biggar

Gap. It is certainly the case that glacial deposition modified the drainage but it is also maintained that river capture was completed before any action by ice.

A Cautionary Note

It cannot be stressed often enough that all rivers are unique and that careful fieldwork and map interpretation are a useful corrective to generalisations. Further, particularly in much of Upland Britain, the impact of ice is important. Rivers are often downcutting through, and transporting, glacial and fluvio-glacial material. Equally, many floodplain features are in fact encountered in the upper courses of valleys which have been shaped by the action of ice into broad-floored glacial troughs.

Water and Life, Global Hydrological Cycle and the Drainage Basin

Read pages 37 to 43.
1 **Explain** the terms **hydrosphere** and **hydrology**.
2 After reading the extract from 'Driving over Lemons', **give examples** of phrases showing the marked seasonal contrasts in climate and river flow in southern Spain.
3 a) Using Figure 3.2 as your guide, **draw** a large long section diagram to show the **hydrological cycle**.
   b) Briefly, **explain** the difference between:
      (i) **evaporation** and **transpiration**
      (ii) **infiltration** and **percolation**
      (iii) **surface runoff** and **throughflow**.
   c) Using the term **closed system**, **write** a paragraph explaining the workings of the hydrological cycle.
4 a) Figure 3.3 shows the main features of a **drainage basin**. **Convert** the information into an annotated sketch map of a drainage basin. Check that it shows: **types of sources**; **tributaries**; **a watershed**; **confluence points**; **trunk river** and **mouth**.
   b) The following are terms (some overlapping) used in discussing the workings of a drainage basin: **precipitation**; **evapotranspiration**; **interception**; **stem flow**; **infiltration**;

throughflow; **leaf drip**; **groundwater flow**; **channel flow**; **percolation**; **evaporation**; **transpiration**; **runoff**; **soil moisture**.
In a table, **group** the terms in four categories: **INPUT**, **OUTPUT**, **STORAGE** and **TRANSFERS**.
   c) Figure 3.5 shows the drainage basin as a systems diagram. Make a copy, using different shading, to show the four main elements of a systems diagram and key.
   d) The **water table** is the upper level of saturated rock/soil. What factors cause its level to vary seasonally?
   e) In what ways is a drainage basin an **open system**?
5 a) Briefly **explain** the term drainage density and suggest how it may be measured.
   b) Using Figure 3.4 as a guide, apply Horton's method of **stream ordering** to the drainage basins in the Loch Etive area shown in the map on page 74. Use tracing paper to help you. **Suggest** possible reasons for the order of basins you find.

Flows and Flood Hydrographs
Read pages 43 to 45.
1 **Explain** the terms **discharge** and **flood (storm) hydrograph**.

2 With the help of Figure 3.8, **explain** the following terms: **peak discharge**; **lag time**; **base flow**; **rising limb**; and **recession (or falling) limb**.

3 **Draw** a large spider-type diagram to summarise the factors that influence the size and shape of a flood hydrograph.

4 Figure 3.9 shows model hydrograph shapes in relation to the character of drainage basins.
  a) Look at hydrographs C and D. Match the land use features of the basins to the appropriate hydrograph.
  b) Explain the shapes of the hydrographs in A and B.
  c) Suggest how **afforestation** and **deforestation** in a drainage basin would affect:
  (i) the rate of interception, **(ii)** the amount of moisture that is lost by transpiration, **(iii)** the volume of water in the trunk river, **(iv)** the potential for flooding in the drainage basin.

5 **a)** Look closely at the map (Figure 3.22) of the Tay and Earn basins. Locate the recording stations at **Killin** and **Kenmore** at the western and eastern ends, respectively, of Loch Tay.
  b) From the 11th January, 1997, heavy rain fell over the Tay Basin. In parts of the basin, farms were alerted. However, there was only minor flooding. Using the data provided in Table 1, **draw a flood hydrograph for Killin** on graph paper. Comment on the relationship between peak discharge and peak rainfall.
  c) On the same frame used for the hydrograph, **superimpose** a discharge graph for the River Tay at Kenmore. What effect did Loch Tay have on the flow at Kenmore?

6 Distinct seasonal contrasts in river discharge are a feature of the Niger Basin in West Africa. Look at figures 3.10–3.14, and read pages 23 to 26 about the seasonal movement of the ITCZ. An atlas is ever-useful.
  a) **Draw** a simple sketch map to outline the size, extent and main features of the River Niger and its basin, including the savanna zones.
  b) What is meant by a **river's regime**? How does a graph of river regime differ from a hydrograph?
  c) Using Figure 3.13 **suggest** why the peak discharge at Mopti lags behind peak rainfall.
  d) Using the graphs in Figure 3.11, **describe and explain** the differences in the discharge of the Niger between:
  (i) Koulikoro and Mopti, and **(ii)** Niamey and Lokoja.
  e) With reference to Figures 3.12 and 3.13, **describe** how farmers, fishers and pastoralists have adapted to the seasonal variations in rainfall and water supply in the savanna lands of West Africa. In your answer, refer to the importance of groundwater.

Table 1

Day	Daily total of rainfall at Killin (mm)	Discharge m³/s at 0900 at Killin	Discharge m³/s at 0900 at Kenmore
01/01/97	0.2	2.0	13.56
02/	1.4	1.84	12.14
03/	0.6	1.84	11.51
04/	0.0	1.76	11.10
05/	0.0	1.62	10.50
06/	0.0	3.41	9.82
07/	0.0	1.60	10.14
08/	0.0	1.49	9.62
09/	0.0	1.43	8.90
10/	0.0	1.41	8.46
11/	17.0	1.38	8.10
12/	5.2	62.37	14.31
13/	2.2	42.98	38.65
14/	3.6	22.24	44.45
15/	0.0	23.45	48.34
16/	0.6	10.58	45.77
17/	4.6	6.83	41.00
18/	5.6	9.21	36.54
19/	0.0	19.19	37.44
20/	0.2	9.57	35.65
21/	0.0	5.78	33.52
22/	0.0	4.57	29.55
23/	1.8	3.71	28.14
24/	1.2	3.54	25.57
25/	0.4	7.73	24.41
26/	0.4	5.66	22.57
27/	0.0	5.41	20.19
28/	0.0	4.52	18.44
29/	0.0	3.84	17.04
30/	0.0	3.37	15.75
31/	0.6	2.78	14.69

Rivers: Profiles and Processes
Read pages 56 to 60

1 Figure 3.22 shows the Tay and Earn basins. Figure 3.20 shows a model **long profile** and those of the Tay, Earn and Isla.

    a) **Describe** the main features of a 'typical' long profile of a river.

    b) Referring to the long profiles of the Tay, Earn and Isla, **suggest** three reasons why they do not conform to the ideal profile.

2 **a)** With the aid of an annotated diagram, **describe** four ways in which a river transports its load.

    b) What factors influence the amount of load that can be transported?

3 **a)** With the aid of an annotated diagram, **describe** four ways in which a river erodes its channel.

    b) What happens to the size and shape of eroded material as it is transported downstream?

4 Under what conditions does a river begin to deposit its load?

5 Figure 3.25 shows **Hjülstrom's Curve** which shows the different velocities necessary to erode, transport and deposit various sizes of load.

    a) What is meant by the **critical erosion velocity curve**?

    b) **Explain** why a greater velocity is required to move B, a small particle of clay, than A, a larger sand particle.

    c) What is the minimum velocity required before the large gravel particles C are transported?

    d) What is meant by the **settling velocity curve**? Explain why clay particles, once in motion, are not deposited even at very low velocities.

    e) Referring to the diagram, briefly **discuss** the relationship between river velocity, erosion, transportation and deposition of varying sizes of particles.

6 **a)** **Copy** Figure 3.42 comparing velocity and discharge between upper and lower courses. **Match** the numbers with the list of items provided.

    b) Briefly **explain** why, in spite of a lower gradient,

river velocity not only stays constant but may increase downstream.

Rivers: Contrasting Landforms – Upper Course to Lower Course
Read pages 60 to 68

1 As you work your way through this section, it is useful to make a large copy of Figure 3.43 overleaf, and complete the appropriate sections as a summary of the main terms.

2 **a)** **Name and describe** five features typical of the **upper course** of a river valley.

    b) With the help of Figure 3.27, **fully explain** how a **'V' shaped valley** is formed.

    c) Make a large copy of Figure 3.29. After matching the numbers with the key, explain how a **waterfall** is formed.

3 **a)** Study Figure 3.30. **Briefly explain** the difference between **vertical erosion** and **horizontal** erosion. Which, vertical or horizontal, becomes more important downstream, and why?

    b) **Copy and complete** the two model diagrams of the upper and middle courses by matching the numbers to the bullet points. Write a paragraph explaining how the valley widens downstream.

4 **a)** With the help of Figure 3.33, **name and describe** five features typical of the **lower course** of a river.

    b) After copying Figure 3.34, **explain** how a **levée** is formed.

    c) Copy Figure 3.35A, Sequence of Meander Formation. With the help of the diagram, **explain** how a **meander** is formed.

    d) **Draw** a cross-section through a meander. Add labels to bring out the differences between the outer and inner sides. How has the **helicoidal flow** of the river contributed to the differences between the two sides?

    e) With the assistance of labelled diagrams (see Figure 3.36), **fully explain** the formation of an **ox-bow lake**.

5 **a)** What is meant by **rejuvenation**? **Suggest two ways** in which a river can be rejuvenated.

    b) What are **river terraces** and how were they formed?

    c) **Explain** how **incised meanders** are formed.

    d) In what ways are rejuvenated river features of value to people?

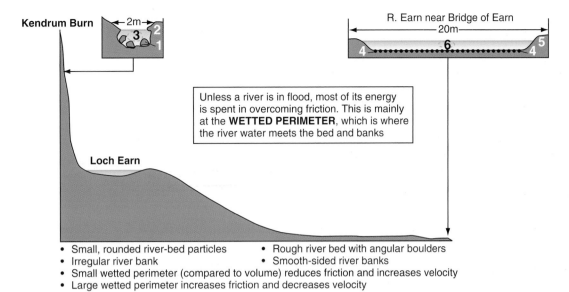

Kendrum Burn
2m
3
2
1
R. Earn near Bridge of Earn
20m
6
4
5
4
Unless a river is in flood, most of its energy is spent in overcoming friction. This is mainly at the **WETTED PERIMETER**, which is where the river water meets the bed and banks
Loch Earn

- Small, rounded river-bed particles
- Irregular river bank
- Rough river bed with angular boulders
- Smooth-sided river banks
- Small wetted perimeter (compared to volume) reduces friction and increases velocity
- Large wetted perimeter increases friction and decreases velocity

Fig 3.42 A comparison of velocity and discharge on the Upper and Lower courses of the R. Earn

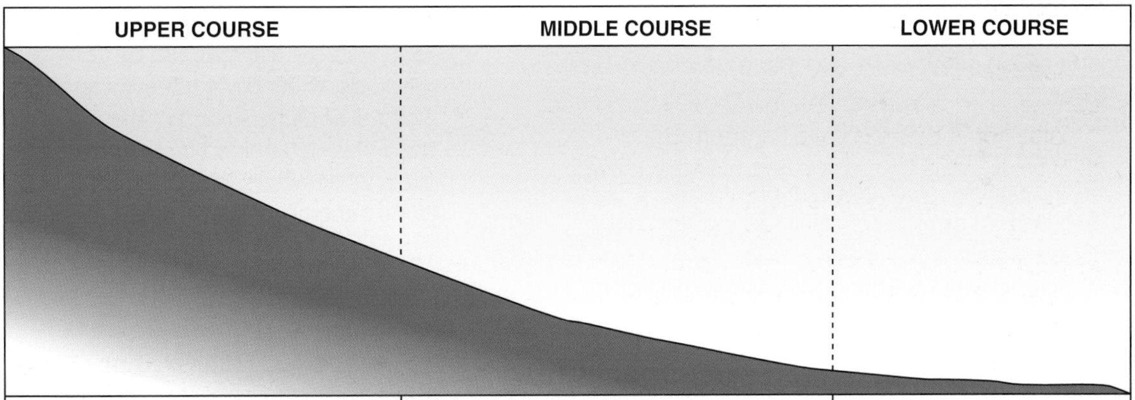

UPPER COURSE
MIDDLE COURSE
LOWER COURSE

Channel Features			
Valley Features			
Gradient			

Fig 3.43 Summary diagram: long profile of a river from Upper to Lower course

OS Map Assignment on Rivers and their Valleys

Two contrasting stretches of the River Earn and its valley can be seen in the two map extracts on the following page

1   Re-write the following table of selected physical features of the Earn and its valley with the six figure references provided.

Six-Figure Reference	Physical Feature
541245	meander
565234	source of River Earn
102200	levée
123196	oxbow lake
176186	upper course stream occupying broad 'V' shaped valley

2   Draw a cross-section across the Kendrum Burn, from 555234 north-east to 565244. Annotate the diagram to show features of relief and drainage.

3   Locate GR 604230. It is part of an alluvial fan. Explain the formation of this landform.

4   When describing a stretch of a river and its valley shown on a topographic map, it is helpful to use a checklist as a guide. Possible topics are shown below. Although there are two separate lists, the information may be blended into a concise answer.

River
◆ Direction of river flow
◆ Width of the river
◆ Tributaries – number, direction of flow
◆ Nature of the course – meandering, flowing straight etc
◆ Special features – ox-bow, waterfall, braiding, levées etc

Valley
◆ Degree of straightness
◆ Depth and Steepness of valley sides
◆ Width and Gradient of valley floor – flood plain, narrow, incised etc
◆ Special features – upper, middle, lower, glaciated valley etc

There is overlap between the checklists but they are a reminder of the main topics which can be discussed from topographic maps. As always with such maps, give accurate evidence by using grid references. You can also use annotated sketch maps to illustrate an answer.

Example 1:   Describe the physical features of the Kendrum Burn and its valley from 541244 to where it flows into Loch Earn.

Example 2:   Describe the physical features of the River Earn and its valley from 090198 to 180184.

5   As part of a fieldwork study along a stretch of the River South Esk and the White Water, one of its tributaries, a senior pupil produced the following table of results based on four sites. Carefully examine the data provided. Compare the river velocity downstream from sample site A to sample site D.

a)   What happens to the speed of the river from site A to site D? Suggest possible reasons for these results.

b)   Why does the bedload get smaller and more rounded from site A to site D?

c)   River speed was measured by using two different methods: either using a flowmeter or timing floats along a 10 metre stretch of the river. The method used was determined by the nature of each of the sites. What factors would likely influence the pupil's choice of measurement technique?

Field work data: white water and river South Esk					
Site	Grid Reference	River Speed	River Gradient	Bedload Size	Bedload Shape
A	237775	0.3 m/sec	40 degrees	30–400 mm	Angular
B	259759	0.4 m/sec	18 degrees	27–394 mm	Sub-angular
C	285756	0.7 m/sec	2 degrees	25–340 mm	Sub-rounded
D	298744	0.9 m/sec	1 degree	21–250 mm	Rounded

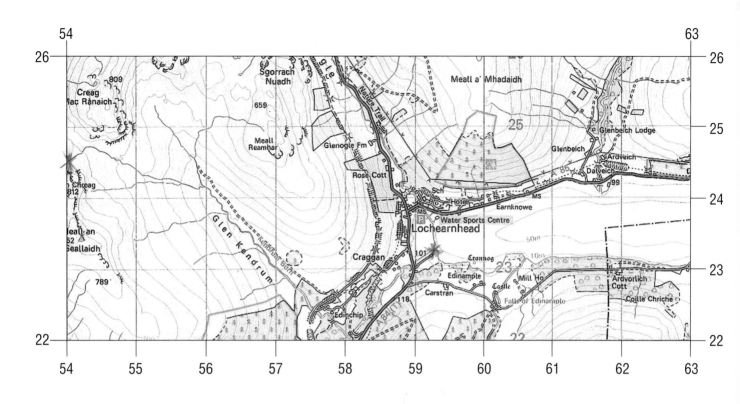

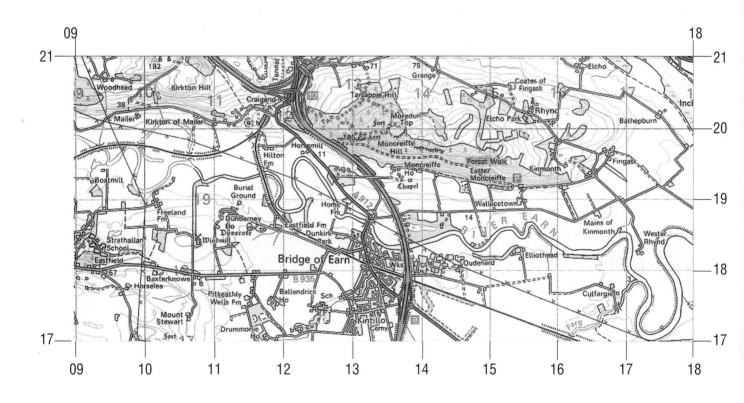

(Top): Upper course of River Earn – Loch Tay. Reduced from 1:50 000. © Crown Copyright

(Bottom): Lower course of River Earn – Perth to Alloa. Reduced from 1:50 000. © Crown Copyright

Extra Assignments

1 With the aid of Figure 3.22, and Figures 3.16–3.19:

a) Outline the main reasons for the January 1993 River Tay flood.

b) Mentioning areas affected, **briefly describe** the impact of the flood.

c) Discuss the measures, both actual and possible, that have been taken to deal with the potential flood hazard in the Tay basin.

2 a) Explain, with the aid of labelled diagrams, the differences between:

(i) radial, dendritic and parallel drainage patterns

(ii) consequent, subsequent and obsequent streams.

b) With reference to the drainage map of the Loch Etive area, Figure 3.44, how many types of drainage patterns can you pick out?

3 With the help of Figure 3.41, **write a paragraph** explaining the past links between the rivers Clyde and Tweed, and the Biggar Gap.

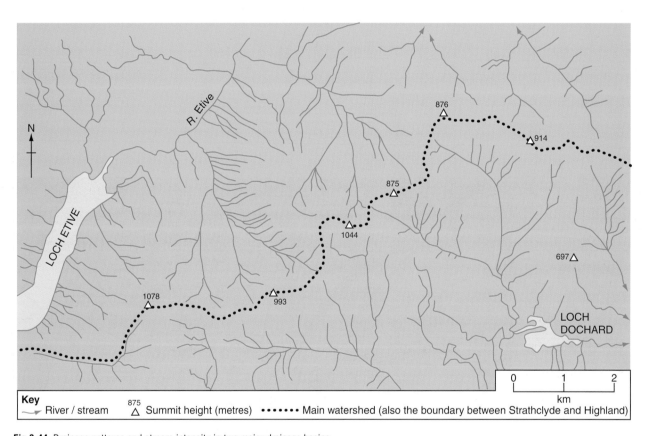

Key

→ River / stream   875 △ Summit height (metres)   •••••• Main watershed (also the boundary between Strathclyde and Highland)

Fig 3.44 Drainage patterns and stream intensity in two major drainage basins

Key terms and concepts

- abrasion (p. 58)
- alluvium (p. 62)
- attrition (p. 59)
- bankful discharge (p. 44)
- base level (p. 56)
- baseflow movement (p. 43)
- braided channels (p. 63)
- channel precipitation (p. 42)
- closed system (p. 39)
- condensation (p. 38)
- corrasion (p. 58)
- corrosion (p. 59)
- delta (p. 65)
- dendritic (p. 66)
- deposit (p. 59)
- discharge (p. 43)
- drainage basin (p. 40)
- drainage density (p. 40)
- drainage pattern (p. 66)
- erosion (p. 58)
- estuary (p. 65)
- eustatic change (p. 65)
- evaporation (p. 38)
- evapotranspiration (p. 41)
- floodplain (p. 62)
- groundwater zone (p. 38)
- helicoidal flow (p. 64)
- Hjülstrom's Curve (p. 59)
- hydraulic action (p. 58)
- hydrograph (p. 42)
- hydrological cycle (p. 38)
- hydrosphere (p. 37)
- impermeable (p. 45)
- incised meander (p. 66)
- interception (p. 41)
- interlocking spurs (p. 61)
- isostatic uplift (p. 65)
- knickpoint (p. 66)
- lag time (p. 43)
- lateral erosion (p. 61)
- levée (p. 62)
- load (p. 57)
- lower course (p. 61)
- meander (p. 63)
- middle course (p. 61)
- obsequent (p. 66)
- open system (p. 41)
- overland flow (p. 42)
- oxbow lakes (p. 65)
- parallel drainage (p. 66)
- peak discharge (p. 43)
- percolation (p. 42)
- permeable (p. 44)
- pervious (p. 45)
- point bar (p. 64)
- pools (p. 63)
- pothole (p. 60)
- precipitation (p. 38)
- profile (p. 56)
- radial drainage (p. 66)
- rejuvenation (p. 65)

- riffle (p. 63)
- river capture (p. 68)
- river terrace (p. 65)
- runoff (p. 39)
- saltation (p. 57)
- slip-off slope (p. 64)
- solution (p. 57)
- stemflow (p. 42)
- storage (p. 39)
- steam ordering (p. 40)
- subsequent (p. 66)

- superimposed (p. 67)
- suspension (p. 57)
- throughflow (p. 39)
- traction (p. 57)
- transpiration (p. 38)
- transportation (p. 57)
- trellised (p. 66)
- trunk river (p. 40)
- upper course (p. 60)
- waterfall (p. 61)
- watershed (p. 40)

Suggested Readings

Broadley, E and Cunningham, R (1991) *Core Themes in Geography: Physical* Oliver & Boyd.

Briggs, D et al (1997) *Fundamentals of the Physical Environment* Routledge.

Guiness, P and Nagle, G (1999) *Advanced Geography: Concepts and Cases* Hodder & Stoughton.

Hocking, JA and Thomson, NR (1979) *Land and Water Resources of West Africa* Moray House College of Education.

McEwen, L (1993) 'Issues in Flood Hazard Management: A Case Study of the River Tay Floods' in *Scottish Geographical Studies* edited by Dawson, AH et al.

Morisawa M (1985) *Rivers: Form and Process* Longman.

Smith, R (1993) *The Great Flood* Perth and Kinross District Council.

Waugh, D (1995) *Geography: An Integrated Approach* (2nd edition) Nelson.

White, ID, Mottershead, DN and Harrison, SJ (1984) *Environmental Systems* George Allen & Unwin.

Whittow, J (1984) *The Penguin Dictionary of Physical Geography* Penguin Books.

Whittow, J (1977) *Geology and Scenery in Scotland* Penguin Books.

Internet Sources

National Climatic Data Centre: www.ncdc.noaa.gov/
Daily flood Summary: www.ca.uky.edu/

4 THE LITHOSPHERE

After working through this chapter, you should be aware that:

◆ the British Isles have a great variety of landforms which are the result of external and internal processes operating over long periods of time

◆ denudation is an important external process and involves weathering, mass movement, erosion, transportation and deposition by rivers, glaciers and the sea

◆ weathering (physical and chemical) and various types of mass movement play an active role in modifying slopes and shaping landforms

◆ the British Isles have varied regional landscapes reflecting the interaction of structure, rock type and the varied processes of denudation. These include glaciated uplands, upland areas of upland carboniferous limestone and the scarpland scenery of south-east England

◆ these distinctive regional landscapes have offered people a variety of economic and social opportunities from the earliest times.

You should also be able to:

◆ identify and label the main landscape features on maps, photographs and sketches

◆ construct and interpret cross-sections and transects.

Understanding the physical landscape is an important part of geography. In this chapter, we examine a critical part of our physical landscape – the lithosphere. Derived from the word 'lithos' (meaning rock), it is essentially the earth's crust and upper mantle. Over millions of years, its surface has interacted with the atmosphere and the hydrosphere. The end result is distinctive landforms produced by various processes. Such landforms are the stage on which so many human activities have taken place from earliest times.

4.1 The Geological Cycle and The British Isles

'From the top of the mountain to the shore of the sea, everything is in a state of change; the rock disolving, breaking and decomposing'

Source: James Hutton

For their size the British Isles have rich and varied physical landscapes. Such landscapes are the result of the interaction between two sets of opposing natural forces which are shown on the diagram of the **geological cycle** (Figure 4.1).

Internal processes: These are powered from the intense heat of the Earth's interior. Essentially, this involves the rise and spread of currents of semi-molten rock which move the giant plates of the Earth's crust. The famous Scottish geologist, James Hutton (1726–97), was the first to realise that such processes have operated throughout millions of years of geological time. Although earthquakes, volcanic activity and folding of mountains no longer occur, they profoundly affected the British Isles during their geological evolution.

External processes: Once the internal processes have created fresh land, it is doomed to change and destruction. The external forces of wind and rain, ice and running water soon start to destroy it. An overall term for this second group of processes is **denudation** and it involves various interrelated operations shown on the diagram below (Figure 4.2).

Processes of denudation

◆ **weathering** involves the breakdown and decomposition of rock into smaller or soluble fragments.

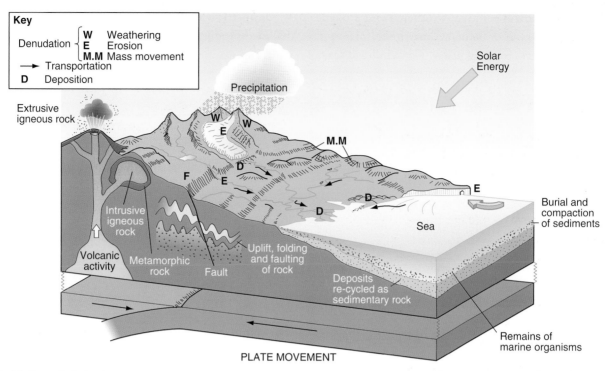

Fig 4.1 The geological cycle

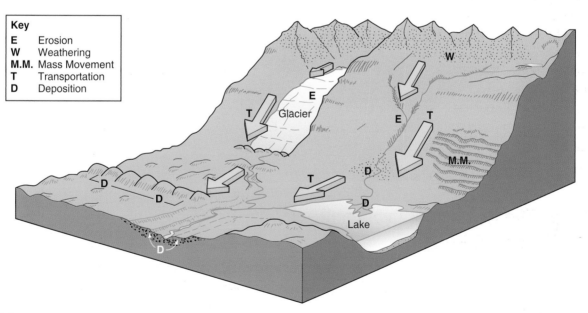

Fig 4.2 Processes of denudation

◆ **mass movement** is the movement of weathered rocks and soil down hill slopes, often assisted by water.

◆ **erosion** involves the wearing away and removal of material. Glaciers, rivers, wind and the sea are the main agents of erosion.

◆ **transportation** removes weathered and eroded material, on a kind of natural conveyor belt, by means of gravity, running water, ice, wind and sea.

◆ **deposition** is the laying down of the various denuded products moved by running water, ice, wind and sea.

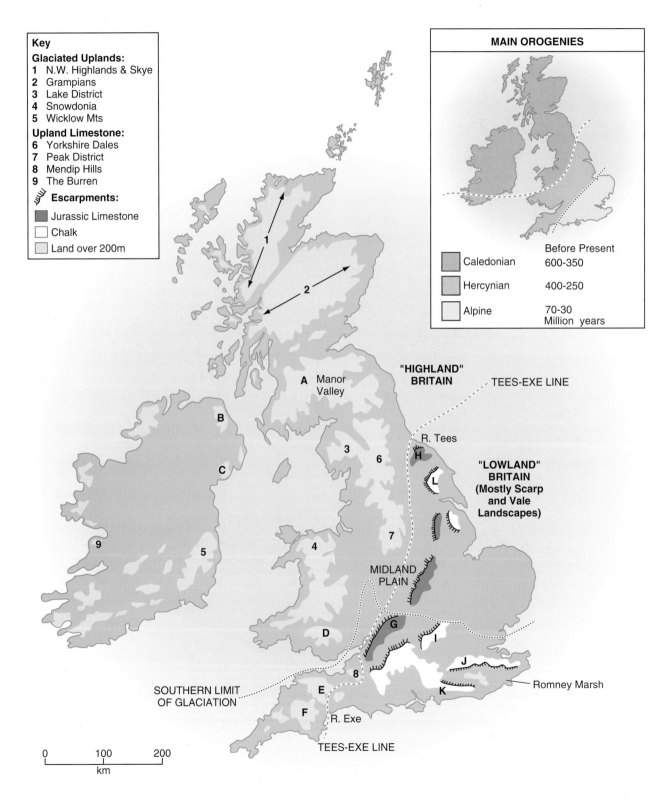

Key

Glaciated Uplands:
1 N.W. Highlands & Skye
2 Grampians
3 Lake District
4 Snowdonia
5 Wicklow Mts

Upland Limestone:
6 Yorkshire Dales
7 Peak District
8 Mendip Hills
9 The Burren

〰 **Escarpments:**
▮ Jurassic Limestone
☐ Chalk
▨ Land over 200m

MAIN OROGENIES

▮ Caledonian   Before Present 600-350
▨ Hercynian   400-250
☐ Alpine   70-30 Million years

"HIGHLAND" BRITAIN

TEES-EXE LINE

A Manor Valley

R. Tees

"LOWLAND" BRITAIN (Mostly Scarp and Vale Landscapes)

MIDLAND PLAIN

Romney Marsh

SOUTHERN LIMIT OF GLACIATION

R. Exe

TEES-EXE LINE

0   100   200
km

Fig 4.3 British Isles: selected landform features

The British Isles: a Physical 'North–South Divide'

A journey from north-west to south-east Britain is like travelling through almost 3000 million years of geological time. During this time, the workings of the geological cycle have had different effects on the landscape. Such a journey through Britain involves:

1   Travelling from the landscapes formed from the oldest and most resistant rocks to the landscapes of the youngest and softest rocks, that is, from the metamorphosed gneiss and schists of the Grampians, the tough granites of the Cairngorm Plateau and the rugged volcanic rocks of the Lake District, past the carboniferous limestones and millstone grits of the Pennines, across the sandstones of the Midland plain to the softer chalk escarpments of south-east England, which were exposed a mere 30 million years ago.

2   Travelling through landscapes formed by three main periods of mounting-building or **orogenies** (see Figure 4.3). During the Caledonian orogeny, masses of sedimentary and volcanic rock were uplifted and buckled into very high mountains. Drastically denuded, these were to be faulted in a NE–SW direction. During the Hercynian orogeny, uplift and folding were responsible for the formation

of the Pennines. Gentle folding, triggered by the Alpine orogeny in Europe, raised and tilted the chalk and clay sediments of south-east England.

3   Travelling from 'Highland Britain' to 'Lowland Britain' across the Tees–Exe line. Identified by the famous Oxford geographer Sir Halford Mackinder (1861–1947), this line broadly separates not only

Fig 4.4 Weathered tombstone at Kinfauns, Perthshire. Notice the way that the sandstone rock has split along layers, and the white lichen covering the inscription.

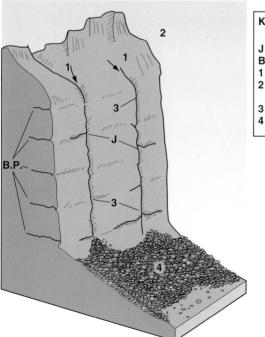

Key	
J	Joint
B.P.	Bedding Plane
1	Water collects in joints and bedding planes
2	Temperature drops below 0˚C. Water in cracks freezes Pressure exerted on cracks
3	Progressive enlargement of cracks
4	Angular rock fragments accumulate as scree

Fig 4.5 Freeze-thaw action has produced the talus slope below the Whin Sill dolerite, Crag Lough, Northern England. Also note Hadrian's Wall, built on top of the igneous escarpment

upland from lowland, but colder, wetter areas of generally poorer soil from warmer, drier areas of generally richer soil. This physical 'North–South' division has been important throughout the peopling, settlement and resource development of the British Isles from the earliest times.

4.2 Physical and Chemical Weathering

Bodies are not the only things that decay in graveyards. Figure 4.4 shows an old headstone whose sandstone layers have split because of weathering.

Weathering is the breakdown and decomposition of rock. There are two main types of weathering: physical and chemical. Both of these processes are aided by weaknesses in the rock such as the fractures called **joints**, which are often vertical, and the **bedding planes** which are the horizontal junctions between layers of sedimentary rock.

Fig 4.6 Sheeting of layers of granite has helped shape Half Dome in Yosemite National Park, California

Physical Weathering

Physical weathering is particularly (but not exclusively) active in colder or drier climates. It involves rocks disintegrating into ever smaller fragments without any change in their chemical composition.

1 **Frost-shattering or freeze–thaw** takes place at high altitudes and in cold humid areas. Water collects in rock fractures. It freezes and expands by about 9%, exerting a great deal of pressure on even the hardest rock. Repeated freeze–thaw action gradually forces the rock to split into large sharp fragments. Masses of such angular rock waste often accumulate below cliffs and mountain tops and are called **scree** (Figure 4.5). In Scotland, most scree-covered slopes built up in the tundra-like conditions immediately after the Ice Age but continued frost action ensures that fresh rock fragments still fall.

2 **Pressure release** occurs when an enormous weight of overlying rock is removed. No longer weighed down, the rock is now free to expand. Fractures develop, sometimes parallel to the surface, and **sheeting** ('pulling off' of rock layers) results in magnificent, massive granite domes such as Half Dome (Figure 4.6).

3 **Exfoliation**, sometimes called 'onion skin' weathering, for obvious reasons, occurs when layers of rock peel off from the core (Figure 4.7). Associated with hot deserts, the traditional

Fig 4.7 Exfoliation – 'onion skin' weathering of rocks in deserts

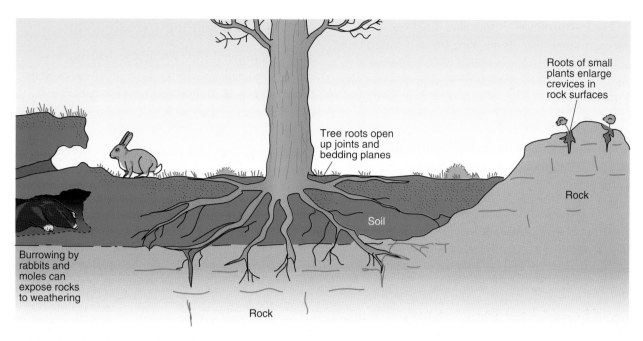

Roots of small plants enlarge crevices in rock surfaces

Tree roots open up joints and bedding planes

Rock

Soil

Burrowing by rabbits and moles can expose rocks to weathering

Rock

Fig 4.8 Biological activity and physical weathering

explanation was that the high diurnal range of temperature resulted in the outer layers of rock first expanding then rapidly contracting. Experiments now suggest that exfoliation is caused by salt crystals expanding as they dry out in rock cracks.

4   Biological activity also plays a part (Figure 4.8). From trees to dandelions, plant roots help to widen joints and bedding planes, a process most easily seen in urban areas when the spread of roots results in the breaking up of paving slabs and the bulging of tarmac surfaces. Even burrowing rabbits and moles help expose fresh material for further weathering.

Chemical Weathering

Chemical weathering is the decomposition of rock by water, oxygen, carbon dioxide and various organic acids. Its processes are especially active under wetter and (usually) warmer conditions. It is therefore most active in humid tropical and subtropical areas, for example, the Yucatan and Malay Peninsulas. It occurs because minerals in freshly exposed rock are unstable and just ripe for a change in chemical composition. Not all minerals are easy prey, however, and can be quite resistant, e.g. quartz in granite.

1   Solution and carbonation is a potent combination, especially in areas of limestone (Figure 4.9). **Solution** occurs when minerals are dissolved by water. Rainwater, however, is a dilute form of carbonic acid (see Figure 4.10 if chemically minded!). Further carbon dioxide is added when rain seeps through soil. **Carbonation** is the dissolving of limestone rock (mainly composed of calcium carbonate) and its removal in solution as calcium hydrogencarbonate. This process is partly responsible for some of the features typical of Karst scenery, such as a limestone pavement. Figure 4.10 shows how carbonation has widened and deepened bedding planes and joints to produce a distinctive landscape of clints and grykes.

Fig 4.9 Limestone pavement, Malham, Yorkshire. You should be able to point out the enlarged fissures (grykes) and paving stones (clints).

2 **Hydration** involves minerals absorbing water and expanding, thereby exerting pressure and causing disintegration. It is often linked with hydrolysis.

3 **Hydrolysis** occurs when minerals such as feldspar and mica are attacked chemically by rainwater. Figure 4.10 shows that these are two of the three constituent minerals of granite and how they are altered to form clay, leaving the chemically resistant quartz in the form of sand.

4 **Oxidation** results when iron compounds, common in many rocks, react with the oxygen in the air and form surface layers of rust.

5 **Biochemical action** involves the release of acids from organic sources such as decaying plants and various bacteria, fungi, lichens and mosses. Chemical weathering is also accelerated by an increase in the acidity of rain with the emission of sulphur dioxide and nitrogen oxide from thermal power stations, vehicles and factories.

General Comments on Weathering

Although physical and chemical weathering have been considered separately, these processes overlap and work together. This can be seen, for example, in the formation of the distinctive, rounded granite tors on the high summits of the Cairngorms and on Dartmoor. Under past warm, humid, tropical climatic conditions, deep chemical weathering of the granite took place. Figure 4.11 shows that the depth of weathering varied. Closely spaced joints allowed more rotting of the granite to take place. The tors, which finally emerged, were mainly formed where the joints were widely spaced. Later physical processes, especially **solifluction**, stripped away the chemically weathered material.

The general name give to the end product of weathering is **regolith**. It is all the **unconsolidated** (loose) material between our feet and the bedrock. It can include all types of rock debris as well as the top layers of soil.

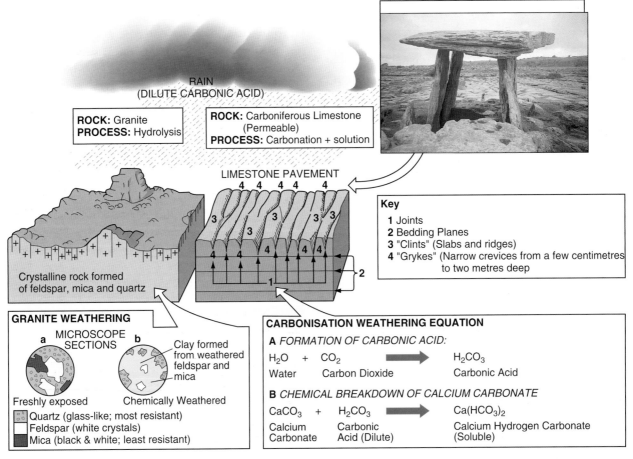

RAIN
(DILUTE CARBONIC ACID)

ROCK: Granite
PROCESS: Hydrolysis

ROCK: Carboniferous Limestone
(Permeable)
PROCESS: Carbonation + solution

LIMESTONE PAVEMENT

Key
1 Joints
2 Bedding Planes
3 "Clints" (Slabs and ridges)
4 "Grykes" (Narrow crevices from a few centimetres to two metres deep

Crystalline rock formed of feldspar, mica and quartz

GRANITE WEATHERING

MICROSCOPE SECTIONS

a
Freshly exposed

b
Clay formed from weathered feldspar and mica

Chemically Weathered

Quartz (glass-like; most resistant)
Feldspar (white crystals)
Mica (black & white; least resistant)

CARBONISATION WEATHERING EQUATION

A *FORMATION OF CARBONIC ACID:*

$$H_2O \;+\; CO_2 \;\longrightarrow\; H_2CO_3$$

Water   Carbon Dioxide   Carbonic Acid

B *CHEMICAL BREAKDOWN OF CALCIUM CARBONATE*

$$CaCO_3 \;+\; H_2CO_3 \;\longrightarrow\; Ca(HCO_3)_2$$

Calcium Carbonate   Carbonic Acid (Dilute)   Calcium Hydrogen Carbonate (Soluble)

Fig 4.10 Chemical weathering examples: The photograph shows limestone pavement in the Burren, Ireland. Part of the vertical stone of the prehistoric tomb was slotted into the east-west gryke

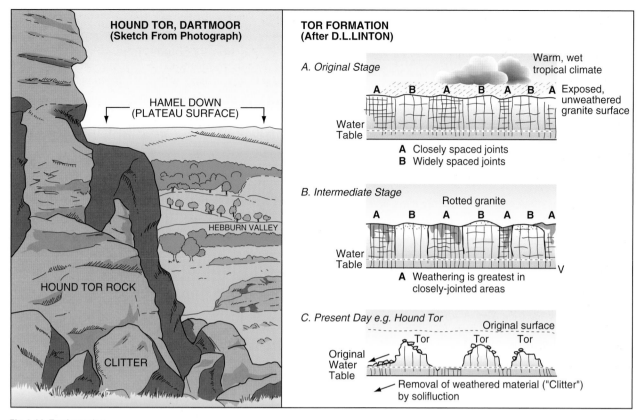

Fig 4.11 Tor formation

4.3 Mass Movement and the Landscape

Descriptions of Mass Movement

A: '. . . shifting material of every kind is objectionable and fraught with peculiar horrors. Up behind Bludenz (Austria) you may see a row of limestone cliffs called Elser Schrofen, whose foot is defended by a 'talus' of rubble which has slowly dropped down from the heights above; and a pretty thing it is, by the way, when you look closely at natural features like this talus, to observe with what flawless accuracy they have been constructed; how these fragments of detritus pass in due order through all gradations of size down the slanting surface, from minute particles like sand at the top to the mighty blocks that form their base. Once, long ago, I conceived the playful project of crossing this rubble-slope from end to end just below the cliffs. I started on its inclined plane . . . it soon struck me as being rather a steep gradient, and not only steep but ominously alive – ready to gallop downhill on a hint from myself; the mere weight of my body could set the whole mass in movement and hurl me along in a rocky flood. While making this sweet reflection I found, with dismay, that it was already too late to turn back; the least additional pressure on one foot might start the mischief; once started, nothing would arrest that deluge; its beginning . . . was going to be my end.

I was in for a ticklish business. Rush down the slope diagonally and evoke the landslide but anticipate its arrival? I preferred to remain in the upper regions, and there finished the long journey, with curious deliberation, on all fours, in order to distribute my weight; and then only by a miracle. It was one of those occasions on which one has ample leisure to look into the eye of death.'

Source: Norman Douglas: *Together*

B: 'Then came such summer rains as had not been known in the hills (Himalayas) for many seasons. Through three good months the valley was wrapped in cloud and soaking mist – steady, unrelenting downfall, breaking off into thunder-shower after thunder-shower . . . Then the sun came out and . . . the hot sunshine lasted for a week, and then the rains gathered together for their last downpour, and the water fell in sheets . . . It was in the black heart of the night that . . . Purun Bhagat heard the sound of something opening with a sigh, and saw two *slabs of the floor draw away from each other . . . he stepped out of the shrine into the desperate night . . . Down the steep, splashy path they poured all together, the Bhagat and his brothers . . . Now they were at the head of the one crooked village street . . . "The hill falls! The hill is falling! Up and out, oh you within." Then the people ran as only hill folk can run, for they knew that in a landslip you must climb for the highest ground across the valley. They fled, splashing across the little river at the bottom, and panted up the terraced fields on the far side . . . they crouched under the pines and waited until the day. When it came they looked across the valley and saw that what had been forest, and terraced field, and track-threaded grazing ground was one raw, red, fan-shaped smear, with a few trees flung head down on the scarp. That red ran high up the hill of their refuge, damming back the little river which had begun to spread into a brick coloured lake. Of the village, of the road to the shrine, of the shrine itself, and the forest behind, there was no trace. For 1.6 km in width and 600 m in sheer depth the mountainside had come away bodily, planed clean from head to heel'.*

Source: Rudyard Kipling: *The Second Jungle Book*

C: *'One of the most horrifying examples of a rapid superficial earth flow is that which devastated the tiny coal-mining town of Aberfan in South Wales in 1966. Just as the townsfolk were starting the day's work on 21 October, the 250 metres high, rain-soaked tip of mining waste which stood on the valley side above the town collapsed and flowed downslope with a thunderous roar, engulfing a row of houses and the school, before coming to rest 800 metres from the tip itself. Under the obscene black sludge of the flow 144 people were suffocated, including 116 children in the school, virtually the entire juvenile population of Aberfan. The grief-stricken parents, joined by thousands of volunteers, including miners from neighbouring coal mines, dug frenziedly in the ruins of the school and the houses, some with their bare hands but to little avail. Apparently, ample warning had been given about the potential hazard, for the tip had moved forward on two previous occasions . . . yet mining debris was still being dumped on the tip when the catastrophe occurred . . . The official investigation disclosed that for thirty years the colliery waste at Aberfan had been tipped on the hillside across a line of natural springs or groundwater seepages. Thus, the foundations of the tip had long been saturated with water, so that, when the granular material*

suddenly failed, high-pore pressures rapidly developed and the entire foundation liquefied like a quickclay.'

Source: John Whittow: *Disasters*

D: *'I saw a great mass of Upper Greensand densely covered with bushes and trees slowly crawling downwards from the highest terrace, while below a river of liquid mud was slipping over the low cliff above the beach . . . the main movement in the night must have been very rapid, as by daylight a huge fan of debris, crested with uprooted trees, had pushed across the beach to beyond low water neap tides.'*

Source: Quoted in D. Brunsden and A. Goudie:
Classic Coastal Landforms of Dorset

Definition and Types of Mass Movement

The four extracts give some idea of the nature of **mass movement**. Mass movement (also called mass wasting) is the downslope movement of the **regolith** (unconsolidated soil, stones and rock) by gravity. Gravity is not the only factor involved: others include the water content of the regolith, the nature of the rock involved, the angle of slope and the presence or absence of covering vegetation. All these factors overcome any natural resistance to movement, and then the force of gravity takes over once it is triggered (Figure 4.12). Actual slope movement may be started in various ways including: (i) heavy rainfall which can increase the weight of material and help to lubricate movement (extracts B, C and D); and (ii) vibrations which can destabilise slope material. Vibrations can be minor in scale (extract A) or be the result of earthquakes.

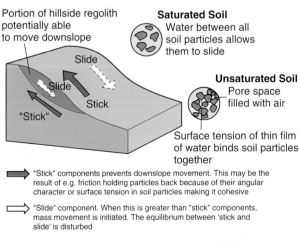

Fig 4.12 The balance between 'stick and slide' on a hillside

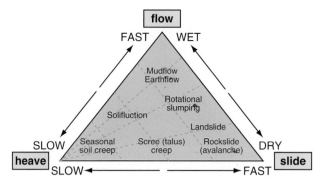

N.B. This is a relatively flexible classification. It reflects the movement of different types of material in very varied conditions

Fig 4.13 A classification of mass movement on slopes (after Carson and Kirkby)

Figure 4.13 shows one possible classification of mass movements based on the speed of flow and type of movement. Inevitably, there is overlap between these different types.

Examples of Mass Movement

1 Rockfalls and scree slopes: Figure 4.13 shows that rockfalls are the most rapid of the mass movements. Material of varying size either falls or rolls from normally bare, almost vertical cliffs. Over the years, the rock debris accumulates at the cliff base and forms a scree slope (also called a **talus slope**). Usually the talus slope consists of (i) mainly an upper straight slope with a gradient of 33°–38° which gives way to (ii) a small concave base (Figure 4.14). The accumulation of the talus is encouraged by physical weathering, particularly freeze–thaw action exploiting weaknesses in any joints and bedding planes on the cliff. Such weathering also explains the varied particle size. Generally, the larger material builds up at the base, while finer grains are washed into the spaces. In Scotland, it is acknowledged that most of the rockfall on the talus slopes accumulated towards the end of the last ice advance. Where conditions are right, for example, at corrie headwalls, fresh material continues to build up, thanks to the colder climate. Older material has often undergone further weathering, vegetation colonisation and soil formation, for example, the thin scree slope at the foot of Salisbury Crags, Edinburgh.

2 Landslides and slumps: These are occasional, rapid movements of large quantities of soil and/or rock. Often, a whole section of a slope gives way when a mass of unsupported, solid rock and/or regolith begins to slide down a slip plane. Sliding is often encouraged by one or more of the following:

◆ movement of rock along bedding planes, often weakened by weathering

◆ saturation of the surface material after prolonged rain, for instance, an overlying permeable rock, making it heavier and liable to slide. Rain can also lubricate the underlying rock, softer clay, for example, making it easier for overlying material to move

◆ undercutting of steep slopes by river erosion or the action of the sea on a cliff.

The factors encouraging sliding are common to both landslides and landslips. Landslips (also called landslumps) are a form of **rotational sliding**. This occurs when a mass of rock slumps along a curved sheer plane. Forming the largest area of landslip topography in Britain are The Storr (719 m) and Quirang in north-east Skye. On a steep scarp face, igneous rocks overlie softer sedimentary strata. After the end of the Ice Age, the sedimentary rocks were unable to support the heavier basalt blocks which slumped in a series of rotational landslips (see Figure 4.15). Other examples can be found on the Dorset coast, east of Lyme Regis, at Cain's Folly where clay rocks underlie stronger limestone and sandstone cappings. The sea undercuts the cliffs, weakens them and the final rotational slide and collapse comes after heavy rain (see Figure 4.16).

3 Flows: Unlike the earlier examples, flows take place gradually, possibly faster at one time of the year than another. Extract D is a good example of a **mudflow** (February 1958), again on the Dorset coast. Known as the Black Venn mudflow, this is the most active British example. The uppermost cliffs have collapsed in a series of rotational slides. At the base of such slips, water and clay particles accumulate. Encouraged by wet winters, mud flows emerge from the base of a rotational slip and slowly spread and pour over earlier, lower slips and flows into the sea.

A: Rockfalls and scree slopes

- Upper part of scree (talus) slope
- Free-fall of frost-shattered material
- Vegetation colonisation of lower scree slopes
- Frost action exploiting weaknesses in joints and bedding planes
- Lower, concave part of scree slope
- Joints and bedding plants
- Grading of scree material upwards from larger to smaller

B: Landslides and slumps:

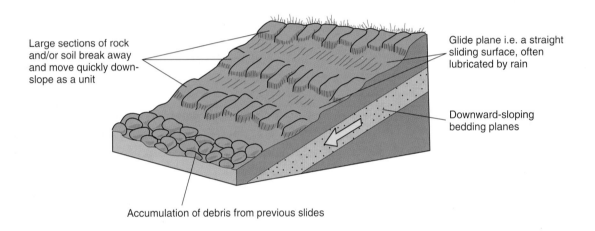

Large sections of rock and/or soil break away and move quickly down-slope as a unit

Glide plane i.e. a straight sliding surface, often lubricated by rain

Downward-sloping bedding planes

Accumulation of debris from previous slides

C: Rotational Sliding (slumps) in N.E. Skye:

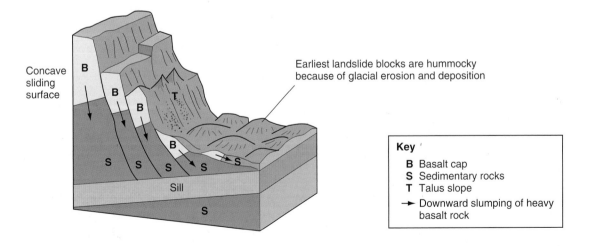

Concave sliding surface

Earliest landslide blocks are hummocky because of glacial erosion and deposition

Key

- **B** Basalt cap
- **S** Sedimentary rocks
- **T** Talus slope
- → Downward slumping of heavy basalt rock

Fig 4.14 Examples of mass movement: landslides and slumps

Fig 4.15 Quirang, north-east Skye: rotational landslips caused by heavy basalt blocks on top of sedimentary rocks form a dramatic landscape

The 1966 Aberfan disaster is another example of a mudflow (Figure 4.17). Extract C describes the tragic events in the Welsh mining village, where a combination of heavy autumn rains and overflowing springs so saturated the hillside slag heap that it behaved like liquid and swept down on the village below (Figure 4.18).

A slower form of flow is **solifluction**. This is the slow, downslope movement of saturated soil on top of **permafrost** (frozen subsoil). Often, the material forms tongues called **solifluction lobes**. These are active during the summer months when the surface layers thaw. The climatic conditions associated with such features are called **periglacial**. Such conditions are found in the Tundra today and, formerly, on the fringes of ice-covered Britain. The smooth, rounded appearance of the chalk uplands, such as the South

Fig 4.16 Rotational landslides and mudflows are responsible for the collapse and retreat of the cliffs

Downs, (see page 123) is thought to be the result of solifluction.

4 **Soil Creep:** With a speed of less than 1 mm/year, soil creep is the slowest and least dramatic of all the downslope movements. Occurring on slopes as gentle as 5°, soil creep is a widespread method of moving extensive amounts of material. Compared to landslides, for example, it is a very slow, unspectacular process but it is still very effective, the upper layers of the regolith moving downslope faster than the lower layers. The consequences of soil creep are shown in Figure 4.19 and include terracettes on the steeper slopes, tilted trees and damaged walls and roads.

Soil creep is the result of repeated expansion and contraction of particles of soil. These processes can take the form of:

◆ alternate freeze and thaw action. Particles are heaved up on freezing and are then pulled downslope by gravity on melting

◆ alternate wetting and drying. Rain encourages soil particles to expand. When drying, the particles contract and settle lower down the slope (Figure 4.20).

The significance of vegetation in ensuring that soil movement lives up to its name and 'creeps', is clearly seen when vegetation is removed. Rapid slopewash is then the result.

5 **Slopewash:** In the strictest sense, slopewash is not a type of mass movement because of the critical role of water in removing soil particles. It occurs on slopes where the amount of rainfall is greater than the capacity of the soil to absorb the water or when the watertable is at the surface of the soil. As the excess water cannot infiltrate the soil, it flows over the surface as **overland flow**. One result is **sheetwash**. This involves whole sheets of soil particles being washed downslope. Sometimes, the overland flow is concentrated into narrow channels or **rills**. These can deepen, particularly when there is little or no vegetation, into gullies which are often eroded down to the underlying bedrock.

Slopewash, particularly on unvegetated slopes, is an important cause of soil erosion. This is a perfectly

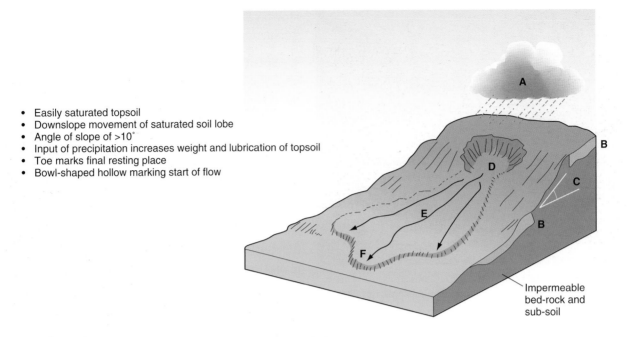

- Easily saturated topsoil
- Downslope movement of saturated soil lobe
- Angle of slope of >10˚
- Input of precipitation increases weight and lubrication of topsoil
- Toe marks final resting place
- Bowl-shaped hollow marking start of flow

ABERFAN 1966: A MUDFLOW WITH TRAGIC CONSEQUENCES>

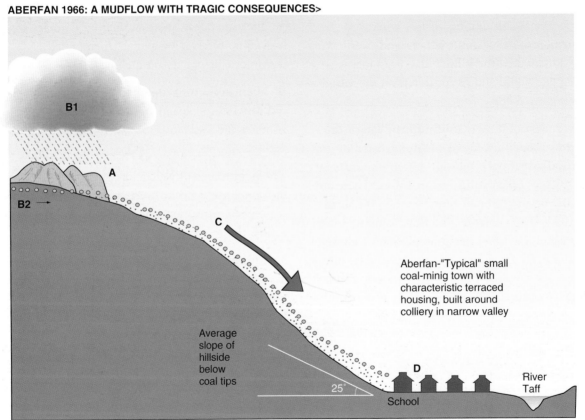

- Tragic death of over 147 people, including 116 children and 5 teachers in the primary school, in a close-knit mining community
- Coal tips, which had accumulated since the 1930s, on valley side above town
- Mudflow, >800m long from tip to toe, consisting of 10000m^3 of material
- Input from (i) heavy rainfall (ii) local spring line which increased the weight and acted as a lubricant

Fig 4.17 Examples of mass movement: mudflow

Fig 4.18 Frenzied rescue work – Aberfan 1966

Fig 4.20 Soil creep terraces can be seen on the right hand side of the photograph. The photo also shows the features of the valley of Riskinhope Burn and is a reminder of the role of gravity in helping to shape the new course of a river valley

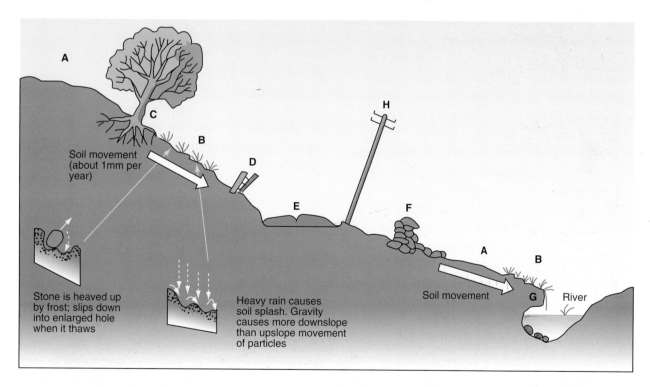

- Soil piles up behind tree trunk, which bends, and roots are exposed on the downslope side
- Overhanging soil and vegetation
- Limited evidence, but still very slow movement on slopes of 5 degrees
- Soil creep terracettes form on sections of steepest slope
- Tension cracks develop on tarmac roads
- Fences bend and break
- Soil piles up behind dry-stane dykes; resulting pressure causes bulges and breaks
- Telegraph pole tilts over

Fig 4.19 Examples of mass movement: soil creep and its consequences

natural phenomenon but increasingly it is initiated by human activities such as overgrazing and deforestation. Other ways in which people can cause mass movement are by road and rail building, quarrying and even large numbers of walkers causing extensive footpath erosion in upland areas.

Mass movement is important because:

◆ it acts in conjunction with other denudational processes, for instance, in shaping the 'V' shaped upper courses of river basins or smoothing the upland slopes of chalk escarpments

◆ mass movement of unconsolidated materials can have tragic consequences, as in Aberfan

◆ it can be partly responsible for spectacular scenery, such as the Quirang, an important part of a country's natural heritage.

To Take You Further

Some of the most spectacular mass movements took place under the North Sea and are called the Storegga slides. These underwater landslides were enormous and occurred off the coast of western Norway (Figure 4.21). Recent research suggests that huge slabs of material were deposited on three separate occasions – the most recent only about 5000 years ago. One slide was 290 km wide and covered an area of seabed equivalent to Scotland. It is suggested that a combination of earthquakes and gas released from decaying methane may have initiated the slides. Such was the scale of these landslides that they, in turn, triggered off enormous ocean waves – tsunamis. These struck the adjoining coastlines of Norway and Scotland with waves up to 10 m above high tide level. Figure 4.21 shows the tsunami deposits related to the second Storegga Slide and the resulting sandy silty deposit.

Speculation about the impact of a repeat occurrence suggests that the loss of life and flooding would be much greater than any storm surges that have so far affected the coastlines of North-West Europe. The chances of a fourth Storegga are so slight, however, that coastal management plans do not make any allowance for such an event.

4.4 Glaciated Upland Scenery

For some 2.4 million years, the landscape of the British Isles was affected by what is popularly called 'the Ice Age'. Known to geologists as the **Pleistocene** period, these years consisted of a series of successive advances and retreats of ice responding rapidly to changes in climate. Periods of extreme arctic cold and glacial activity called **stadials** were separated by shorter, warmer periods called **interglacials**. Our present interglacial dates from around 8000 BC when the last traces of glacier ice vanished.

Types of Glaciers and the Loch Lomond Stadial

All of Scotland was covered by ice at certain stages during the Pleistocene. Indeed, all of the British Isles, as far as a line between the Thames estuary and the Bristol Channel, was glaciated (Figure 4.3); to the south, the non-glaciated areas experienced tundra conditions similar to northern Russia and Canada today. Thanks to the processes of erosion, transportation and deposition, glaciers and their related meltwater rivers shaped the landscape, often sculpting distinctive landforms. Some of the freshest of these date from the 'last fling' by glaciers known as the **Loch Lomond Stadial** which lasted from 11 000–10 000 years ago. A sample landscape reconstruction from that time appears in Figure 4.22 and shows ice coverage on Skye with glaciers occupying the Cuillin mountains. These glaciers may be grouped into three types, according to their size and shape:

1 The main mass of ice, covering an area of 150 km², formed a small **ice cap** and buried most of the Cuillins. Some summits, with alpine-like peaks sharpened by freeze–thaw action, protruded above the ice. Such peaks are known as **nunataks**, an Inuit word.

2 **Valley glaciers** (slow moving 'rivers of ice') flowed from the ice cap and reached inlets of the sea, e.g. Loch Scavaig.

3 Smaller **corrie glaciers** occupied armchair-shaped hollows in the mountains.

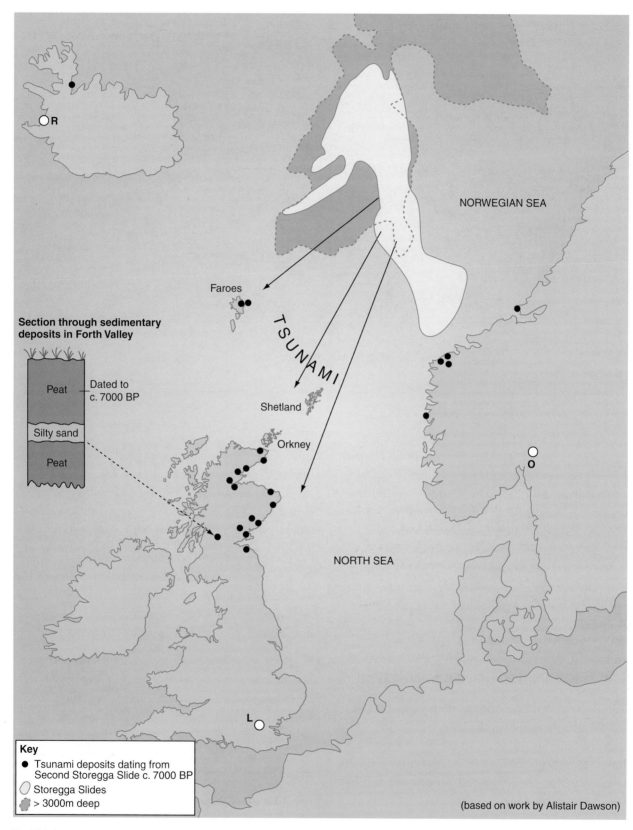

Fig 4.21 Subsea mass movement: Storegga Slides

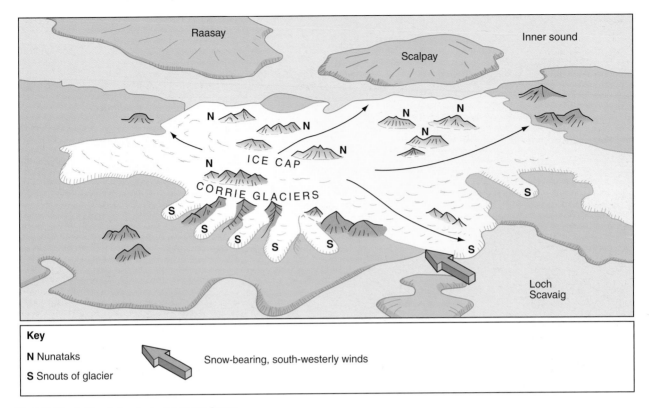

Key

N Nunataks

S Snouts of glacier

Snow-bearing, south-westerly winds

Fig 4.22 The Cuillins during the Loch Lomond Stadial

The Loch Lomond Stadial began when the climate abruptly changed in response to a slowing down and cooling of the North Atlantic Drift, whose relative warmth is normally critical for Scotland, which is on the same latitude as Labrador. Heavy snow was brought by moist south-westerly winds blowing over North Atlantic water, then some 7°–11°C colder than now. Such climatic conditions encouraged the rapid build up of ice, particularly over the Western Highlands.

Glacial Ice and the Glacier as a System

Glacial deposits in the Loch Lomond area suggest that ice was up to 600 m thick during the last stadial. Such ice built up and decayed rapidly. Figure 4.23 shows (a) how glacial ice formed and (b) the way in which glacial ice builds up and decays, which is best thought of as a system.

(a) Formation of glacial ice: Snowfall passes through several stages before becoming mature ice able to radically alter the landscape.

◆ In upland areas, snow builds up above the **snow line**. This is the lowest level of permanent snow and, during the Pleistocene, the heavy winter falls far exceeded any summer melting, particularly at higher altitudes and on north facing slopes with a shadier aspect. At this stage snow is fluffy – each snow flake is mainly air and has a delicate lace-like pattern.

◆ Over several years, the snow piles up, the air is squeezed out and granules of crystallised snow called **névé** (French) or **firn** (German) accumulate. (A literal translation of '**firn**' is snow 'from last year').

◆ After some 40–50 years, with ever more snow accumulating and by further compacting of the névé/firn, glacier ice forms. Dense, crystalline, impermeable and blue in colour, the solid glacial ice continues to accumulate. Once an adequate thickness of ice has built up, the glacier moves downslope under the influence of gravity.

(b) The glacier as a system: Figure 4.23 shows a model glacier and it is a reminder (see page 41) that, like a river, a glacier functions as a system with inputs and outputs. **Input** mainly takes place in the **zone of accumulation** – usually at the upper part of the glacier where precipitation (mainly snow but

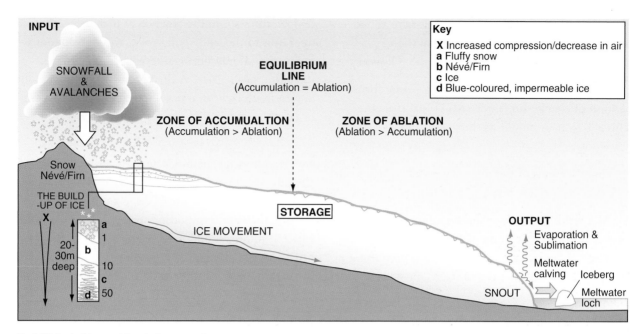

Fig 4.23 Ice build-up and the glacier as a system

also hail and rain) is highest and where avalanching from the valley sides can occur. **Storage** is the glacier itself, predominantly in the form of ice crystals. **Output** takes place in the **zone of ablation** – in the lower part where melting occurs. The various outputs include: (mainly in summer) meltwater streams (flowing from the inside of the glacier, its base and surface), evaporation and iceberg calving (icebergs breaking off) into the sea and meltwater lakes that may form at the front or **snout** of the glacier.

The overall size and ability of the glacier to shape the landscape depends on the **glacial budget**. This is a balance between the total accumulation (input) and total ablation (output) for the glacier for a year. If a glacier is advancing with its snout moving downvalley, then accumulation is greater than ablation and this is referred to as positive net balance. If the glacier is retreating, then ablation exceeds accumulation and the loss is called a negative net balance.

Processes of Glacial Transportation and Erosion

Glacial Transportation

Figure 4.24 show the distance that ice is capable of transporting material. An **erratic** is a rock which is not found locally but has been moved by glacial ice (see page 108). Erratics of rock such as the Loch Doon granite help trace patterns of glacial movement, sometimes over substantial distances. Glaciers, therefore, like rivers, carry out the dual function of transportation and erosion. Effectively, therefore, glaciers function as conveyor belts transporting eroded material. Unless eroded material is moved, the erosion process will come to a stop.

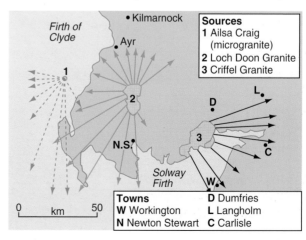

Fig 4.24 Transportation of erratics in south-west Scotland

How does solid ice actually flow? Ideas are varied but, depending on the temperature of the ice, movement involves one or more of the following:

1   Ice over 30 m thick can flow in a plastic form because of changes in its internal crystal structure. This allows it to mould its shape as it moves downslope.

2   Large sections of the upper layers of a glacier can slide past each other along fault lines.

3   As a result of pressure, a layer of meltwater forms upstream from rocks in the pathway of the ice. This lubricates the movement of the glacier, permitting it to slip downwards over the obstacle and then re-freeze.

The pace at which a glacier moves is slow, ranging from a few metres to several kilometres a year. Much depends on whether they are warm-based or cold-based glaciers. **Warm-based glaciers** erode more efficiently because they put on a turn of speed, glaciologically speaking! Not unlike the skater gliding across the ice on a thin film of 'meltwater' caused by the pressure of the skate, the warm-based glacier typically has a summer temperature of around 0 °C at its base and therefore lubricating meltwater is released. **Cold-based** glaciers have lower base

temperatures so, in the absence of meltwater, they freeze on to the underlying rock. Any movement is minimal and only occurs internally. Less erosion, therefore, takes place. A final point about movement: due to friction, both types of glacier move faster on or near the surface than at depth, and in the middle than at the sides.

As a transporting agent, glaciers can be very effective. Visually, this explains why the surface of a glacier is anything but 'snow white'. Glaciers can be grimy because they are transporting material denuded from adjacent hillslopes and it can cover much of the surface of the glacier. The overall name for the debris is **moraine** and it can be found on the surface, at the base and within the glacier (Figure 4.25).

Glacial Erosion Processes

It is accepted that glaciers are effective agents of erosion yet glacier ice is slow moving and softer than rock. Basically, glaciers pick up rock fragments and use them as erosional tools. Two main interacting processes are involved:

1   **Plucking:**   ice freezes on to bedrock, fractures it and incorporates it into the base and side of the glacier. It is really a form of glacial quarrying. The

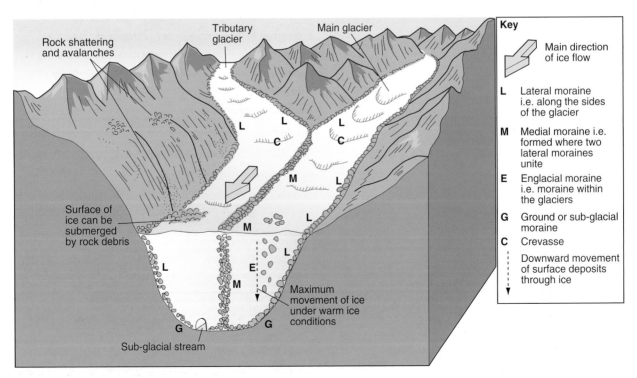

Fig 4.25 Glacial transport and moraine movement

Fig 4.26 Striations scored by rocks embedded in a glacier

process is made easier if frost action has already helped to loosen the material and the rock is well-jointed. Also, as the rock is removed under the ice, the rock beneath will expand and further break up. Surfaces affected by plucking tend to be jagged (Figure 4.37).

2 Abrasion: tends to produce smooth landforms and results from angular pieces of rock grinding away at the valley sides and floor. One indication of this process is the presence of **striations** cut into bare rock surfaces. These are grooves gouged out by the rocks embedded in the glacier's base and ranging in size from a few centimetres to a metre in depth and up to 100 m long. Examination of glacial striations can suggest the direction of past ice movement (Figure 4.26).

The effectiveness of these processes depends on various factors, including: whether they are warm-based or cold-based glaciers; the degree of resistance from the bedrock; the gradient of the land being eroded; the input from rockfalls and frost-shattering landing on the glacier and, finally, the amount of pre-glacial weathering of bedrock.

Although discussed later (see page 105), it needs to be remembered that glacial meltwater is also a very important agent of erosion and transportation in its own right. Often flowing on the surface, sometimes disappearing down openings and **crevasses** (deep fissures in the glacier caused by stresses in the ice), meltwater is capable both of assisting glaciers and also shaping its own landforms.

Landforms Resulting from Glacial Erosion

Glacial erosion dramatically modified the main pre-glacial landforms and drainage patterns of Scotland. It must be stressed that impressive, 'classic' erosional features, such as corries and glacial troughs are the end product of several periods of glacial activity. It is possible to group the erosional landforms by size.

Large-scale Erosional Landforms:

Corries

Corries (from the gaelic 'coire') are also known as cirques [French] and cwms [Welsh]). Corries are very distinctive landforms. Make use of Figure 4.27 as you read this section. Usually, the 'classic' corrie is a large, deep, rock hollow or basin (not unlike a bucket-shaped armchair) cut into a mountainside or the edge of a plateau such as the Cairngorms. Coire an Lochain (9803) and Coire an t-Sneachda (9903) are located on the northern edge of the Cairngorms. The map shows that their semi-circular, steep backwalls are topped by rugged cliffs. Most Scottish corries do not contain corrie lochs but the Cairngorms have the highest examples, with seven out of nine such lochans in the area over 900 m. A good example is the moraine-dammed Loch Coire an Lochain (9400), north-west of Braeriach (Figure 4.28), Scotland's highest corrie loch (see extract below). The map shows that the corries already mentioned are orientated toward the north. This is not true of all corries but a northerly aspect meant greater protection from the sun. Snow, therefore, was able to heap up and gradually initiate the various processes of corrie formation. These processes are outlined below and summarised in Figures 4.29A–C:

◆ corries began life as snowpatch hollows. These formed when patches of snow built up in high level, mountain side hollows, mainly oriented between north to north-east.

◆ the hollow gradually deepened into a nivation hollow as névé built up under more snow. Then ice and yet more ice built up. Ice accumulation was assisted by spring–summer thawing. During a thaw, water penetrated cracks in the base rocks of the hollow. Slow, steady freeze–thaw action repeated over the years caused more rock to break up. Thanks

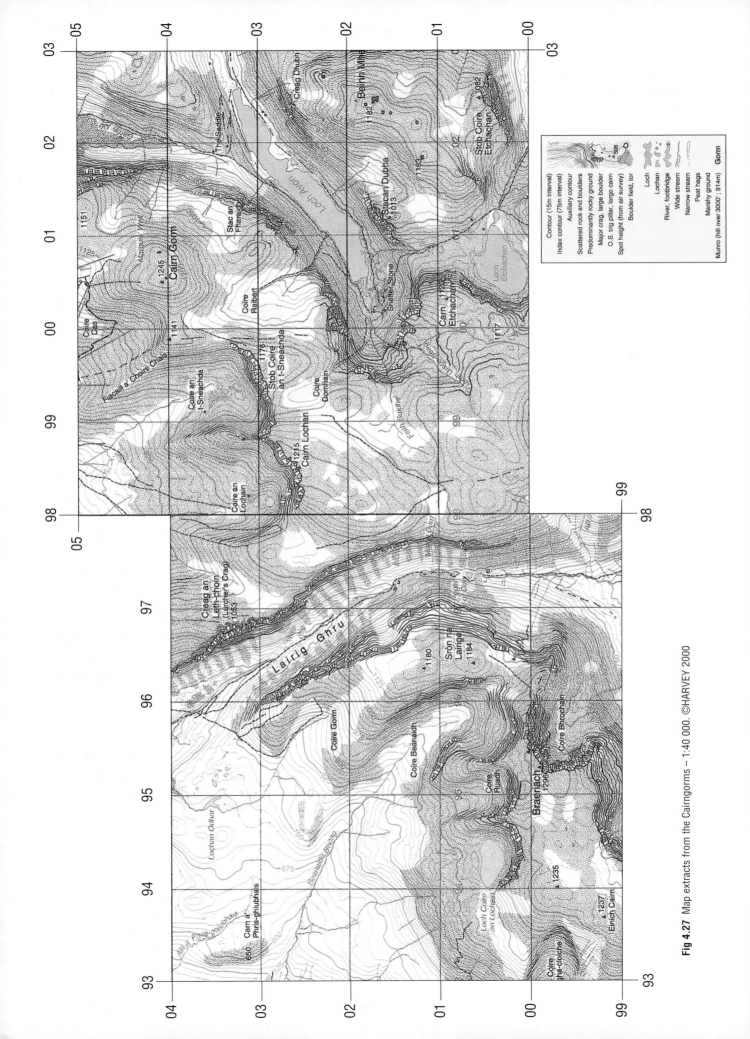

Fig 4.27 Map extracts from the Cairngorms – 1:40 000. ©HARVEY 2000

to meltwater streams, the disintegrated material was removed, the hollow was enlarged and a steep backwall formed.

◆ continued accumulation of snow–névé–ice produced a corrie glacier. Pressure from fresh, annual inputs of ice caused the glacier to flow by **rotational sliding** which resulted in backwards and downwards erosion. Backward erosion was assisted by plucking at the corrie backwall and the resulting embedded debris abraded the corrie base. Figure 4.29B shows that freeze–thaw action on the backwall also supplied further abrasion 'tools' as loosened debris fell down the bergschrund. The bergschrund also allowed summer meltwater penetration which helped lubricate rotational movement. (The **bergschrund** is a large crevasse at the back of a corrie glacier and separates flowing ice from ice attached to the backwall).

◆ rotational movement meant that erosion at the corrie edge was not so powerful. A **sill** or **rock lip** developed at the corrie entrance. Often this was covered with moraine deposited during the most recent phase of glaciation – the Loch Lomond Stadial. In some corries a deep, rounded lochan (tarn in the Lake District) formed from remaining ice, decaying as the climate warmed. Frost action encouraged fans of scree (talus) to build up below the headwall, often extending to the corrie floor.

Climb as often as you will, Loch Coire an Lochain remains incredible. It cannot be seen until one stands almost on its lip, but only height hides it . . . it is not shut into the mountain but lies on an outer flank, its hollow ranged daily by all the eyes that look at the Cairngorms from the Spey . . . Two cataracts, the one that feeds it, falling from the brim of the plateau over rock, and the one that drains it, show as white threads on the mountain. Having scrambled up the bed of the latter
. . . one expects to be near the corrie, but no it is still a long way off. And on one toils into the hill. Black scatter of rock, pieces as large as a house, pieces edged like a grater. A bit of tough going. And there at last is the loch, held back tight against the precipice. Yet as I turned, that September day, and looked back through the clear air I could see out to ranges of distant hills. And that astonished me. To be so open and yet so secret . . . the Loch of the Corrie of the Loch . . . this distillation of loveliness!
Source: From *The Living Mountain* by Nan Shepherd

Aretes and Pyramid Peaks

Corries are not solitary landforms. Usually they occur in groups, often directly adjoining one another. Dividing one corrie from another are the 'arms' of the 'armchair'. These are called **aretes** and are steep sided ridges which formed as the hollow deepened to form the corries. Frost-shattering further sharpened the ridges leaving them rocky, jagged and often with a partial cover of scree. One named example on the map is Facaill a' Choire Chais (9904). Associated with aretes are **pyramid peaks**. These steep-sided, frost-shattered, isolated summits formed when the backwalls of three or more corries converged, thanks mainly to plucking action. The 'classic' example is the Matterhorn but Scotland has a few pyramid peaks – the best group is found in the Cuillins (Figures 4.30/31). Instead, many Scottish peaks often tend to be particularly steep on one or two sides only.

Glacial Troughs and Ribbon Lochs

Figures 4.27 and 4.32 show one of the most impressive **glacial troughs** in Scotland. Glen Avon/A'an has been cut deeply into the Cairngorm plateau (see extract below) thanks to glacial erosion which widened, over-deepened and straightened the former pre-glacial, 'V' shaped valley. The resulting landform is a steep-sided glacial trough. Troughs vary, depending on rock hardness, glacier size and intensity of erosion but the following features are common:

◆ Although they are often called 'U' shaped valleys, the actual cross section is generally parabolic in shape. This partly reflects later deposition of moraine, scree and alluvial material which masks the lower slopes of the valley (Figure 4.33).

◆ typical upland pre-glacial valleys frequently had interlocking spurs. Glaciers, unlike rivers, are just not manoeuvrable. As valley glaciers advanced, therefore, they abraded these protrusions leaving **truncated spurs** and steepened the sides of the trough (Figure 4.34).

◆ steepening is also pronounced at the **trough head**. This marks the descent of glacial ice into the valley, usually as corrie glaciers descending or from an ice cap. (In the case of Glen Avon/A'an the ice flowed from an ice cap, flowing across the Cairngorm plateau in a north-easterly direction). The

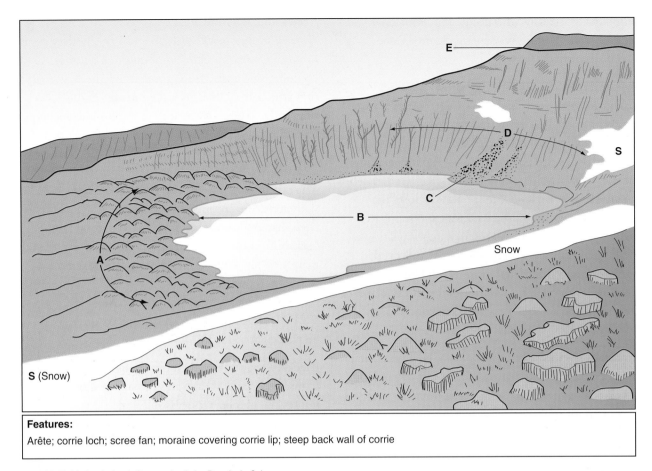

Features:
Arête; corrie loch; scree fan; moraine covering corrie lip; steep back wall of corrie

Fig 4.28 Field sketch: Loch Core an Lochain, Braeriach, Cairngorms

trough head drops some 300 m from the plateau to the trough below

◆ compared to the usually smooth cross section, a trough frequently has an **irregular long profile**. This is the result of selective over-deepening, particularly by abrasion. Abrasion can be very active in certain stretches of a trough, as a result of various factors including: rotational sliding by the glacier; the addition of a tributary glacier, giving more erosional power; zones of weaker rocks along fault lines; and squeezing the glacier between harder rock constrictions, forcing further downcutting. The result is one or more elongated **rock basins**, deeper than the next section of the trough floor. A rock basin may become the site of a long, narrow, post-glacial **ribbon loch** (e.g. Loch Avon/A'an). Britain's deepest ribbon loch is Loch Morar, 315 m deep, yet the loch's surface is merely 15 m above sea level. It only just missed becoming a sea loch or fiord. Typical of Scotland's fretted west coast, a sea loch is the 'drowned' end of a glacial trough, flooded as a result of the post-glacial rise in sea level

◆ present day rivers draining glacial troughs are small in relation to the trough. These rivers are called **misfit rivers** because they lack the energy to erode such large troughs.

I had climbed all six of the major summits, before clambering down the mountain trough that holds Loch Avon. This loch lies at an altitude of 690 m, but its banks soar up for another 450 m. Indeed farther, for Cairn Gorm and Ben MacDhui may be said to be its banks. From the lower end of this 2.5 km gash in the rock, exit is easy but very long ... But higher up the loch there is no way out, save by scrambling up one or other of the burns that tumble from the heights.

... I first saw it (Loch Avon) on a cloudless day of early July ... when the noonday sun penetrated directly into the water, we bathed. The clear water was at our knees and at our thighs ... We waded on into the brightness ... then I looked down ... we were standing on the edge of a shelf that ran some yards into the loch before plunging down to the pit that is the true bottom.
Source: From *The Living Mountain* by Nan Shepherd.

A (on a smaller scale than b and c)

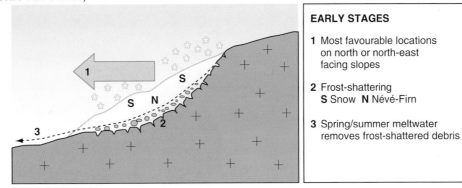

EARLY STAGES

1 Most favourable locations on north or north-east facing slopes

2 Frost-shattering
   S Snow **N** Névé-Firn

3 Spring/summer meltwater removes frost-shattered debris

B DURING PHASE OF ACTIVE GLACIER OCCUPATION

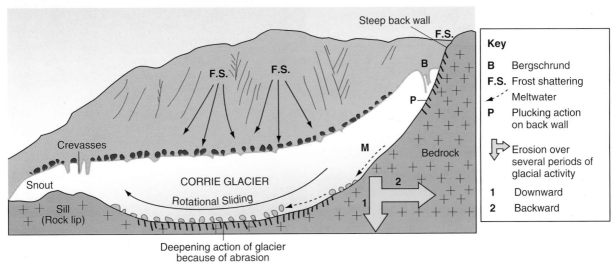

Deepening action of glacier because of abrasion

Key

B Bergschrund

F.S. Frost shattering

- - - Meltwater

P Plucking action on back wall

⇨ Erosion over several periods of glacial activity

1 Downward

2 Backward

C CORRIE(& LOCHAN) TODAY

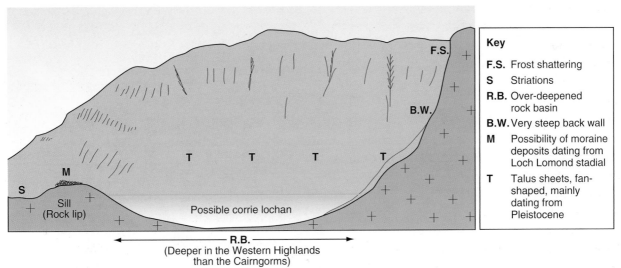

Key

F.S. Frost shattering

S Striations

R.B. Over-deepened rock basin

B.W. Very steep back wall

M Possibility of moraine deposits dating from Loch Lomond stadial

T Talus sheets, fan-shaped, mainly dating from Pleistocene

Fig 4.29 Corrie formation

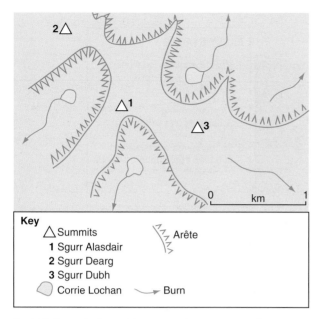

Key
△ Summits
1 Sgurr Alasdair
2 Sgurr Dearg
3 Sgurr Dubh
◯ Corrie Lochan   → Burn
〈〈〈 Arête

Fig 4.30 Sgurr Alasdair and the surrounding corries in the Cuillins, Skye

Hanging Valleys

Glaciers following pre-glacial river valleys repeat the pattern of a main trunk glacier fed by smaller tributaries. Size matters: tributary glaciers have less erosional power and are not able to excavate vertically as efficiently as the main glacier. After ablation, the result is a hanging valley, that is the smaller, tributary trough, ends abruptly above the lower trunk valley. This can be an ideal site for

Fig 4.31 The pyramid peaks of the Black Cuillin, the closest in Britain to Alpine scenery

Fig 4.32 Loch Avon/Loch A'An. Match this with the map extract. Try to work out the direction of the photographer was facing. What features can be clearly picked out on both the map and photograph? Note the recreational use of such areas, with geomorphological impact.

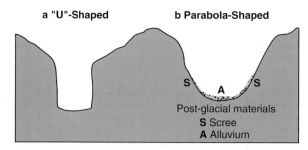

Fig 4.33 Glacial troughs: model cross-sections

waterfalls to develop. Figure 4.35 shows the Grey Mare's Tail, the famous Dumfriesshire waterfall. Draining from Loch Scene, a corrie lochan, the Tail Burn drops from its tributary valley down to Moffat Dale, a long, straight, fault-guided trough.

Taking Scotland as a whole, by far the greatest concentration of major erosional features, that is, corries, troughs and rock basins, is in the western Highlands. This reflects the overall higher altitude and the heavier precipitation, past and present, of the area and therefore the best conditions for widespread severe erosion. In the east, glacial erosion was more selective but locally was severe and in the Cairngorms the plateau tops were little affected by cold-based ice, while faster moving, warm-based ice over-deepened the troughs below.

Small-Scale Erosional Landscapes

Roches Moutonnées

Roches moutonnées are resistant, rocky hillocks (from roughly ten to a hundred metres long and from several to a hundred meters high in Scotland) commonly found in glacially eroded terrain. Figure 4.36 shows one of a series near Grantown on Spey. Their shape and character reflects the nature of the rock and a combination of abrasion and plucking erosion. They also demonstrate the direction of glacier flow. Figure 4.37 shows that the streamlined appearance of the upstream part, or **stoss**, is the product of abrasion by ice embedded stones. As it is worn away, striations are frequently incised into its surface. The downstream part, or **lee side** is craggy and is the result of plucking. This pulls away blocks of rock, a process encouraged by the weakening effect of jointing in the rocks.

(The name roches moutonnées was first coined in 1787 by the geologist and mountaineer De Saussure

because their shape was similar to *moutonnées*, fashionable, wavy wigs).

Crags-and-Tails

In looking at glacial erosion in upland areas, objections could be made to any consideration of crags-and-tails at this stage in the text. They are included because:

◆ there are good examples in Central Scotland including Dundee Law, North Berwick Law and Traprain Law. Obviously these are not upland landforms and neither are Edinburgh's crags-and-tails. Its largest 'classic' crag-and-tail provided the site for the nucleus from which Edinburgh grew

◆ it is often stated that crags-and-tails are a depositional feature but decide for yourself to what extent erosion played the critical role in the formation of Edinburgh's main crag-and-tail. Figure 4.38 shows that Edinburgh Castle stands on the crag. This is a volcanic plug formed of very hard rock (basalt), the much denuded remains of a former

a. Pre-Glacial Landscape

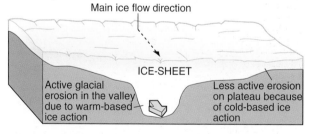

1 "V"-shaped cross section of river valley   2 Inter-locking spurs

b. During Glaciation

Main ice flow direction

ICE-SHEET

Active glacial erosion in the valley due to warm-based ice action

Less active erosion on plateau because of cold-based ice action

a. Post-Glacial Landscape

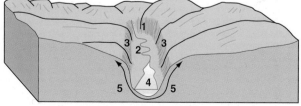

1 Steep trough head	2 Misfit river
3 Truncated spurs	4 Ribbon loch occupying rock basin
5 Glacial trough	

Fig 4.34 Glacial trough evolution in the Cairngorms (very simplified)

Fig 4.35 The Grey Mare's Tale, a famous Dumfriesshire waterfall, draining from a hanging valley

Fig 4.36 This shows a Roche Moutonnée near Grantown on Spey. The display board reminds us that sites of geomorphological importance are part of our landscape heritage.

volcano which erupted some 350 million years ago. The crag's hard rock resisted erosion more effectively than the tail. The tail is formed from softer sedimentary rock which once completely surrounded the plug. Ice advancing from the west was diverted over and around the crag, scooping out and over-deepening the sedimentary rocks at the front and sides of the crag by some 60 m. The sedimentary rocks, immediately downstream in the lee of the crag, were sheltered and formed a 'tail'. This tapering tail dips north-eastwards and was the site for Edinburgh's Old Town. Other Edinburgh examples include Calton, Craiglockhart and Blackford Hills.

To Take You Further

Other Glacially Eroded Features

1 Diffluence channels: these are glaciated valleys that formed when excess ice from one valley overflowed a col (a depression in a mountain range) into a neighbouring valley. One Cairngorm example is The Saddle, shaped by ice overflowing from Glen Avon/A'an northwards into Strath Nethy (see Figure 4.27).

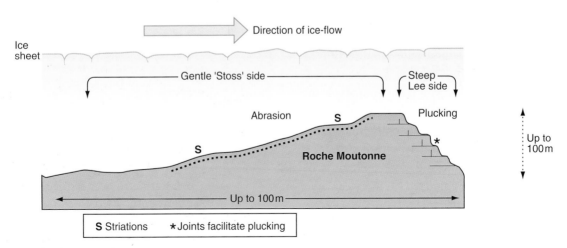

Direction of ice-flow

Ice sheet

Gentle 'Stoss' side

Steep Lee side

Abrasion   S   Plucking

S

Roche Moutonne

Up to 100m

Up to 100m

S Striations   ∗Joints facilitate plucking

Fig 4.37 Formation of roche moutonnée

2 Knock and lochan topography: Although technically a low lying feature (average height of 100 m) this glaciated landscape is found in the North-West Highlands and the Outer Hebrides. Figure 4.39 shows a lowland of irregular relief consisting of rough bare bedrock shaped from Lewissian gneiss. Ice has scoured the surface leaving roches moutonnées, rocky knolls (knocks) and small lochans. Rising above such a landscape are residual mountains (Canisp, Suilven and Stac Pollaidh are examples), formed from Torridonian Sandstone.

3 Meltwater channels: Glacier meltwater is a very potent erosional agent and is capable of eroding channels such as that shown in Figure 4.40 on the edge of the Pentland Hills. As the ice melted at the end of the Loch Lomond Stadial, the lower slopes of uplands emerged above the ice. Figure 4.41 shows meltwater flowing at the edge or just under the ablating ice. One particular feature of subglacial flow is that water flows at high pressure and can flow both uphill and downhill. Such meltwater flow has created a complex network of channels dissecting the lower

Fig 4.39 Suilven rises above a platform of Lewissian Gneiss that has been glacially eroded into knock and lochan topography

slopes of the Cairngorms, the Pentlands and the Southern Uplands.

Landforms Resulting from Glacial Deposition

Around 10 000 years ago, the end of the last phase of ice advance (the Loch Lomond Stadial) was marked by a rapid rise in temperature. Deglaciation was under way and it gradually revealed a variety of landforms resulting from glacial deposition. Broadly speaking, these landforms are very widespread, first, across the lowlands of central, eastern and southern Scotland and, second, on the valley floors of upland

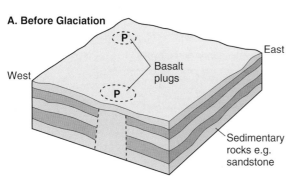

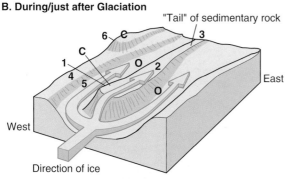

Fig 4.38 Formation of crag and tail landscapes in Central Edinburgh

Fig 4.40 Carlops meltwater channel on the edge of the Pentland Hills. Any comments on the two large mounds on the left of the picture?

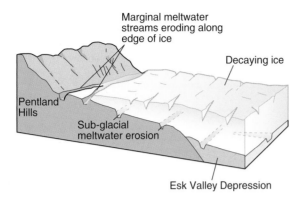

Fig 4.41 Formation of meltwater channels (see Figure 4.40)

areas. Other glacial deposits now lie offshore, for example, the 'Wee Bankie' moraine (Figure 4.42), drowned after the post-glacial rise in sea level. An overall name for glacial deposits is **drift**. Most drift was deposited in the ablation zone, particularly near the snout. Drift can be divided into two groups: (i) sediments which are directly deposited by the ice and called **till**; (ii) sediments that are deposited by the vast quantities of meltwater found at the margins of decaying ice and are called **fluvioglacial deposits**.

Till Deposits

Subglacial Till Deposits
These are an intriguing mixture of material that has been directly carried and deposited under the ice (subglacially). Typically, till is **unsorted**, **unstratified** (non-layered) and consists of varying sizes of **rocks**, **clays**, **sands** and **silts**. Normally, the **rocks are fairly angular** in shape, not rounded like those on a river bed. As the glacier melted, the till spread out, covering the bedrock to form till sheets. Such sheets are the main parent material for most of Scotland's soils.

Drumlins
This name (from Gaelic, **druim**, a mound) is given to streamlined, oval hillocks mainly formed from thick till, sometimes with a core of rock. Frequently, they occur in groups or **swarms**. Figure 4.42 shows their distribution in Scotland, especially in the Glasgow area (including the city), the Merse and lowland Galloway. They tend to come in various shapes and sizes but Figure 4.43 shows some characteristic features: a steeper, blunt end or **stoss**, facing upstream towards advancing ice; a gentler, downstream **lee slope**; an

average height of 50 m in lowland areas, a breadth of some 500 m and a long axis length of 1 km or more. Figure 4.43 shows sample drumlins in the Merse where they form a rolling terrain. The map shows that the long axes of the drumlins run parallel to each other and to the direction of glacier flow. Ideas about drumlin formation vary: one possibility is that previously deposited till has been dragged along and moulded subglacially into its elongated form by later, fast-flowing ice; another is that drumlins are erosional features, carved from the till by ice movement. Whatever the cause, the result, as you look out over a drumlin swarm, is a distinctive 'basket of eggs' topography.

Ice-marginal Moraines
As well as forming sheets, till forms ridge-like

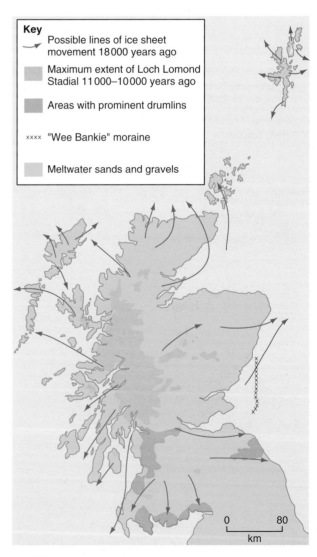

Fig 4.42 Aspects of glaciation in Scotland

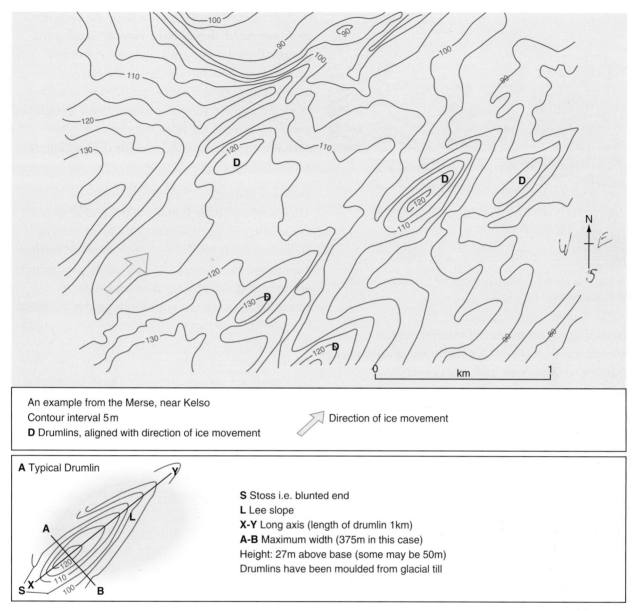

An example from the Merse, near Kelso
Contour interval 5 m
D Drumlins, aligned with direction of ice movement

Direction of ice movement

A Typical Drumlin

S Stoss i.e. blunted end
L Lee slope
X-Y Long axis (length of drumlin 1km)
A-B Maximum width (375m in this case)
Height: 27m above base (some may be 50m)
Drumlins have been moulded from glacial till

Fig 4.43 A drumlin swarm ('basket of eggs' topography)

features and hummocks along the sides of a glacier. These are moraines and are mainly the result of:
(i) the dumping of debris from the ablating ice, and
(ii) the pushing of loose, ground material by the ice. Several types of moraine can be distinguished:

1 Terminal (or End) moraines mark the maximum extent that a valley or corrie glacier or ice cap has advanced. In upland areas, a terminal moraine is formed from conspicuous mounds of poorly sorted till that runs across a valley, or sometimes a corrie mouth, at right angles to the direction of ice flow. Their size varies according to

age and the amount of ice; some examples are up to 30 m high and often littered with large, protruding angular rocks. Certain terminal moraines hold back lochs and corrie lochans. Finally, it is not unknown for glaciers to re-advance after a period of ice standstill. If that happens, previously deposited moraine may be bulldozed into large, mound-shaped **push** moraines.

2 Lateral moraines develop along the edges of an advancing glacier. These form from debris eroded by the advancing ice and especially from frost-shattered material loosened at the valley sides. As the

Fig 4.44 This shows hummocky moraine in the Creag Meagaidh National Nature Reserve near Loch Laggan

supporting ice ablates, lateral moraines start to collapse but they can still form distinctive landforms today, sometimes with fresh scree covering.

3   Medial moraines form on active glaciers where two lateral moraines merge when two glaciers flow together. Often they extend downward to the base of the enlarged glacier.

4   Hummocky moraine can also be seen in the Torridon area at Coire nan Ceud Cnoc ('the Valley of a Hundred Hills') and along the floors of many other Highland glens. It consists of a large number of seemingly irregular mounds, closely heaped up together. Current thinking suggests that they are really closely spaced push moraines, formed during a series of small but repeated re-advances of active glaciers (Figure 4.44).

Erratics

These are glacially transported blocks of rock (see page 95) which, like drumlins and moraines, are good clues when attempting reconstructions of the former flow and activity of ice sheets (Figure 4.45).

Fluvioglacial Deposits

During deglaciation, vast amounts of meltwater were released, often at high velocity. It flowed beneath, through and over the decaying ice, transporting considerable amounts of material. Fluvioglacial material differs from till in that it is **sorted**, **stratified** (layered) and, because it was transported

by water over longer distances than ice, consists of more **rounded deposits of mainly sand and gravel** (Figure 4.46). (NB: fluvioglacial deposits can include large boulders, such was the force of the water). Basically, the meltwater deposits reworked morainic material in two places (i) in the **proglacial zone** beyond the ice front and, (ii) in the **ice-contact zone** immediately beside the ice (Figure 4.47).

Outwash plains

These are important features of the proglacial zone. Often, these were deposited at valley mouths by meltwater rivers much larger than those occupying the present day channel. Because of the vast amount of outwash deposits, river flow frequently was braided as it sorted out sands from gravels. South of Blairgowrie is an extensive outwash plain spread by the glacial ancestor of the River Ericht. The heavier, coarser gravels are closest to the uplands while the finer sands are farthest away from the former ice front. Nearby are Fingask and Stormont lochs, two small water-filled depressions called **kettle holes**. These depressions were formed by the burial of a remnant piece of ice left behind as the glacier retreated. The ice slowly thawed, the covering gravel collapsed and a depression remains. If the depression is deep enough to tap the water table, a kettle loch forms. Kettle holes do not always contain lochs. If they do, the lochs need not be small: Loch Leven, Kinross-shire, is perhaps the best known example.

Landforms typical of the ice contact zone are shown in Figure 4.47. Many lower slopes of upland valleys

Fig 4.45 Several erratics sit on the top of this ice-smoothed rock

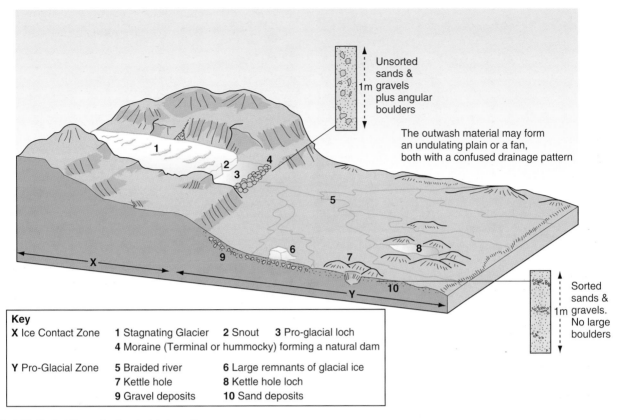

Unsorted sands & gravels plus angular boulders

1m

The outwash material may form an undulating plain or a fan, both with a confused drainage pattern

Sorted sands & gravels. No large boulders

1m

Key

X Ice Contact Zone   **1** Stagnating Glacier   **2** Snout   **3** Pro-glacial loch
       4 Moraine (Terminal or hummocky) forming a natural dam

Y Pro-Glacial Zone   **5** Braided river   **6** Large remnants of glacial ice
       7 Kettle hole   **8** Kettle hole loch
       9 Gravel deposits   **10** Sand deposits

Fig 4.46 Deglaciation and outwash plain formation

are masked by such features. It shows that meltwater streams, transporting sands and gravels, flowed in tunnels through the ice, along the sides of the glacier and in channels on the glacier surface. Gradually, the sands and gravels choked the tunnels and new routes were found. Eventually, when the ice melted, the sand and gravel formed the following landforms.

Eskers (Figure 4.48A)

These look rather like railway embankments and are long, narrow, winding, steep-sided ridges of varying size, composed of stratified sands and gravels. Like the channels of a braided river, they sometimes form a series of linked ridges. Among the best Scottish examples are the Flemington eskers. These are around 5–10 m in height and wind almost continuously north-eastwards across the countryside for 10 km from Culloden towards Nairn. Unlike many other eskers, they have not been altered by large-scale quarrying for sand and gravel. Eskers formed during deglaciation as sand and gravel deposits built up in subglacial tunnels fed by meltwater streams.

A Stagnant Ice During Closing Stages of Loch Lomond Stadial

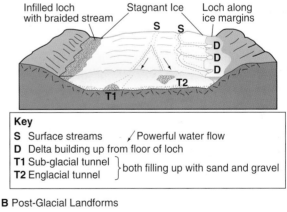

Infilled loch with braided stream   Stagnant Ice   Loch along ice margins

Key

S Surface streams   / Powerful water flow
D Delta building up from floor of loch
T1 Sub-glacial tunnel } both filling up with sand and gravel
T2 Englacial tunnel

B Post-Glacial Landforms

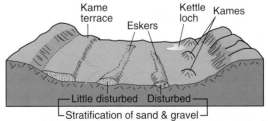

Kame terrace   Eskers   Kettle loch   Kames

Little disturbed   Disturbed
Stratification of sand & gravel

Fig 4.47 Deglaciation and ice contact landforms

Fig 4.48A Eskers are fluvio-glacial features, formed from stratified sand and gravel

Fig 4.48B The two small hills in the middle are kames and are fluvio-glacial in origin. This was taken north-west of Kirriemuir.

Kames (Figure 4.48B)

Formed closer to the ice margin, these are small, mound-shaped hillocks consisting of alternating layers of sand and gravel. Their size and steepness varies according to how they formed. Figure 4.47 shows that some form as small deltas built up by meltwater streams as they deposit material in small lochs along the margin of the ice. Others develop in crevasses choked by debris from streams on the glacier's surface. Kames composed mainly of gravel tend to have steeper sides while those formed mainly of sand are gentler and more rounded. Frequently associated with these mounds are kettle holes. Some contain lochs but many of these hollows are floored with peat deposits.

Kame Terraces

These are fluvioglacial benches with level or gently sloping tops. Often, they were built up by meltwater streams depositing sand and gravel between the ice margin and the ice-free slopes. Sometimes, the sand and gravel sediment was discharged into a narrow loch at the ice margin which eventually filled. As the ice melted, the steeper ice-contact side of the terrace often slumped down, disturbing the stratified sand and gravel sediments. Examples are found between Dunkeld and Pitlochry, acting as sites for villages (e.g. Ballinluig) and farms. The 'classic' examples (designated as a Site of Special Scientific Interest) are at the mouth of Loch Etive, where they descend in a step-like fashion down the valley side.

Finally, it is important to remember that Nature is never so clear cut as suggested by the text and diagrams. Hummocky moraines can be found mixed up with kames. Even a single mound may be formed from unsorted till and sorted fluvioglacial sand and gravel.

To Take You Further

Periglacial Landforms

Periglacial landforms are the result of prolonged winter freezing and brief summer thawing. Such climate conditions and associated landforms are typical of the present day Tundra regions. Before and after main glacial stadials, however, periglacial ('around the ice sheet') conditions prevailed in upland Britain. Areas south of the Thames Valley–Bristol Channel line, also experienced periglacial conditions when uplands to the north were ice covered.

One of the key features of periglacial areas is **permafrost**. This is a subsoil layer that is permanently frozen and potentially many metres thick. Above it, the ground thaws in the short summer, varying in depth from a few centimetres to several metres, and this is called the **active layer**. It is in this layer that periglacial processes are able to affect the landscape. Sample periglacial processes and landforms are illustrated in Figure 4.49. The following processes can be briefly mentioned:

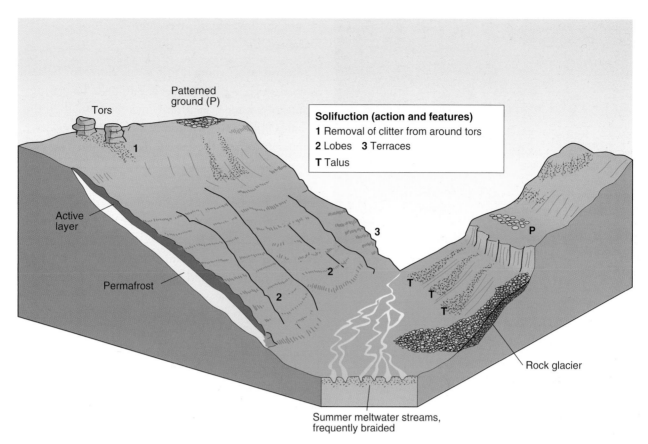

Fig 4.49 Sample of periglacial landforms

◆ **Solifluction** is the most widespread activity. When the summer thaw commences, the active layer becomes saturated with meltwater and slowly flows downhill upon the frozen layer below. Rates vary but speeds of 10 cm/year on 10° slopes have been measured. Solifluction takes the form of sheets, terraces and lobes (tongues) of this seasonal sludge. Although their movement is confined to a limited season, solifluction sheets and lobes have a distinct effect on relief. In the chalk downs of south-east England, for example, solifluction has helped to round dip-slope surfaces, fill in valleys and encouraged the formation of coombes (see page 127).

◆ **Freeze–thaw activity** was particularly pronounced and was responsible for large features such as **talus slopes** (see page 82) and **rock glaciers**. Rock glaciers are lobes of frost-shattered rocks which extend from the base of talus slopes. The largest Scottish example is in Strath Nethy and is some 300 m long. Smaller features include **patterned ground** which can take the form of stripes, circles and polygons. Water freezing below

loose surface material causes it to expand. Such frost heaving sorts out varying sizes of stone into patterned ground.

The Parallel Roads of Glen Roy

Figure 4.50 shows the three 'parallel roads' of Glen Roy, arguably one of Scotland's best known landforms, and even noted on the 1:50 000 O.S. map (Sheet 41). By choking up the lower ends of valleys, glaciers obstructed the flow of streams and produced lochs. During the Loch Lomond Stadial, such proglacial (ice-dammed) lochs formed in Glen Roy and its neighbouring valleys, north-east of Spean Bridge. The three Glen Roy 'Roads' are 350 m, 325 m and 260 m high and represent three former loch levels. As the glacier front ablated, the loch was slowly lowered from 350 m. Each of these heights coincides with cols which acted as outlets for each of the loch levels, ensuring that the loch shorelines, (the 'Roads') had sufficient time to develop. Considering that at its maximum the Glen Roy ice-dammed loch was 16 km long, up to 200 m deep and that the shorelines are between 9 m and 11 m wide, these

Fig 4.50 How many of the Parallel 'Roads' of Glen Roy can you pick out?

landforms must have developed in a relatively short period of time (Figure 4.51).

The Atlantic Conveyor System

It is always interesting to speculate about the next Ice Age. Understandably, popular interest centres on global warming. Yet current thinking links global warming with the possible return of Ice Age conditions to Britain and may explain past Ice Ages. The key lies in the North Atlantic Drift. It transports enormous quantities of relatively warm salty water to wash the shores of North-West Europe. As it warms the prevailing air masses travelling eastwards to Britain, the salty waters of the NAD cool and sink to return southwards at lower depths, part of a large oceanic 'conveyor belt' system.

Climatologists maintain that if this conveyor belt, so critical in giving North-West Europe a 'staggering bonus of winter warmth', is 'switched off' then the result will be a sudden and dramatic drop in temperatures and the onset of an Ice Age. The 'switch off' mechanism, it is believed, will result from vast quantities of fresh water swamping and diluting the salty waters of the NAD. Because the waters of the NAD will no longer be dense enough to sink, the conveyor belt will weaken. Evidence from deep ice cores obtained in Greenland suggests that the Loch Lomond Stadial may have resulted from icebergs released into the Atlantic from North America and 'switching off' the North Atlantic Conveyor. Today it is 'switched on'. Some scientists maintain that global warming will cause (i) melting of ice in Greenland, and (ii) increased rainfall over the northern Atlantic, resulting in excess fresh water released into the ocean waters and the onset of another Ice Age as the conveyor belt is 'switched off'.

Glaciated Areas: Difficulties and Opportunities

A glance at any OS map of a glaciated upland shows that, in general, the population density is low and settlement is concentrated at the margins and in the

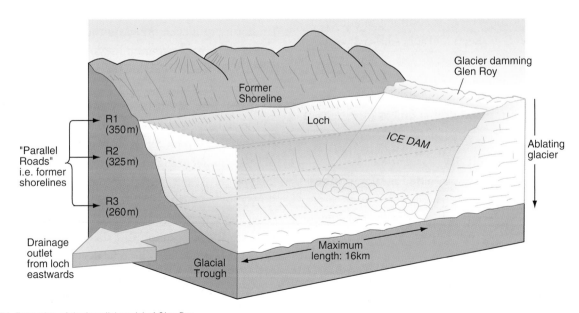

Fig 4.51 Formation of the 'parallel roads' of Glen Roy

Key

1 Slopes too steep for cultivation
2 Blanket-bog: waterlogged, acid soils
3 Very heavy rainfall, resulting in leaching of soil minerals and a high drainage density
4 Hard, impermeable rock, e.g. gneiss, promotes rapid run-off
5 Talus on steep slopes
6 Very low average temperatures and very short growing season because of high altitude
7 Poor fluvio-glacial soils
8 Thin, or no, soil because of glacial erosion
9 Very high wind speeds e.g. on Cairngorm Plateau
10 Marshy, waterlogged deltas where rivers enter lochs
11 High drainage density i.e. a large water surplus

Fig 4.52 Environmental factors limiting human activity in upland areas

valleys. Yet in summer and at weekends, the resident population can be swollen by incursions of visitors from densely populated lowlands and abroad. From ancient field patterns to funicular railways, and from the North-West Highlands to the Brecon Beacons, the resource-base of glaciated uplands has offered varied economic and social opportunities since prehistory, in spite of environmental constraints.

Environmental Factors Limiting Human Activity in Upland Glaciated Areas

Figure 4.52 shows a variety of environmental difficulties which can be found in areas such as the Cairngorms. Not all are directly the result of glaciation, some are climatic difficulties, others are the result of geology, drainage and soils but they all interrelate.

Economic and Social Opportunities of Glaciated Uplands

Tourism, Recreation and Conservation

Peoples' tastes change: glaciated uplands were not always seen as areas of high scenic quality. Attitudes to the Scottish Highlands and the Lake District changed as they were popularised from the 19th Century onwards by the writings of Sir Walter Scott and William Wordsworth, the visits of Queen Victoria and Prince Albert and improved transport links as the railway network extended. To 'active recreations', such as hunting, shooting and fishing, have been added climbing, hill walking, skiing and orienteering, as well as the 'passive recreation' of car-based tourism. Golf courses may be sited in fluvioglacial deposits (Gleneagles) or drumlin swarms (Lenzie). Related to tourism is the scenic significance of upland glaciated areas. Glaciation has helped shape the scenic resources of National Parks in England and Wales – the Lake District, Snowdonia and the Brecon Beacons – as well as of the Loch Lomond/Trossachs area and the Cairngorms, proposed as Scottish National Parks. Research by the Scottish Tourist Board shows that most visitors to Scotland are attracted by its dramatic scenery.

Farming and forestry

In spite of the environmental constraints of slope, soil, altitude and climate, hill farmers raising sheep, cattle and red deer use the lower slopes for grazing and the valley bottoms for winter feeding and arable in-bye land. Since 1919 and the establishment of the Forestry Commission, extensive reafforestation has covered former grassland and moorland with exotic plantations of trees such as Sitka Spruce. A landscape of industrial forestry, typified by deeper ploughing technology, new drainage techniques, genetically-improved tree stock and high fertiliser input, has transformed the sides of many glacial troughs but not without criticism of its visual impact. Certain farms and estates practise integrated land use linking farming, forestry, tourism and conservation of the remnants of the former Caledonian Forest.

Sand and Gravel Extraction

Kames and Eskers can be exploited for sand and gravel used by the building industry. Quarries exploiting these important resources are quite common in areas of fluvioglacial deposition.

Hydropower and Water Supply

Glaciated uplands, especially the Scottish Highlands and North Wales, have the best conditions in Britain for hydropower generation. These include a combination of steep valleys able to be dammed, a large head of water and extensive catchment areas. Although conditions are not so ideal as in Norway, for example, civil engineering skills have very effectively linked main troughs and hanging valleys, corrie lochans and ribbon lochs through giant 'plumbing' schemes to generate electricity. Glacial lochs are also major sources of water supply, such as that from Loch Katrine to Glasgow.

4.5 Carboniferous Limestone Scenery

There's not enough water here to drown a man, nor enough trees from which to hang him, nor enough earth to bury him.

17th Century description of the Burren by one of Cromwell's generals

Figure 4.53 shows part of the Burren, a distinctive area of County Clare. Here, as suggested in the above quote, rocks dominate the landscape. The bare, exposed rocks are the surface expression of massive beds of Carboniferous limestone – up to 780 m thick in this part of west Ireland. Originally formed some 340 million years ago in the warm, shallow waters of the Carboniferous ocean, this

Fig 4.53 The Burren in County Clare, west Ireland

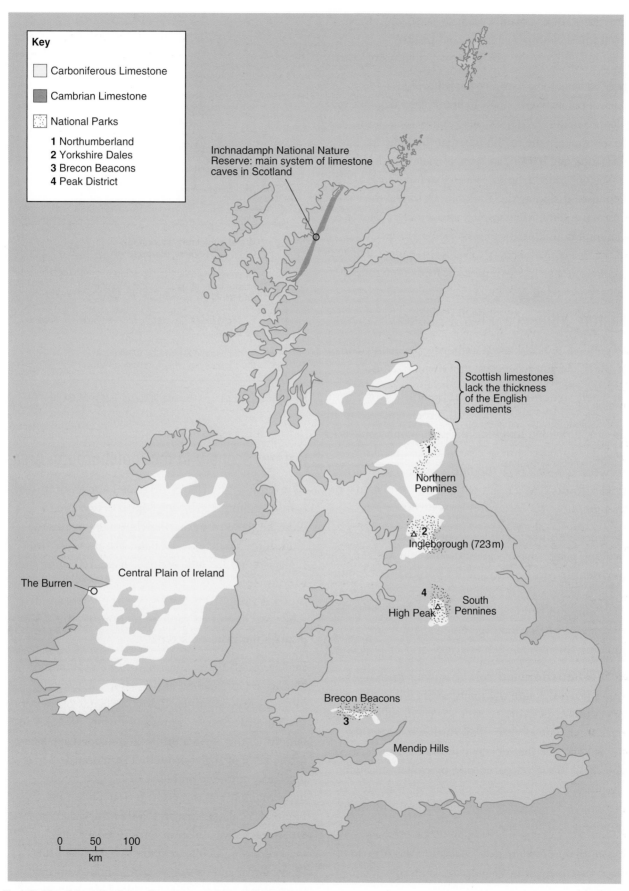

Key

☐ Carboniferous Limestone

▨ Cambrian Limestone

▨ National Parks

1 Northumberland
2 Yorkshire Dales
3 Brecon Beacons
4 Peak District

Inchnadamph National Nature Reserve: main system of limestone caves in Scotland

Scottish limestones lack the thickness of the English sediments

1
Northern Pennines

△ **2**
Ingleborough (723 m)

4
South Pennines
High Peak △

Central Plain of Ireland

The Burren

Brecon Beacons

3

Mendip Hills

0   50   100
km

Fig 4.54 Map of the carboniferous limestone areas in the British Isles

limestone rock is mainly calcium carbonate, derived from the skeletons of animals and plants accumulating on the sea floor over millions of years.

Look at a geological map of the British Isles. Appropriately, Carboniferous limestone is depicted in blue – a reminder of the shallow, blue tropical seas. It covers more of the British Isles than any other rock formation and is found in central Ireland, the Midland Valley of Scotland, South Wales, extensive areas of the Pennines (including the Yorkshire Dales and the Peak District) and the Mendips of Somerset (Figure 4.54). Carboniferous limestone outcrops can produce a distinctive type of scenery called Karst, named after such a limestone area in Slovenia in the former Yugoslavia.

Factors Influencing Karst Scenery

In general, typical karstic landforms are mainly the result of the interplay of the following factors.

◆ **rock structure:** carboniferous limestone is a hard, resistant, usually grey rock. As this rock emerged from the sea, it divided into blocks (Figure 4.55) thanks to a well developed system of clearly marked breaks between the rock beds (**bedding planes**) and vertical cracks (**joints**). As a result of the rock's structural strength, it is able to support large caves.

◆ **permeability:** such a system of joints and bedding planes makes carboniferous limestone a **permeable** rock. Although not porous, permeability permits water to pass through the rock by following the joints and bedding planes. It also means little in the way of surface drainage unless there is an impermeable cover, such as glacial drift or peat bogs.

◆ **Carbonation and solution** (see page 83): as water flows through the joints and bedding planes, the complex network of routes is chemically enlarged over thousands of years. This further increases the permeability of the rock and allows the limestone to absorb all the locally occurring precipitation. Such chemical action occurs because rainfall is dilute carbonic acid which acts on the calcium carbonate, removing it in solution.

(Note that in certain areas, glaciation was also an important process in shaping the landscape – see below)

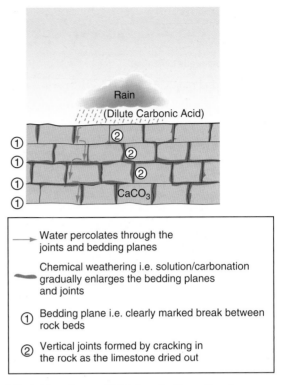

Water percolates through the joints and bedding planes

Chemical weathering i.e. solution/carbonation gradually enlarges the bedding planes and joints

① Bedding plane i.e. clearly marked break between rock beds

② Vertical joints formed by cracking in the rock as the limestone dried out

Fig 4.55 Bedding planes and joints in carboniferous limestone

Karstic Scenery in the Ingleborough Area

As a result of the factors described above, distinctive karstic landforms may develop. It is possible to illustrate these from the Ingleborough area in the Yorkshire Dales National Park. Rising to 723 m, Ingleborough has a distinctive shape, reflecting the various types of rock. The sketch (Figure 4.56) shows it rising from a plateau in the foreground while the simplified geology section (Figure 4.57) displays the main rock types.

Ingleborough's crest is formed from a capping of Millstone Grit, which, as the name suggests, is a tough, gritty sandstone. Below lie bands of rocks of the Yoredale Series. Consisting mainly of shales and sandstone, they give a distinctive stepped appearance to the sides of Ingleborough. The Yoredale Series gives way to the main block of Carboniferous limestone known as the Great Scar Limestone. Figure 4.57 shows that this 200 m thick slab of limestone forms a plateau with its edges often marked by high cliffs or **scars**. These fall to the River Greta and its valley, whose floor is formed from older, harder, impermeable shale deposits.

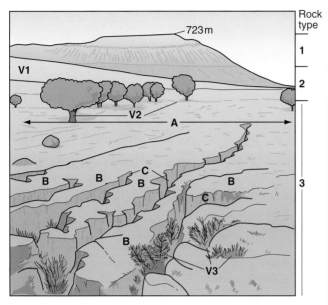

Fig 4.56 Field sketch of limestone scenery looking towards Ingleborough

Karstic (and related) landforms can be seen in the massive Great Scar Limestone and may be divided into four main groups. These are illustrated on the OS map extract (Figure 4.58), block diagram (Figure 4.59) and in photographs (Figures 4.61 to 4.64).

Surface Solution Features

Originating as horizontal bedding planes, limestone pavements consist of natural paving stones or **clints**. With their rough surfaces, these are divided into generally rectangular blocks of varying size and separated by **grykes**, which are joints widened and deepened by the processes of carbonation and solution. Figure 4.60 shows how the horizontally bedded pavement initially formed under a cover of soil and glacial drift. This cover gave a greater input of CO_2 which speeded up the process of solution. Later glacial advance exposed the pavement to continued weathering (both physical and chemical)

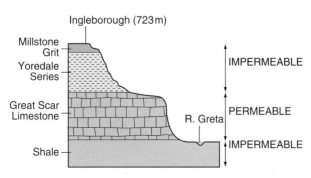

Fig 4.57 Very simplified geological section from Ingleborough to the River Greta

Photo of limestone scenery looking towards Ingleborough

further deepening and widening the grykes. Weathering has allowed the shady, humid grykes to harbour a rich and varied plant life, as suggested in the following quote:

> *Sometimes a mere knife-edge of rock is all that remains of the block between the fissures. On the surface no living thing appears save here and there a gnarled and wind-haggled hawthorn bush, but in the clefts are luxurious growths of harts-tongue fern and other shade-loving plants – wood sorrel, wood-garlic, geranium, anemone, rue and enchanter's nightshade.*
>
> Source: *Geology of Yorkshire* by Kendall and Groot

Unfortunately, human activity in recent years has helped denude these pavements through walking and the use of limestone for garden rockeries. Ultimately, solution and even the action of running water will destroy the surface of the pavement.

Solution is also responsible for closed hollows of varying size. Parts of the limestone plateau are pitted by hollows called 'dolines'. In the Ingleborough area the funnel-shaped dolines are known as **shake holes** and are up to 11 m in diameter and 3 m deep. Figure 4.60 shows that they form where glacial and alluvial

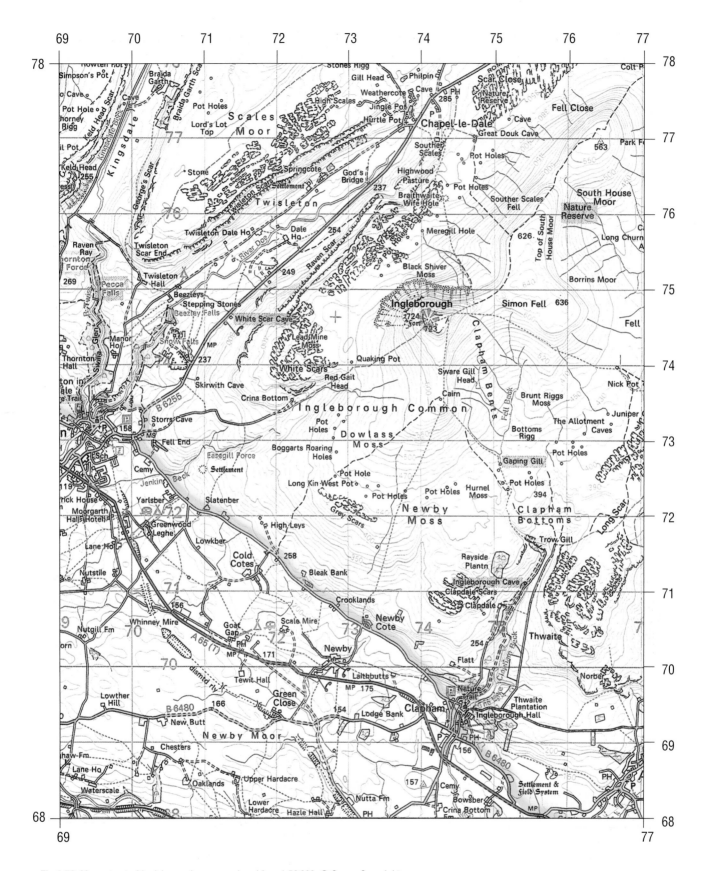

Fig 4.58 Map extract of Ingleborough area – reduced from 1:50 000. © Crown Copyright

deposits have fallen into joints enlarged and widened by solution. The debris blocking such features may be eroded by river action. This results in the formation of **swallow holes**, also known as **potholes**, the most spectacular example of which is Gaping Gill (Figure 4.61). Its formation has been aided by the strong vertical jointing of the limestone, the heavy rainfall and past glacial meltwater action (see section on glaciokarst).

Drainage Features

Fell Beck is one of the many surface streams draining the impermeable upper slopes of Ingleborough. When Fell Beck reaches the limestone outcrop, it flows over the oval-shaped lip (20 m × 10 m) of Gaping Gill – the second highest pothole in Britain – to plunge dramatically as a waterfall some 110 m into the largest single cave room in Britain. This is just one of many subterranean streams flowing underground via potholes, chemically dissolving and widening by erosion the many fissures into systems of narrow passages and caves. At times, the flow can be spectacular, particularly after heavy rain permits the water to thunder over subterranean waterfalls and into rock pools. Typically, the water produces a series of tunnels and caves in a step-like sequence.

Once the underground streams reach impermeable rock, for example, shale, they emerge from a spring or **resurgence** (see Figure 4.59). Fell Beck travels through 11 km of subterranean passages before emerging at Clapham Beck Head.

Dripstone Deposits

As water seeps into the caves, it again comes into contact with air and some of the CO_2 is lost. The water is, therefore, less acidic and holds less of the calcium hydrogencarbonate. This allows deposits of a whitish mineral called calcite slowly to build up into the following formations:

◆ **Stalactites** are 'icicles' of calcite formed by the dripping of water from cave roofs. Their shape, size and thickness vary: straw stalactites are thin, hollow and fragile (and therefore easily vandalised); bulbous stalactites are thicker; while some take the form of curtains hanging from cave roofs. The rate of growth varies but 7 mm every year is an average figure for stalactites in the Ingleborough caves

◆ **Stalagmites** are rounded, not pointed and jut upwards from the cave floor. They grow as drips land on the floor

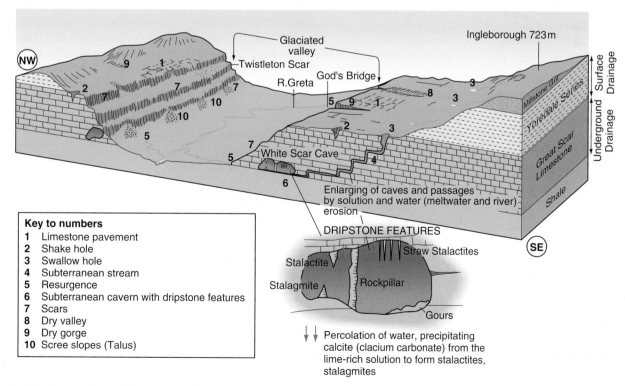

Key to numbers
1 Limestone pavement
2 Shake hole
3 Swallow hole
4 Subterranean stream
5 Resurgence
6 Subterranean cavern with dripstone features
7 Scars
8 Dry valley
9 Dry gorge
10 Scree slopes (Talus)

Fig 4.59 Sample features of Karst scenery in Ingleborough area

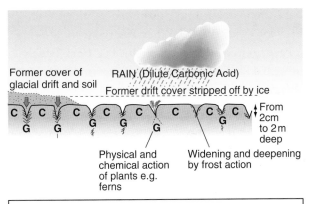

Key

⬇ Increased CO_2 from soil speeds up solution

≋ Carbonation and solution deepen and widen joints

G Grykes

C Clints

Fig 4.60 Limestone pavement formation

◆ **Rock Pillars** form where stalactites and stalagmites meet

◆ **Gours** are calcite ridges formed on cave and passage floors as CO_2 is lost from subterranean streams. Passage sides are also lined in places with flowstone ledges formed through the seepage of groundwater.

Glaciokarstic features

The distinctive Karst landscape is essentially the result of solution, which best operates in warm, interglacial conditions. The Ingleborough area, however, was affected by past cold climates and the action of glacial ice and meltwater. Certain features of this area, therefore, are best labelled **glaciokarst**. As well as stripping off topsoil from the pavements and widening and deepening valleys, ice plucked and abraded the bare, cliff-like **scars**, e.g. Twistleton Scars and Raven Scar, which outcrop above the valley of the River Doe (Figure 4.62). A feature of the final glacial advance is the presen.ce of erratics: the Norber stones (Figure 4.63), east of Ingleborough, stand on top of a limestone pavement. The erratics, formed of Yoredale or Millstone Grit, are much less soluble compared to the underlying pavement, which they protect from the rain. Consequently, they are perched on pedestals up to 30 cm high. Vast quantities of meltwater, released

Fig 4.61 Gaping Gill Hole – located at the junction between impermeable shale and carboniferous limestone. Fell Beck drains from the middle right of the picture (just below the walker). Note the soil creep and the well-used footpaths

Fig 4.62 Twistleton Scars can be seen in the background. River Doe flows in the foreground – the valley bottom is formed from impermeable sandstones

Fig 4.63 Norber Stones: these are found south-east of Ingleborough and are good examples of glacial erratics. They are composed of Millstone Grit or Yoredale Shales and are perched on pedestals of protected Carboniferous Limestone

Fig 4.64 A footpath to Gaping Gill Hole passes through the steep sided dry valley (gorge) called Trow Gill, probably eroded by glacial meltwater exploiting the weak line of a major joint

from stationary ice, flowed over frozen, therefore, impermeable, limestone and were particularly significant in eroding present day **dry valleys** and **gorges** such as the steep-sided, upper part of Trow Gill Gorge (Figure 4.64). Once the climate warmed, such valleys and gorges became dry, as water again was able to sink underground. Meltwater also further enlarged subterranean passages, shafts and caves. Subsequently, post-glacial tundra conditions and freeze–thaw action have shattered limestone along joints and bedding planes, allowing **scree slopes** to accumulate at the foot of scars such as Twistleton.

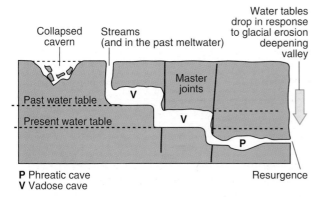

P Phreatic cave
V Vadose cave

Fig 4.65 Caves and varying water tables

Overall, karstic scenery evolved slowly: from its accumulation as shell detritus on the ocean floor over millions of years, to the slow chemical action of surface solution and then, most recently, geologically speaking, the various denudational processes hastened by the effects of the Ice Ages.

To Take You Further

Caves and Gorge Formation

Thanks to the exploration by enthusiastic cavers of many kilometres of passages, shafts and caves in the Yorkshire Dales, various levels of caves can be identified. Normally, these levels are related to the water table because this marks the greatest amount of lateral flow. Water tables, however, migrate, as shown in Figure 4.65. Many caves were completely filled with water in the early stages of formation. These are called **phreatic** because they were formed below the water table. Typically, they are tubular in cross-section. Some caves formed along the water table while others formed when it was higher. These are **vadose** caves because they are above the present water table. In such caves, thanks to the steady percolation of water from above, the characteristic dripstone deposits are able to form.

Through time, portions of cave roofs collapse, as shown by heaps of limestone blocks littering their floors. It has been argued that massive collapse of cavern roofs was responsible for the formation of limestone gorges. Examples of such gorges are Winnats Pass in the Peak District, Cheddar Gorge in the Mendips and Goredale Scar near Malham in the Yorkshire Dales. While the southern end of Goredale Scar shows some sign of collapse, it has recently been suggested that powerful meltwater erosion played a critical role in the formation of the scar.

Carboniferous Limestone: Economic and Social Opportunities

Areas of Carboniferous limestone, and related impermeable rocks, were first settled and utilised by small communities of prehistoric hunters and gatherers, as shown by artefacts from caves such as the Victoria cave at Settle. Today, there are varied categories of land use. Sometimes these are in

Fig 4.66 Photograph shows the impact of quarrying in the Yorkshire Dales. The quarry lies 1.5 km north-east of Ingleton.

competition because karstic scenery is valued for both its scenic and its mineral resources (Figure 4.66).

Farming

The traditional farmscape of the Yorkshire Dales is characterised by: flower-rich meadows cut for hay, the rearing of dairy cattle at lower levels; hardy hill sheep (e.g. Herdwick) on the thinner, upland soils of the fell tops; farm and field boundaries of dry-stone walls; and stone built barns (called laithes) sited in the lower fields so that hay and manure did not have to be carried far. Modern farming techniques (cutting of grass for silage, use of fertilisers rather than dung, mechanisation and modern livestock sheds) and field amalgamation (resulting in deterioration of defunct dykes) have altered the traditional farmscape. Designation of part of the Dales as an ESA (Environmentally Sensitive Area) encourages interested farmers, through various grants, to maintain the traditional farmscape.

Quarrying

Limestone is quarried for building stone, steel making and cement. Land use conflict can result between those opposing the visual impact of a quarry, the noise of the extraction processes and the heavy traffic generated, and those welcoming the employment opportunities and economic benefits triggered by the exploitation of a useful raw material.

Recreation and Tourism

Karst areas provide excellent opportunities for various forms of outdoor recreation. Demand from neighbouring cities and towns can be heavy, particularly at summer weekends, and is reflected in eroded footpaths, congested roads and overcrowded 'honeypots'. Active 'surface' recreation includes: fellwalking on the Three Peaks – Ingleborough, Pen-y-ghent and Whernside; rock climbing on scars – Twistleton; visiting waterfalls (appropriately called 'forces' locally) – Thornton Force; and conducting geography and biology fieldwork in an area with 'classic' landforms and rich plantlife. Below ground level, there is the appeal of its subterranean geography for tourists, cavers and speleologists. Many 'passive' recreationists and their cars are pulled by the characteristic scenery with distinctively shaped peaks, bold, naked scars, stone farmsteads with their never-ending succession of stone dykes snaking their way up to the fell tops and attractive villages like Clapham and Ingleton. Initially, such villages were sited close to resurgences and functioned, in past times, as market centres and/or weaving villages. Today, those settlements which are especially accessible function as dormitory villages. Careful planning usually allows new housing to blend with the old – often characterised by limestone walls and Yoredale roof tiles. Poor planning, however, may result in yet more conservation disputes.

4.6 Scarpland Scenery of South and East England

The landscape shown on this photograph (Figure 4.67) is quite typical of the **scarplands** of south and east England. Such scarplands extend from Devon in the south to Yorkshire in the north. Generally speaking, two main elements make up such a landscape.

1 Hills, such as the Cotswolds, Chilterns and the North and South Downs, which are called **escarpments** (cuesta is an alternative name). Among their characteristic features are

◆ a steep slope which rises from the lowland and can form a distinctive landmark. This is known as the **scarp slope**

◆ a gentle slope dipping away from the **crest** of the scarp slope. This is known as the **dip slope**.

Fig 4.67 The photograph is taken from the Devil's Dyke car park and is looking west along the scarp face of the South Downs. Identify the crest, scarp and dip slopes. Note the land use contrasts between escarpment and clay vale as well as the hang-gliders.

Figure 4.68 shows the main escarpments and the direction faced by their steep, scarp faces. It also distinguishes the two main rock types forming such escarpments: Jurassic limestone and Chalk. Jurassic limestone escarpments, such as the Cotswolds, are older, harder and generally higher than those formed from chalk. They are briefly discussed on page 129.

2 Between the escarpments are low lying, broad **vales** such as the Vales of Oxford and Sussex. These are mainly formed from clays and sandstones.

The Formation of Chalk Escarpments and Clay Vales

The formation of typical chalk scarpland scenery (often described as 'scarp and vale') is the result of the type of rock, earth movements and differential erosion by rivers.

Like Carboniferous limestone, chalk is a **permeable** rock. Unlike Carboniferous limestone, however, chalk is **porous**. This permits water to pass through the pore spaces between the fine chalk grains. Of the main types of limestone, chalk is the whitest and chemically the purest. Dating from some 95 to 65 million years ago, during the later Cretaceous period, the chalk was formed from fine-grained calcium carbonate shells deposited in a once extensive, clear

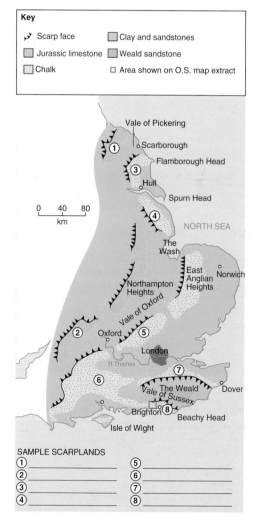

Fig 4.68 Scarplands of South and East England

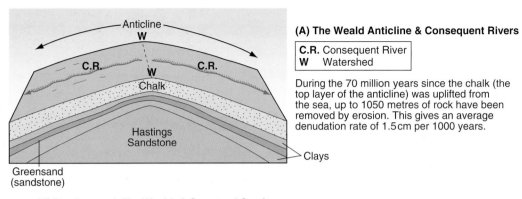

(A) The Weald Anticline & Consequent Rivers

C.R. Consequent River
W Watershed

During the 70 million years since the chalk (the top layer of the anticline) was uplifted from the sea, up to 1050 metres of rock have been removed by erosion. This gives an average denudation rate of 1.5 cm per 1000 years.

(B) The Downs & The Weald: A Structural Section

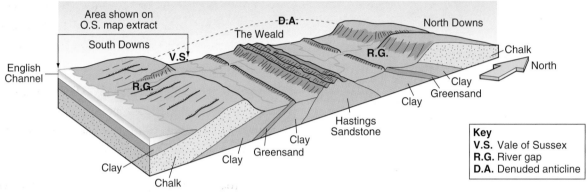

Key
V.S. Vale of Sussex
R.G. River gap
D.A. Denuded anticline

Fig 4.69 (A): The Weald anticline and consequent rivers **(B):** The Downs and the Weald: a structural section

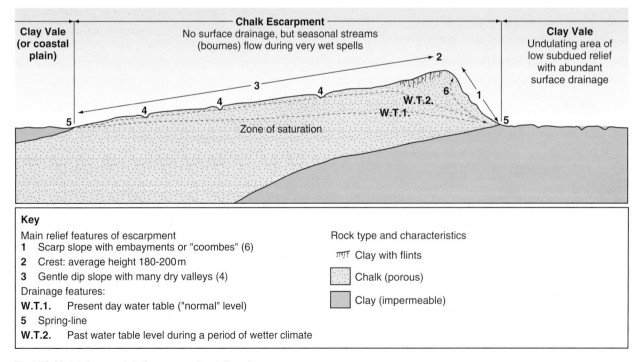

Key

Main relief features of escarpment
1 Scarp slope with embayments or "coombes" (6)
2 Crest: average height 180–200 m
3 Gentle dip slope with many dry valleys (4)
Drainage features:
W.T.1. Present day water table ("normal" level)
5 Spring-line
W.T.2. Past water table level during a period of wetter climate

Rock type and characteristics
Clay with flints
Chalk (porous)
Clay (impermeable)

Fig 4.70 Model diagram of chalk escarpment and clay vale

and warm sea. Clay, on the other hand, is a softer rock and, once saturated, is **impermeable**.

At one time chalk, along with clays and Jurassic limestones, formed horizontal sediments on the floor of a shallow sea. Later, these sediments were uplifted to form dry land and then tilted. Tilting took place during the Alpine orogeny (mountain building period), some 30 million years ago, when the rocks of southern Europe were dramatically buckled by the collision of the African and Eurasian plates. Away on the northern fringes, minor geological ripples gently tilted the sedimentary layers of what is now south and east England into a series of upfolds (called **anticlines**) and downfolds (called **synclines**).

Having been tilted, the different sedimentary rocks were then affected by **differential erosion**. In particular, river erosion exploited the differences between the sedimentary rocks in terms of hardness and permeability. **Consequent** rivers initiated the erosion, gradually cutting through the chalk anticline to the clay. Then came the more rapid erosion (Figure 4.69A) and lowering of the clay compared to the chalk, because of its softer nature and its impermeability. This was especially aided by the development of tributary, **subsequent** streams as they eroded the outcrops of clay. The porous chalk, on the other hand (Figure 4.69B), by allowing rainwater to sink through, was eroded at a much slower rate, and stands out as bold escarpments (see page 123).

The South Downs and the Vale of Sussex: Relief and Drainage

Features typical of chalk scarpland scenery can be illustrated from the South Downs and the Vale of Sussex immediately to the north (see Figure 4.70). It is very useful to keep looking at the OS map extract (Figure 4.71) and the photograph (Figure 4.67) to appreciate the key features of relief and drainage.

The South Downs: A Chalk Escarpment

Figure 4.69B shows that the South and the North Downs are the eroded remnants of a sedimentary upfold called the Weald anticline. It also shows that the scarp slope of the South Downs faces north while the dip slope gently drops southwards to the coastal

plain. Overall, the South Downs extend a distance of some 90 km from Beachy Head in the east to Butser Hill in the west. South flowing rivers such as the Ouse and the Adur divide the escarpment into a series of 'blocks'. The OS map (Figure 4.71) shows the flood plain of the River Adur occupying one of these low-lying alluvial gaps between the 'blocks'. It also shows that the steep, north slope of hills such as Truleigh (2210) and Edburton (2310) make up part of the scarp face and that their summits form part of the South Downs crest line rising to over 200 m. The land then gently dips southwards with a typically smooth, gently rounded topography, broken only by the presence of **dry valleys**, such as Bushy Bottom (2209).

Dry valleys and related features called **coombes** diversify the scenery. Coombes sometimes take the form of embayments (semi-circular hollows) in the scarp face. Steyning Bowl (1609) is a good example. Some dry valleys (Bushy Bottom (2209)) are quite long, broad and shallow; others, notably the Devil's Dyke (2611), are shorter and steep sided (Figure 4.72). Valleys such as these display many features associated with normal river valleys but are dry. (Some ideas about the origins of dry valleys and coombes are given below).

Surface drainage is generally absent from the escarpment but temporary streams called **bournes** can flow in the usually dry valleys, especially during wet winter spells. This happens because the water table (the level of saturation) is high. Figure 4.70 shows that its level can fluctuate. It also shows that springs occur at the interface of the porous chalk and the impermeable clay. Two lines of springs are found, one at the foot of the scarp slope and the other on the dip slope. These springs have been one of the factors in the growth and development of spring line villages such as Fulking (2411) in Figure 4.74 and Edburton (2311).

In places, there are deposits of clay-with-flints overlying the chalk. Thicker, damper soils develop, often supporting copses of deciduous woodland.

The Vale of Sussex: A Clay Vale

Extending north from the scarp face, the Vale of Sussex is a gently undulating lowland of clay and

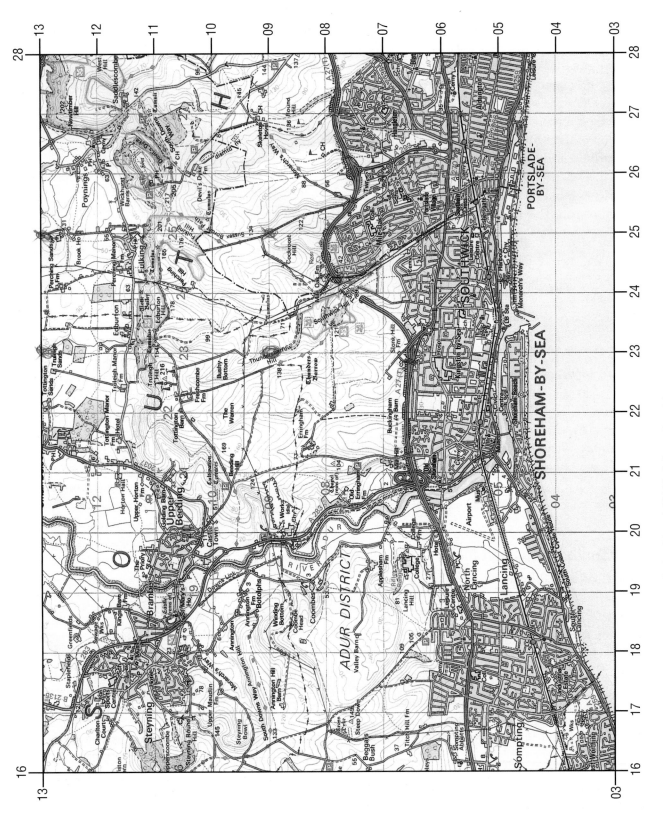

Fig 4.71 Map extract – Brighton and Lewes. Reduced from 1:50 000. © Crown Copyright

Fig 4.72 Devil's Dyke Dry Valley – imagine a river flowing through this valley

sandstone, around 40 m high on average. The underlying clay is a soft, fine-grained rock which, once saturated, supports abundant surface drainage. Small streams flow northwards from the scarp foot springs and feed into the trellised drainage network (see page 66) which eventually serves the River Adur in its floodplain. Once covered with thick, deciduous woodland and marsh areas, these lands were gradually cleared and drained. Man-made ditches (e.g. 2011) are a means of improving drainage in these clay lowlands.

To Take You Further

Dry Valleys and Coombes: Some Ideas About Origins

Dry Valleys

Bearing in mind the porosity of the chalk, how were such valleys formed? In medieval times, it was believed that the Devil's Dyke was dug by the Devil in an attempt to flood the Vale of Sussex. Today's geomorphologists have somewhat different views.

◆ One suggestion is that dry valleys were formed at a time when precipitation was greater than today. A higher water table could have supported permanent surface drainage

◆ Another suggestion is that they were formed at a time of Tundra-type climate. Under such conditions, the ground would be affected by permafrost. Pore spaces would have been frozen and, therefore,

impermeable. Any surface water, formed in the summer from melting snow, would have flowed quite energetically and helped erode such valleys.

Coombes

◆ Such features are most likely to be the result of past freeze–thaw action which loosened chalk particles on the scarp face and subsequent solifluction processes which caused this material to slowly slump downslope. This also took place at a time of past tundra-type climate (see page 111).

◆ Another suggestion is spring sapping. The chalk around the springs was constantly saturated and the flow of water through the springs weakened the surrounding chalk at the base of the scarp face. As a result of stream erosion the slope behind steepened and collapsed to form the hollow coombes.

Settlement and Land Use: An Outline

Map evidence, including field systems, tumuli (burial chambers) and Iron Age forts on the OS map, points to prehistoric settlement on the crest and dip slope. This setting offered: (i) protection (see 262112); (ii) elm and lime woodland that was easier to fell; and (iii) thinner, chalky soils easier to plough than the clays of the vale. By medieval times, however, villages had developed along the spring line. Such a site offered: (i) a water supply for domestic purposes and power for mills; (ii) level, well-drained land, and (iii) access to both the chalk uplands (whose dip slopes supported sheep grazing) and the mixed farming land of the clay vales.

Today, chalk escarpments such as the South Downs and neighbouring clay vales afford a variety of economic and social opportunities. These are summarised in Figure 4.73.

Scarp and Vale Landscapes: Economic and Social Opportunities

Distinctive scarp and vale landscapes, such as the South Downs, have been settled and utilised since prehistoric times. Various (overlapping) categories of human use of this area can be distinguished.

Building Materials

Flint, a silica impurity, is found in beds in the chalk

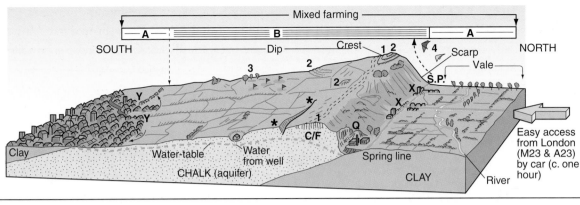

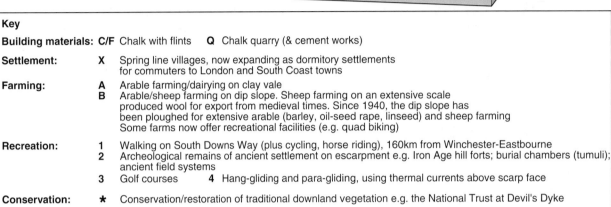

Key

Building materials:	**C/F**	Chalk with flints **Q** Chalk quarry (& cement works)
Settlement:	**X**	Spring line villages, now expanding as dormitory settlements for commuters to London and South Coast towns
Farming:	**A**	Arable farming/dairying on clay vale
	B	Arable/sheep farming on dip slope. Sheep farming on an extensive scale produced wool for export from medieval times. Since 1940, the dip slope has been ploughed for extensive arable (barley, oil-seed rape, linseed) and sheep farming. Some farms now offer recreational facilities (e.g. quad biking)
Recreation:	**1**	Walking on South Downs Way (plus cycling, horse riding), 160km from Winchester-Eastbourne
	2	Archeological remains of ancient settlement on escarpment e.g. Iron Age hill forts; burial chambers (tumuli); ancient field systems
	3	Golf courses **4** Hang-gliding and para-gliding, using thermal currents above scarp face
Conservation:	✱	Conservation/restoration of traditional downland vegetation e.g. the National Trust at Devil's Dyke

Fig 4.73 Economic and social opportunities for scarplands: the South Downs

and was used to make early tools and weapons. Because chalk is quite a soft rock, flint frequently was used for buildings and can be seen in the spring line villages, for instance, in the churches. Cement production uses chalk as the main raw material but there are objections to the quarries and the processing works. On the vales, brick works were sited beside suitable clay deposits and sandstones have been exploited for building stone.

Farming

Sheep rearing on the escarpment was the main source of wealth in medieval times. Mixed farming developed on the vale, aided by clay's ability to retain moisture. Since World War Two much of the dip slope has been ploughed. Wheat, barley and oilseed rape are now grown in usually large agribusiness farms. Many of these apply herbicides and fertilisers which are potentially harmful to the traditional downland ecology formerly maintained by extensive sheep flocks.

Settlement

Originally, settlement took place on the crest and dip slope of the escarpment. Evidence includes: iron age

forts, ancient field systems and burial sites, e.g. tumuli. By medieval times, settlement centred on the spring line villages. Larger villages and towns developed at gaps in the chalk (e.g. Steyning). Spring line villages grew because of the water supply and because the villages lay at the interface of upland sheep rearing and clay vale arable farming. Attractive villages and small towns are within easy commuting distance of London and also appeal to retired people if they can afford the house prices.

Recreation

This is varied but can include: hang gliding; walking along the South Downs Way; riding, using the many bridle paths; golfing and cycling. Easy access from Greater London and the coastal towns puts a great deal of pressure on the Downs, especially at weekends and in summer.

Conservation

The South Downs are designated as an Area of Outstanding Natural Beauty (AONB). Pressure on the landscape (from tourism and recreation, new farming techniques and population pressure) has led many to press for National Park status and this was finally acknowledged in 1999. Every year the South Downs receive some 30 million visitors – more than any other National Park – yet housing, roads and industrial development are nibbling away at the edges. Equally, intensive farming is affecting the rare plants, insects and butterflies. Conservation projects, such as those carried out by the National Trust at the Devil's Dyke, are restoring the traditional downland turf, rich in herbs, orchids and gentians.

Jurassic Limestone: Very Generalised Features

The Cotswolds, like the South Downs, are an escarpment with its scarp face overlooking the lowlands of the Severn valley and its dip slope gently extending south-eastwards into the Vale of Oxford. With their distinctive yellow-brown stone built villages and drystone dykes, the Cotswolds are also an AONB (Area of Outstanding Natural Beauty). They differ, however, in that their underlying Jurassic limestone has features almost half way between chalk and Carboniferous limestone. Unlike chalk, there are rock outcrops (with bedding planes and joints) and the slopes are not so smooth. Because the Jurassic limestone is harder than chalk, the Cotswolds are higher (reaching 300 m in a few places) but, as it is softer than Carboniferous limestone, it has few subterranean features and does not support typical Karst surface features.

ASSIGNMENTS

Geological cycle, Weathering and Mass Movement
Read pages 78 to 92.

1 a) Using Figure 4.1 **explain** the difference between the external and internal processes in the geological cycle.

b) Refer to Figure 4.2. Carefully **distinguish** between the following:
   (i) weathering and mass movement
   (ii) erosion and denudation.

2 a) Study Figure 4.3. With the aid of an atlas, **name** the upland areas A to F.

b) Identify the definitions shown below. Select from: **orogeny**; **sedimentary rocks**; **metamorphic rocks** and **igneous rocks**
   (i) formed by the cooling of molten magma
   (ii) major period of mountain building
   (iii) formed by deposits of old rock material, plant remains or animal remains
   (iv) formed from other rocks altered by great pressure and intense heat.

c) **Name** the three **orogenies** and indicate how they have affected the British Isles.

d) What is the **Tees–Exe line**? Why is it a significant feature of the geography of Britain?

3 a) What is meant by weathering?

b) Clearly **distinguish** between:
   (i) weathering and erosion.
   (ii) physical weathering and chemical weathering.

c) With the help of a diagram, **describe and explain** the processes involved in freeze–thaw weathering. Suggest ways in which this process is assisted by rock weaknesses (see Figure 4.5).

d) **Explain** the terms **pressure release** and **exfoliation**.

e) Study Figure 4.8. In what ways does biological activity assist physical weathering?

4 a) Study Figure 4.10. **Describe and explain** the chemical weathering processes that take place in the development of a limestone pavement.

b) **Suggest** the effects on weathering rates if the climate becomes: **(i)** colder **(ii)** drier.

c) **Describe** the effect of hydrolysis on granite. Would you expect the rate of hydrolysis to be faster or slower in summer?

d) What is **regolith**?

5 a) Give a **definition** of **mass movement**.

b) Look at Figure 4.12. Distinguish between the 'stick' component and the 'slide' component on a hillside.

c) Suggest various factors that might disturb the balance between 'stick' and 'slide' and trigger mass movements.

d) Read extracts A to D. Pick out phrases that show:
   (i) various trigger mechanisms
   (ii) the character of the mass movement.

6 a) Make a large copy of Figure 4.14. **Rockfalls and scree slopes**. **Match** the letters with the bullet points. **Describe and explain** the formation of the talus/scree slope.

b) Study Figures 4.14 and 4.15. **Describe and explain** the formation of the Quirang.

c) Copy the diagrams of mass movement and the Aberfan mudflow (see Figures 4.17 and 4.18). Match the bullet points with the letters on the

Fig 4.74 Village of Fulking from the South Downs

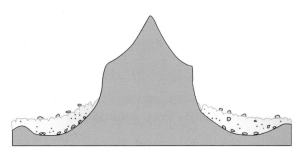

Fig 4.75 Assignment – Glaciated scenery

diagrams. Describe and explain the Aberfan mudflow.

d) Make a large copy of Figure 4.19, Aspects of Soil Creep. **Match** the letters with the bullet points. **Outline** the main processes involved in soil creep.
How does the impact of soil creep compare with that of landslides?

Glaciated Scenery (with reference to Erosion)
Read pages 92 to 106.

1 a) **Explain** the terms **Pleistocene**, **stadials** and **interglacials**.

b) What, and when, was the **Loch Lomond Stadial**? **Explain** the link between the Loch Lomond Stadial and the North Atlantic Drift.

c) From Figure 4.42, identify the areas of Scotland affected by the Loch Lomond Stadial.

2 Study Figure 4.23.

a) Briefly **explain** how glacial ice forms.

b) **Name** two inputs in the **zone of accumulation** and two outputs in the **zone of ablation**.

c) What happens to a glacier's **snout** when: **(i)** accumulation exceeds ablation, and **(ii)** ablation exceeds accumulation?

3 a) What are **erratics**? How do they assist geographers in reconstructing ice movement? (See Figure 4.24.)

b) Briefly **explain** how a glacier (i.e. a large solid mass of ice) is able to move.

c) Which erode more efficiently: **warm-based glaciers** or **cold-based glaciers**? Explain your answer.

d) Look at Figure 4.25. What is **moraine**? Carefully distinguish between:
(i) lateral and medial moraine
(ii) englacial and ground moraine.

e) **Describe** two processes by which glaciers erode.

f) With the aid of a diagram, **explain** the formation of a roche moutonnée (see Figures 4.26, 4.36 and 4.37).

4 a) What is a **corrie** and what are its main physical features?

b) **Copy** the field sketch of Loch Coire an Lochain (Figure 4.28). **Annotate** your sketch, using the named features provided.

c) (i) With the help of one or more diagrams, **explain**, as fully as possible, the formation of a corrie.

(ii) In what ways did the following assist in corrie formation?
◆ a northerly aspect
◆ frost shattering
◆ meltwater

d) Figure 4.75 above lacks any explanatory labels. When complete it should help explain the formation of **aretes**. Possible annotations might include: **corrie glacier; steep backwall; frost shattering of exposed ridge and peak; pyramidal peak. Copy and complete** the diagram. Write brief notes to explain the formation of an arete.

5 Study Figures 4.32, 4.33 and 4.44.

a) Which is the better term, 'U' shaped valley or glacial trough? Explain your answer.

b) With the aid of Figures 4.34A, B and C, **draw** clearly labelled diagrams to explain the development of a **glacial trough**. You should discuss the landforms and processes before, during and after glaciation.

c) Carefully study the photograph of Loch A'an/Avon. Read the extract from 'The Living Mountain' and study the map extract (Figure 4.27). Identify as many landforms as you can and

draw a fully labelled sketch based on the photograph.

d) **Explain** the formation of a hanging valley. Draw either an annotated sketch or a diagram to illustrate your answer.

6 After referring to Figure 4.38, **explain** the formation of the **Crag and Tail landscapes** of central Edinburgh.

Glaciated Scenery (with reference to deposition and fluvio-glacial deposition)
Read pages 95 to 114.

1 a) With the help of the two section diagrams shown on Figure 4.46, carefully **distinguish** between **glacial till** and **fluvioglacial deposits**.

b) Study Figures 4.42 and 4.43. Write a short account about **drumlins**. You should mention their location, their characteristic features and how they might have been formed.

c) What are the main features of **terminal moraine**? How does it differ from **hummocky moraine**?

2 Study Figures 4.46, 4.47 and 4.48.

a) Fluvioglacial landforms are varied. Match the landform in Table 1 with the appropriate description. The descriptions are not in the correct order.

b) Carefully **distinguish** between the formation of **eskers** and **kames**.

c) Look carefully at Figures 4.47A and 4.47B. Explain the difference in stratification between the two eskers.

Carboniferous Limestone Scenery
Read pages 114 to 122. (Also revise the earlier work on chemical weathering, page 83)

1 a) Refer to Figure 4.54 and an atlas. Name four upland areas of carboniferous scenery in the British Isles.

b) Why is the term 'karstic' applied to carboniferous limestone scenery?

c) On a large piece of paper outline, with the help of simple diagrams, the way in which Karst scenery is influenced by **rock structure, permeability** and **solution**.

2 a) Describe the location of the Ingleborough area (see atlas and Figure 4.54).

b) Copy Figure 4.56. With the aid of Figure 4.57, identify the following features and match them with the letters and numbers on the sketch: **flat topped summit of Ingleborough; capping of impermeable Millstone Grit; stepped profile of impermeable Yoredale Series; permeable Carboniferous limestone; extensive area of limestone pavement with chocolate bar appearance; rectangular blocks or clints; deep grooves or grykes (joints widened and deepened by solution); ferns occupying grykes; small copse of hawthorn; rough grazing on upper slopes**.

Table 1

Landforms	Description
Esker	Mound-shaped hillock of fluvioglacial sands and gravel.
Kame	An enclosed depression in an area of outwash sands and gravels. If the water table is high enough a loch may form.
Kettle/Kettle Loch	A long, narrow, winding steep-sided ridge of sand and gravel.
Outwash plain	A sand and gravel bench with fairly level tops found along valley margins.
Kame Terrace	Area of sorted gravels and sands often crossed by braided streams.

3 'Karstic scenery, both surface and underground, is unique.'

a) Study Figure 4.59. In a table, **group** the various features of the Karst scenery in four categories: **surface solution features; drainage features; dripstone deposits; glacio-karstic features**.

b) With the help of a diagram, **describe and explain** the typical features associated with a limestone pavement (see Figure 4.60).

c) Look at Figure 4.59. **Copy** the section of the diagram showing dripstone features. **Describe and explain** these underground features typical of karstic scenery.

d) In what ways have glaciers and freeze–thaw action contributed to karstic scenery?

Scarpland Scenery of South and East England
Read pages 122 to 128. (Also revise earlier work on drainage patterns. See page 66)

1 a) Study Figure 4.68. With the help of an atlas, **name** scarplands 1–8. Write your answers in two columns to distinguish between Jurassic Limestone and Chalk escarpments.

b) A 'typical' escarpment has two contrasting slopes. Name them and state how they differ.

c) In which direction do the scarp slopes of the following escarpments face:
(i) Cotswolds, (ii) North Downs, (iii) South Downs?

d) What are the main differences between chalk and clay?

2 Write the heading: The Formation of Chalk Escarpments and Clay Vales.

After studying Figures 4.69 and 4.70, rearrange the following statements **in the correct order**:

◆ During the Alpine orogeny (about 30 million years ago), the Jurassic Limestone, Chalk and Clay sediments were tilted into a series of upfolds (anticlines) and downfolds (synclines).

◆ Sedimentary rocks such as Jurassic Limestone, Chalk and Clay were deposited in horizontal beds on the floor of a shallow sea. About 70 million years ago they were uplifted to form dry land.

◆ Today, the more resistant chalk forms escarpments (e.g. the South Downs) with its distinctive steep scarp slope and gentler dip slope. The less resistant clay forms the lower lying clay vales (e.g. the Vale of Sussex).

◆ Differential erosion by rivers has exploited the differences between the rocks. Initiated by consequents (e.g. on the Wealden Anticline), clay (soft and impermeable) has been eroded into plains. Chalk, however, is harder and porous and has eroded at a slower rate.

3 a) On a large piece of paper, make a copy of Figure 4.70 – a model diagram of a chalk escarpment and a clay vale.

b) Explain the difference in level between the present day and past water tables.

c) Study Figure 4.72. It shows the Devil's Dyke – a classic example of a **dry valley**. **Draw** a sketch to show its main features. Suggest how such a feature was formed.

d) What is a **bourne**? Consult an atlas and give two place names associated with the term 'bourne'.

e) Where is the spring line located? **Explain** why it has developed there.

f) Why is the spring line of significance for settlement?

g) What are coombes? What are possible explanations for their origins?

4 a) Briefly, in what ways has rock type influenced the presence or absence of surface drainage in an area of scarp and vale scenery?

b) Study Figure 3.40. **Describe** the development of a trellised drainage pattern.

5 Study the photograph (Figure 4.67). **Draw** an annotated sketch to show as many features of the physical and human geography as you can distinguish.

6 Briefly **suggest** ways in which a Chalk escarpment differs from a Jurassic Limestone escarpment.

Assignment on Cairngorm 1:40 000 map extract

1 Construct a table to match the following physical features – **steep backwall of a corrie, corrie lochan, small hanging valley drained by stream, arete, ribbon loch occupying glacial basin; truncated spur** – with the following six

figure grid references: 943004, 008008, 015025, 015020, 995031 and 947006

2 a) Draw a cross-section from the summit of Cairngorm 1245 m south-east to Beinn Mheadhoin 1182 m. (NB. Loch A'an is around 10 m deep along the line of section).

b) Annotate the diagram to show the main features of relief.

3 Draw a large annotated sketch map of the area bounded by Eastings 98 and 00, and Northings 02 and 05, to show the main features resulting from glacial erosion.

4 When describing the relief of an area shown on a topographic map, it is useful to use a check list as a guide. Possible topics, appropriate to a glaciated upland include:

◆ **Heights** – maximum, minimum, range of altitude?

◆ **Slopes** – steepness, concave, convex, even, scree cover?

◆ **Landforms** – rounded summit-areas, steep-sided valley, rocky ridges, corrie, aretes?

◆ **Possible type of rock**

At all times, locate examples by grid references. Attempt simple, annotated sketch-maps to illustrate your answer. With help of a simple annotated sketch map, describe the relief of the area in the six grid squares covering Eastings 95 to 97 and Northings 00 to 03.

5 Is the Lairig Ghru a 'typical' glacial trough? Justify your answer with map evidence.

6 Using map evidence, outline the opportunities and difficulties for human activities in this part of the Cairngorms.

Assignment on Ingleborough 1:50 000 map extract

1 Re-write the following table to match the following landscape features with the six figure references provided.

Six-Figure Reference	Landscape Feature
730757	steep-sided, narrow dry valley
727758	limestone pavement
745707	summit formed from millstone grit
751727	resurgence
742746	scar
755716	swallow hole

2 Using tracing paper, and on the same scale as the map extract:
 a) draw an outline map of the Easting and Northings;
 b) carefully trace the 250 metre and the 550 metre contours;
 c) mark on all the streams and rivers shown.

3 Give map evidence to show the presence of underground cave systems in the Ingleborough area.

4 Draw a simple annotated cross-section along a line from 713780 to 760723. You should attempt to name as many landforms and drainage features as possible.

5 Describe the River Doe and its valley between 735766 and 694733. You should cover: direction of valley and river, valley width, valley depth, steepness of sides, river width, nature of river course, tributaries and special features eg waterfalls.

6 Giving map evidence, describe the various land uses which you can identify, suggesting any potential for land use conflict in the area.

Assignment on South Downs 1:50 000 map extract

1 Match the following physical and human features – **escarpment crest, prehistoric burial mound (tumulus) on a dip slope, dry valley, spring line settlement, north-facing scarp slope** – with these six figure grid references: 265111, 245111, 226108, 253101, 246114.

2 **a)** Two contrasting areas of land are shown in the grid squares below. On a larger copy of the grids, draw a suitable contour line (50m?70m?) to divide the area into a upland area and a lowland area. Label the areas A and B.
 b) Describe the relief of the land in each of areas A and B.

Use the following check list to assist you.
◆ **Heights** – maximum, minimum, range of altitude
◆ **Slopes** – steepness, concave, convex, gentle etc.
◆ **Landforms** – scarp face, crest, dip slope, undulating lowland etc.
◆ **Possible type of rock**

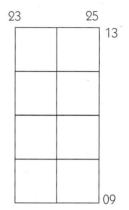

3 **a)** What evidence is there of prehistoric settlement in the area shown on the map extract? Give reasons to explain the distribution of prehistoric settlement.
 b) What factors may have influenced the siting and growth of settlements such as Edburton (2311), Fulking (2411) and Poynings (2612)?

4 Draw an annotated sketch to show the main physical features of the River Adur from 203130 to the coast.

5 Giving grid references, outline the various land uses you can identify north of grid line 07. Comment on any potential for land use conflict in the area.

Extra Assignments

1 With reference to the formation of Tors (Figure 4.11), **discuss** the view that both physical and chemical weathering interact.

2 What are the Parallel Roads of Glen Roy? How were they formed? (See Figures 4.50 and 4.51).

3 What landforms are the result of periglaciation? Suggest how they were formed.

4 **a)** What are the difficulties limiting human activities in either glaciated uplands (Figure 4.52) or upland Carboniferous limestone areas?

 b) For one of the following, outline the economic and social opportunities that each offers: **glaciated uplands; Carboniferous limestone uplands; scarplands**. (see Figure 4.73).

Key terms and concepts

- abrasion (p. 97)
- anticline (p. 125)
- arete (p. 99)
- bedding plane (p. 82)
- bergschrund (p. 99)
- biochemical action (p. 84)
- bournes (p. 125)
- carbonation (p. 83)
- chemical weathering (p. 83)
- coombes (p. 125)
- corrie (p. 97)
- corrie glacier (p. 92)
- crag and tail (p. 103)
- crest (p. 122)
- crevasses (p. 97)
- denudation (p. 78)
- deposition (p. 80)
- diffluence channels (p. 104)
- dip slope (p. 122)
- drift (p. 106)
- drumlin (p. 106)
- dry valley (p. 121)
- erosion (p. 79)
- erratic (p. 95)
- escarpment (p. 122)
- esker (p. 109)
- exfoliation (p. 82)
- fluvioglacial (p. 108)
- frost-shattering or freeze–thaw (p. 82)
- geological cycle (p. 78)

- glacio-karstic (p. 120)
- glacial trough (p. 99)
- gorge (p. 121)
- gours (p. 120)
- hanging valley (p. 102)
- hydration (p. 84)
- hydrolysis (p. 84)
- impermeable (p. 125)
- interglacials (p. 92)
- joints (p. 82)
- kame (p. 110)
- kame terrace (p. 110)
- kettle hole (p. 108)
- knock and lochan topography (p. 105)
- Loch Lomond stadial (p. 92)
- mass movement (p. 86)
- meltwater channel (p. 105)
- moraine (p. 106)
- mudflow (p. 87)
- nunataks (p. 92)
- orogenies (p. 81)
- overland flow (p. 89)
- oxidation (p. 84)
- periglacial (p. 89)
- permafrost (p. 110)
- permeability (p. 116)
- physical weathering (p. 82)
- Pleistocene (p. 92)
- porous (p. 123)
- pothole (p. 119)
- plucking (p. 96)

- pressure release (p. 82)
- pyramid peaks (p. 99)
- regolith (p. 84)
- ribbon loch (p. 100)
- rills (p. 89)
- roche moutonnées (p. 103)
- rock basin (p. 100)
- rock glacier (p. 111)
- rock lip (p. 99)
- rock pillar (p. 120)
- rotational sliding (p. 87)
- scar (p. 116)
- scarp slope (p. 122)
- shake holes (p. 117)
- sheetwash (p. 89)
- sill (p. 99)
- slopewash (p. 89)
- snout (p. 95)
- soil creep (p. 89)
- solifluction (p. 89)
- solifluction lobes (p. 89)
- stalactites (p. 119)
- stalagmites (p. 119)
- Storegga slides (p. 92)
- striations (p. 97)
- swallow holes (p. 119)
- syncline (p. 125)
- talus/scree slope (p. 82)
- till (p. 106)
- truncated spur (p. 99)
- vale (p. 123)
- valley glacier (p. 92)
- weathering (p. 82)
- zone of ablation (p. 95)
- zone of accumulation (p. 94)

Suggested Reading

Broadley, E and Cunningham, R (1991) *Core Themes in Geography: Physical* Oliver & Boyd.

Briggs, D et al (1997) *Fundamentals of the Physical Environment* (2nd Edition) Routledge.

Brunsden, D et al (1988) *Landshapes* David & Charles Channel Four Television Company.

Gordon, JE (ed) (1997) *Reflections on the Ice Age in Scotland* SAGT/SNH.

Guiness, P and Nagle, G (1999) *Advanced Geography: Concepts and Cases* Hodder & Stoughton.

Sissons, JB (1967) *The Evolution of Scotland's Scenery* Oliver & Boyd.

Waugh, D (1995) *Geography: An Integrated Approach* (2nd edition) Nelson.

Whittow, J (1977) *Geology and Scenery in Scotland* Penguin Books.

Internet Sources

British Geological Survey: www.bgs.ac.uk

Earth Science Resources: www-hpcc.astro.washington.edu/scied/earth.html

US Geological Survey: www.usgs.gov/network/science/earth/earth/.html

Scottish National Heritage: www.snh.org.uk

Staffordshire Learning Network: www.sln.org.uk/geography

5 THE BIOSPHERE

After working through this chapter, you should be aware that:

◆ after the last glaciers vanished, plants colonised Scotland in a series of stages until most of the country was covered with various woodland communities

◆ various vegetation communities – oak woodland, heather moorland and pine forests – developed as ecosystems in response to various factors, including the influence of people from prehistoric times

◆ ecosystems develop through various stages from pioneer until climatic climax is reached. This can be illustrated from coastal dune belts, for example, Tentsmuir and different forms of derelict land, such as urban wasteground and pit bings

◆ coastal dune and derelict land ecosystems are important biotic and recreational resources

◆ soil is a basic resource and that there are different types (e.g. podsol, brown earth and gley) which have distinctive horizons, colour, texture and drainage characteristics

◆ soil profiles are influenced by various processes, including podsolisation, gleying, organic and nutrient movement

◆ soils have been profoundly influenced by human activity since prehistoric times

◆ climate, vegetation and soils interact to form major biomes, for example, coniferous and temperate deciduous forests, which have been affected variously by human activity from prehistoric times onwards

You should also be able to:

◆ interpret data from vegetation surveys and explain the pattern of vegetation succession

◆ recognise and analyse soil profiles and data from field surveys.

5.1 Introducing Vegetation: Scottish Wildwoods

As the last glaciers vanished from Scotland, summers steadily became warmer, winters milder and the growing season lengthened. These conditions allowed plants to gradually colonise what was an uninviting, treeless landscape, often with extensive areas of bare rock. Several simplified stages of vegetation development can be picked out (Figure 5.1).

◆ Plants typical of present day areas of Tundra were the first colonists. Grey or yellow coloured lichens spread over bare rock surfaces. Mosses (e.g. sphagnum), grasses, sedges, ferns, hardy heathland, and scrub juniper spread over the least eroded areas. Such low-lying, open vegetation was well adapted to the harsh subarctic conditions.

◆ On lower ground, as the weather grew warmer, the winds lessened and soils further developed, conditions became suitable for tree growth. Species that had retreated when the ice originally advanced, now mainly migrated northwards in stages, colonising the treeless landscape. Dwarf and silver birch, followed by hazel, aspen, rowan and willow, made up these pioneer species.

◆ Scots pine, alder and oak came next, invading the earlier birch and hazel woodlands, often outshading them. Taller than the birch, the Scots pine, with its distinctive crown, came to predominate over large areas of the Grampians, growing up to 800 m on well-drained soils and forming what was later called the Caledonian Forest. Spreading around the lower flanks of the Grampians came broadleaved deciduous species, especially oak, elm and alder. Slowly, these trees migrated north, oak eventually reaching Skye in about 4000 BC.

As a result of all these vegetation changes, Scotland, by about 3000 BC, was a predominantly wooded

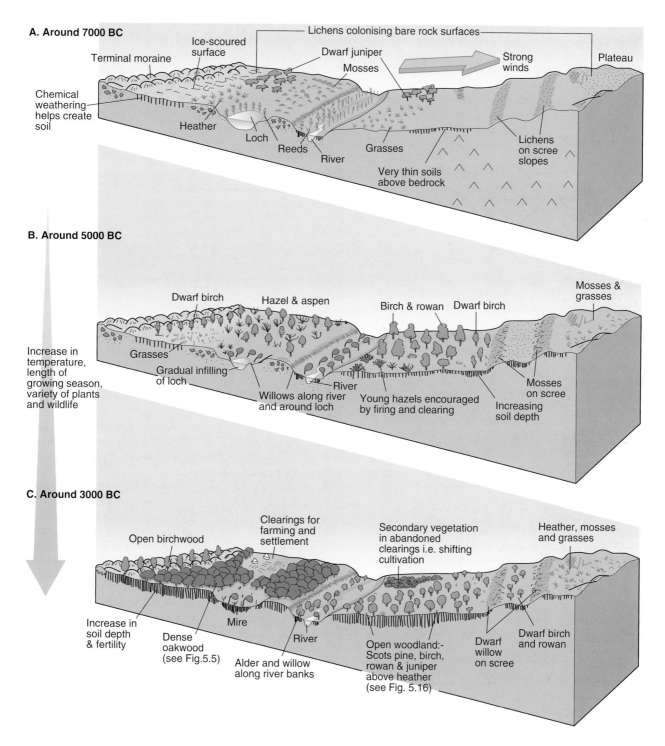

Fig 5.1 Vegetation succession in Scotland

country. Figure 5.2 shows the broad pattern of the main types of woodland at that time. But it was not a land of blanket forest coverage. Although never so rich in terms of species as before the Ice Ages, Scotland had a varied plant cover. At the coast, sand dunes and salt marshes had their own distinctive vegetation, while the wind-swept Western and

Northern Isles supported open scrub woodland. Inland, above the tree line, tundra-type plants prevailed on the highest, exposed mountain slopes and the Cairngorm summit plateau. Valley bottoms, often damp and soggy, were best suited to water tolerant trees such as the alder and willow. Well-drained, south-facing areas often sustained a varied

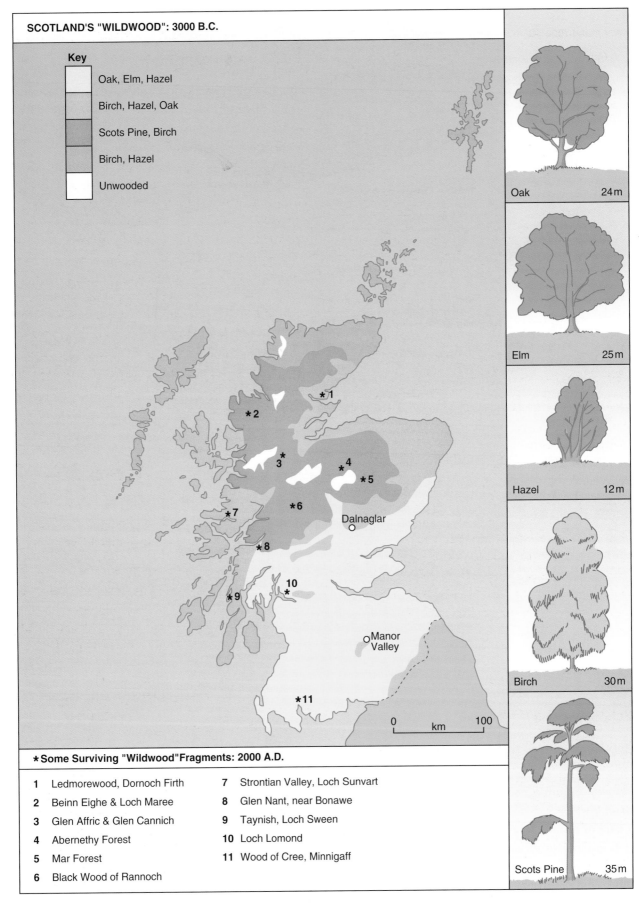

SCOTLAND'S "WILDWOOD": 3000 B.C.

Key
- Oak, Elm, Hazel
- Birch, Hazel, Oak
- Scots Pine, Birch
- Birch, Hazel
- Unwooded

Oak — 24 m

Elm — 25 m

Hazel — 12 m

Birch — 30 m

Scots Pine — 35 m

Dalnaglar

Manor Valley

0 — km — 100

★Some Surviving "Wildwood" Fragments: 2000 A.D.

1	Ledmorewood, Dornoch Firth	**7**	Strontian Valley, Loch Sunvart
2	Beinn Eighe & Loch Maree	**8**	Glen Nant, near Bonawe
3	Glen Affric & Glen Cannich	**9**	Taynish, Loch Sween
4	Abernethy Forest	**10**	Loch Lomond
5	Mar Forest	**11**	Wood of Cree, Minnigaff
6	Black Wood of Rannoch		

Fig 5.2 Scotland's 'Wildwood' 3000 BC

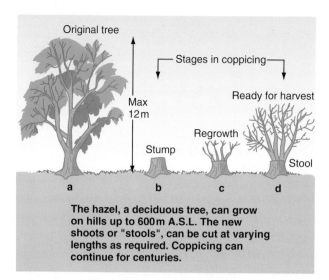

Fig 5.3 Coppicing hazel

The hazel, a deciduous tree, can grow on hills up to 600 m A.S.L. The new shoots or "stools", can be cut at varying lengths as required. Coppicing can continue for centuries.

cover of mixed woodland and shrubs, offering the early (Mesolithic) hunters and gatherers a wide range of edible berries, nuts, fungi and herbs as well as tools and shelter.

Overall, the pattern of vegetation that emerged during these post-glacial years was a response to many factors: climate, soils, drainage, relief, aspect, competition between plants, and the role of early people. Archaeological evidence from Colonsay suggests that the growth of hazel trees (a very important source of easily-stored food) may have been deliberately encouraged by hunters and gatherers who burnt out clearings and, possibly, coppiced the trees (Figure 5.3). By 3000 BC early farmers were making an impact by felling. Assisted by the spread of disease, elm trees, for example, had declined markedly.

To Take You Further

Pollen Analysis (Figure 5.4)

Various methods are used to reconstruct vegetational history. All flowering plants have pollen and pollen analysis involves gathering pollen grains preserved in peat or loch sediments. By analysing how frequently the grains occur in a core, the pollen provides clues about changes in the vegetation. By plotting such fluctuations on a diagram, a reasonable picture of changing vegetation emerges.

5.2 Woodlands, Moorlands and Ecosystems

The dominant wild woodlands — oak in the south, Scots pine in the east and birch in the north — that mainly covered Scotland by 3000 BC (Figure 5.2) are examples of **ecosystems**. An ecosystem is a unit which links living organisms — plants, animals, people — with each other, and with their physical environment — rock, soil, air and water. Ecosystems vary in type and size. Figure 5.5 shows an oak forest ecosystem. Such a forest includes smaller ecosystems: a raindrop held in a leaf can support micro-organisms, as can a heap of dung or even a dead tree. Equally, the oak forest is part of a world-wide ecosystem (called a **biome**) of temperate deciduous forest. In turn, all the varied biomes make up the **biosphere**, the thin life-bearing zone of the earth. Studying such an oak forest — its structure and how it functions — can reveal the workings of an ecosystem.

An Oakwood Ecosystem

The Oakwood Ecosystem and its Stratified Structure

The plants of an oakwood community are often stratified into four fairly clearly defined layers:

1 The **dominant** plants are the oaks whose crowns form, in most places, an almost complete canopy. These oaks create their own microclimate by reducing the amount of light which penetrates, and modifying forest temperature and humidity.

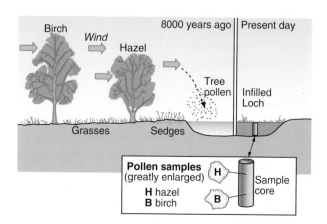

Fig 5.4 Pollen analysis

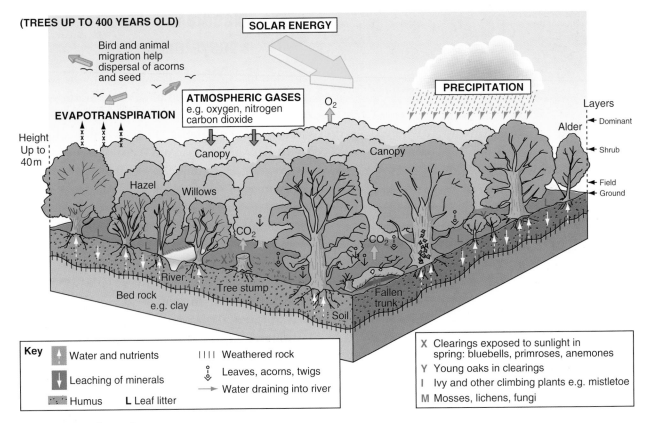

Fig 5.5 An oakwood ecosystem

2 The **shrub layer**, consists mainly of hazels and possibly willows in wet places, as well as younger oaks.

3 Below is a **field layer** whose composition partly depends on the shrubs and dominants above. Shade-tolerant brackens, grasses and ferns are common while carpets of bluebell burst into flower in spring, taking advantage of light penetrating the seasonally leafless canopy and shrub layer.

4 On the soil the **ground layer** is usually dominated by simpler plant forms, notably lichens, mosses and fungi.

Within this community, therefore, all the plants compete for light, water and nutrients. Below the ground they compete for root space. They survive because each species has different requirements from the forest environment. Each occupies its own micro-habitat or **ecological niche**. A mature oak woodland has a rich diversity of species, both plant and animal, reflecting a large and varied number of niches.

The Oakwood Ecosystem and Energy

All oakwood plants and animals are ultimately dependent on **solar energy** to 'drive' the ecosystem. Critical is **photosynthesis** which involves (i) the capture of **light energy** by the green pigment **chlorophyll** in leaf cells, and (ii) its conversion into **food energy** in the form of **carbohydrates**. Photosynthesis, therefore, allows energy to pass through the **food chain** and enables the whole ecosystem to function.

Each of the stages in the chain is called a **trophic level**. Plants are the first level – called **primary producers** – and all the succeeding organisms in the food chain are **consumers**. Figure 5.6 shows the various types of consumer in the chain: in basic terms, the green plant passes through the body of a herbivore (plant-eater) and then through a carnivore (animal-eater) or omnivore (eats both plants and animals). Plants and animals that die are chemically decomposed by various soil organisms which recycle nutrients back into the system. Energy, however, is lost as it passes through each **trophic level**. This explains the pyramidal nature of Figure 5.7. Most

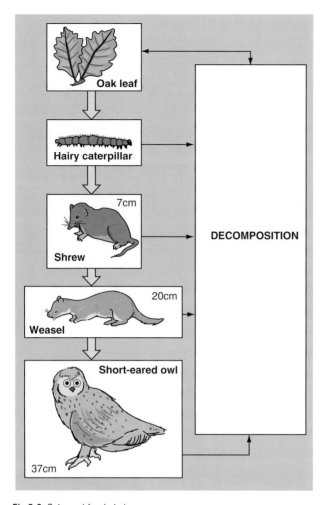

Fig 5.6 Oakwood food chain

energy is spent sustaining life at each of the levels, thereby limiting the amount transferred upwards. At each trophic level, only about 10% of the energy received by the organism is passed on to the next consumer in the chain. As energy is effectively only moving in one direction, such an ecosystem is an open system. A final point: given the diversity of species in an oakwood and the complexity of nature, it is better to think of many food chains forming an interdependent **food web**. Few animals feed on just one other kind – it is risky to be overdependent. Energy transfer can be a very complicated matter!

The Oakwood Ecosystem and Cycling of Materials

We have seen how an ecosystem such as the oakwood is maintained by energy flow. Equally important, and related to energy flow, is the cycling of important chemical elements in the environment. Various natural cycles make sure that there is an input of water, air and nutrients essential to sustain life. Three examples of these are the hydrological, nutrient and carbon cycles, as shown in Figure 5.8. The figure shows how an ecosystem such as the oakwood, part of the biosphere, interacts with the atmosphere, hydrosphere and lithosphere.

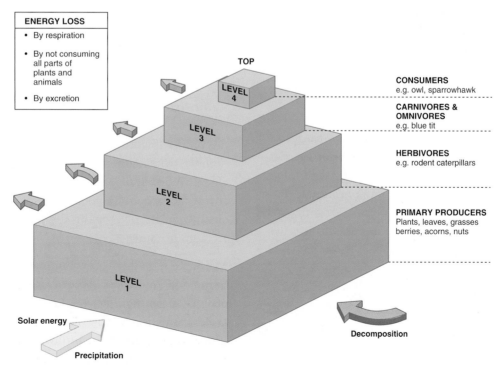

Fig 5.7 Trophic levels

The Oakwood Ecosystem and People

Figure 5.9 shows coppiced oaks in Mugdock Wood, now an SSSI (Site of Special Scientific Interest) and part of Mugdock Country Park, north of Glasgow. It is a mixed wood, with oak the dominant tree on the steeper, drier ground, while alder and willow predominate on the lower, damper areas.

The photograph shows that the oak has been coppiced, part of a long history of woodland management that may date back for at least four hundred years, judging by the size of the coppiced stools and evidence from old maps. Timber was produced commercially while parts of the wood were let for grazing. Actions such as coppicing alter light intensity and influence humidity while grazing can encourage the spread of grasses and consume acorns.

Little of Scotland's broadleaved woods can now be described as 'wildwood'. From the earliest hunter-gatherers onwards, people have been influencing such ecosystems directly or indirectly, thereby altering the balance of nature. Careful management of old oakwoods such as Mugdock or at Glen Nant, Argyllshire, and on Loch Lomond-side, however, can ensure that they retain much of their ecological heritage and allow people to enjoy them.

Heather Moorland Vegetation: A Managed Ecosystem

From Pinewood to Heather Moorland

Arguably, an even better example of a managed ecosystem is heather moorland, often incorrectly assumed to be Scotland's 'natural' vegetation. Figure 5.10 shows the distribution of the main types of moorland. Compare this map with that for 3000 BC (Figure 5.2) and you can see that the changes in vegetation are quite pronounced. Forest has given way to heather moorland, and this is further confirmed by Figure 5.11 – a pollen diagram for Dalnaglar, Perthshire. It shows an increase in heather and grass pollen, and a decline in tree pollen. Since prehistory, a combination of felling, burning and grazing has ensured the removal of extensive stands of Scots pine and birch from the uplands. Heather

moorland spread and became the dominant vegetation in some upland areas. It was then seen as a resource to be managed for hill farming, deer stalking, grouse shooting and other recreations (restricted during the shooting season) as well as being recognised for its landscape value. Over the years, heather management has involved grazing and burning:

◆ grazing of any surviving trees prevented any natural regeneration. Unless fenced off, damage by red deer in particular prevented the survival of young saplings. Over the years, heather has provided essential nutrition for red deer, goats, cattle, red grouse and sheep, especially after the introduction of large-scale commercial farming of Cheviots and Blackface from the late eighteenth century

◆ management has meant periodical burning of the heather, called '**muirburn**'. This aims to provide young, fresh growth for grazing animals and became particularly important with the exploitation of grouse moors from the mid-nineteenth century.

Heather, Grouse and the Landscape

Originally, the red grouse inhabited the Scots pine-and-birch woods of upland Scotland but, like the red deer, it has adapted to the open moorland ecosystem. A hardy bird, it is mainly found 250–700 m above sea level and depends on the green shoots, buds and seeds of common heather for the bulk of its diet. Other food comes from related moorland plants, some of which appear in the profile (Figure 5.10), and midges (countless in number) which are a useful source of protein for chicks. Additionally, effective digestion of such a coarse diet requires a good supply of grit in the gizzard: quartz-rich granite and sandstone uplands are a good source.

Managing a heather ecosystem involves careful burning to ensure that there are different ages of the plant near at hand. Figure 5.12 shows that the life cycle of a typical heather shrub involves four growth-phases: **pioneer**, **building**, **mature** and **degenerate**. Careful staging of muirburn allows grouse to eat fresh buds from the pioneer and

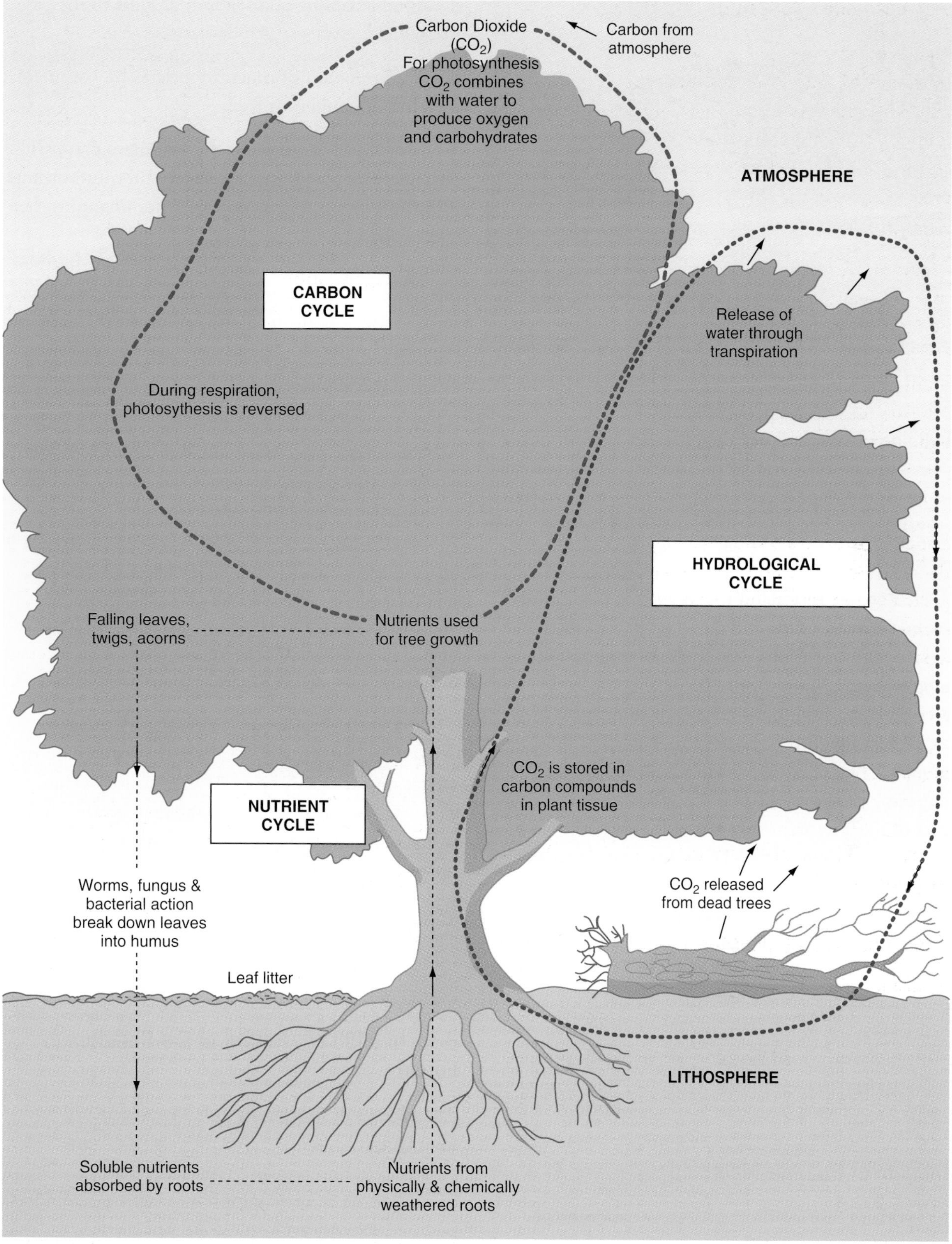
Carbon Dioxide
(CO_2)
For photosynthesis
CO_2 combines
with water to
produce oxygen
and carbohydrates
Carbon from
atmosphere
ATMOSPHERE
CARBON
CYCLE
Release of
water through
transpiration
During respiration,
photosythesis is reversed
HYDROLOGICAL
CYCLE
Falling leaves,
twigs, acorns
Nutrients used
for tree growth
CO_2 is stored in
carbon compounds
in plant tissue
NUTRIENT
CYCLE
CO_2 released
from dead trees
Worms, fungus &
bacterial action
break down leaves
into humus
Leaf litter
LITHOSPHERE
Soluble nutrients
absorbed by roots
Nutrients from
physically & chemically
weathered roots

Fig 5.8 Recycling of materials

Fig 5.9 Coppiced Oaks at Mugdock Country Park, near Glasgow

building phases, and use the older, taller heather for cover and nesting. Spring muirburn of small patches of heather (0.5–2.0 ha in size), in 50 m wide strips, therefore, explains the striped nature of heather-clad upland slopes (Figure 5.13).

Such a grouse moorland is every bit as people-made as the related features which accompanied the development of grouse shooting: large shooting lodges, coach houses, gamekeepers' cottages, lines of turf-built shooting butts across the moors and new estate roads giving access for those able to afford such a 'sport' (Figure 5.14) which, by the 1880s was disposing of about 500 000 red grouse a year. It was part of the Victorian disposition for exterminating wildlife on a huge scale. To ensure that there were large numbers of red grouse, deer, hare and capercaille for 'sport', gamekeepers were encouraged to hunt out the top predators in the food web – hen harriers, golden eagles, otters, wild cats, foxes and pine martens (Figure 5.15). Such virtual extermination, along with the clearing of what remained of the 'wildwood', effectively meant that these Victorian landowners destroyed one ecosystem while establishing another.

Decline of Heather Moorland

In Scotland, during the past fifty years, both the extent of heather and the number of grouse have declined for various related reasons. Heather loss is the result of:

◆ reclamation of lower hill land by liming,

draining, fertilising and seeding of grass to increase stock yields

◆ reafforestation of much moorland by commercial planting of coniferous trees

◆ poor muirburning. Careful timing and control of temperatures is required for successful muirburning: if burning is too fierce, normally dominant heather takes a long time to recover, allowing coarse grass and bracken to invade. Also, in the wetter heather moors of the north and west, and on steeper slopes, soil erosion – both sheet and gulley – can result, especially if already being overgrazed by hill sheep and red deer.

Reduction of the heather coverage means a reduced habitat for grouse. Today, the density of birds per square kilometre is half that of the 1930s. Additional factors include disease (often caused by a parasitic ring worm); poor spring weather; loss because certain birds of prey (e.g. the hen harrier) are protected species (but can be killed illegally, often by gamekeepers); plant damage caused by the heather beetle and the fact that there are fewer gamekeepers nowadays for effective management.

To conclude: heather moorland is a managed ecosystem. Effectively, heather is as much a crop as wheat or barley. Because of its resource value, centuries of burning has meant that heather has been the dominant plant over extensive areas of upland, with varied food web links (Figure 5.15). Its maintenance depends on effective management and that ultimately responds to prevailing economic and political factors, not forgetting land ownership.

Back to 3000 BC: Return of the Caledonian Forest

Across areas of heather moorland, particularly underlain by extensive areas of peat, it is quite common to find preserved, bleached roots and stumps of ancient Scots pine and birch. They are reminders of the once extensive Great Wood of Caledon, of which less than 1% remains, now scattered in 85 small clumps (Figure 5.2). Direct descendants of those that colonized post-glacial Scotland (Figure 5.1), their survival is partly because some trees were too remote

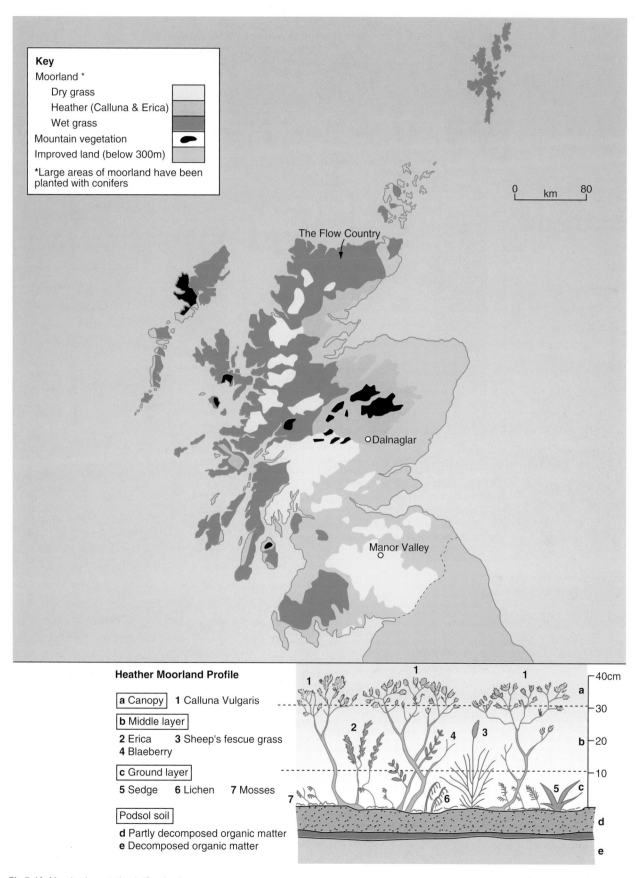

Key

Moorland *

Dry grass

Heather (Calluna & Erica)

Wet grass

Mountain vegetation

Improved land (below 300m)

*Large areas of moorland have been planted with conifers

0   km   80

The Flow Country

○ Dalnaglar

Manor Valley
○

Heather Moorland Profile

a Canopy   **1** Calluna Vulgaris

b Middle layer

2 Erica   **3** Sheep's fescue grass
4 Blaeberry

c Ground layer

5 Sedge   **6** Lichen   **7** Mosses

Podsol soil

d Partly decomposed organic matter
e Decomposed organic matter

Fig 5.10 Moorland vegetation in Scotland

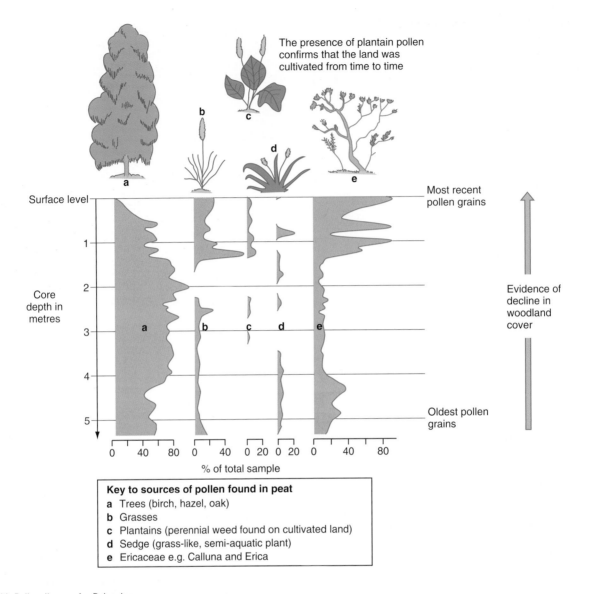

The presence of plantain pollen confirms that the land was cultivated from time to time

Fig 5.11 Pollen diagram for Dalnaglar

for exploitation. These botanical 'fossils', therefore, were spared the deforestation that affected Scotland from prehistoric times.

Now, various conservation groups and enlightened private landowners are attempting to protect and expand the remnants of native forest (see Figure 5.2). Such woods vary but can include a mixture of Scots pine and birch as dominants, junipers and rowans, which often form a shrub layer, while heather, blaeberry, ferns, grass and flowering herbs form the field layer (Figure 5.16). As open woodland they were once home to elk, brown bear, wild boar, beaver, lynx and wolf – all hunted to extinction by the eighteenth century. Some distinctive forms of

wildlife, however, do survive: notably the pine marten, wild cat, red deer, capercaillie (Scotland's answer to the turkey and reintroduced in 1837) and the pine-needle dependent crossbill which is unique to Scotland.

Various conservation and regeneration measures have been put into practice and others, perhaps a little optimistically, are proposed to help recreate this most distinctive ecosystem.

◆ **Extensive deer fencing** helps to keep out the red deer, with their appetite for young saplings. In Glen Affric, for example, Forest Enterprise is creating a number of fenced reserves downwind from existing seed sources of rowan, willow and birch, while pine

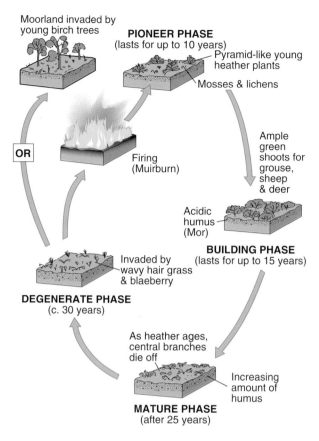

Moorland invaded by young birch trees

PIONEER PHASE
(lasts for up to 10 years)

Pyramid-like young heather plants

Mosses & lichens

OR

Firing (Muirburn)

Ample green shoots for grouse, sheep & deer

Acidic humus (Mor)

BUILDING PHASE
(lasts for up to 15 years)

Invaded by wavy hair grass & blaeberry

DEGENERATE PHASE
(c. 30 years)

As heather ages, central branches die off

Increasing amount of humus

MATURE PHASE
(after 25 years)

Fig 5.12 Life cycle of heather and muirburn

Fig 5.13 Heather moorland, Northern Perthshire, with its distinctive patchwork quilt effect as a result of muirburn

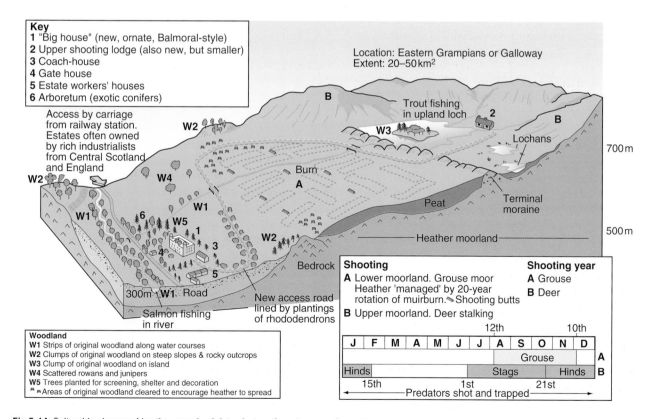

Key
1 "Big house" (new, ornate, Balmoral-style)
2 Upper shooting lodge (also new, but smaller)
3 Coach-house
4 Gate house
5 Estate workers' houses
6 Arboretum (exotic conifers)

Location: Eastern Grampians or Galloway
Extent: 20–50 km²

Access by carriage from railway station. Estates often owned by rich industrialists from Central Scotland and England

Trout fishing in upland loch

Lochans

700 m

Burn

Peat

Terminal moraine

500 m

Heather moorland

Bedrock

New access road lined by plantings of rhododendrons

300m   Road

Salmon fishing in river

Woodland
W1 Strips of original woodland along water courses
W2 Clumps of original woodland on steep slopes & rocky outcrops
W3 Clump of original woodland on island
W4 Scattered rowans and junipers
W5 Trees planted for screening, shelter and decoration
Areas of original woodland cleared to encourage heather to spread

Shooting
A Lower moorland. Grouse moor Heather 'managed' by 20-year rotation of muirburn. Shooting butts
B Upper moorland. Deer stalking

Shooting year
A Grouse
B Deer

J	F	M	A	M	J	J	A	S	O	N	D	
								Grouse				A
Hinds							Stags			Hinds		B

12th   10th

15th   1st   21st
Predators shot and trapped

Fig 5.14 Cultural landscape of heather moorland: late nineteenth century sporting estate

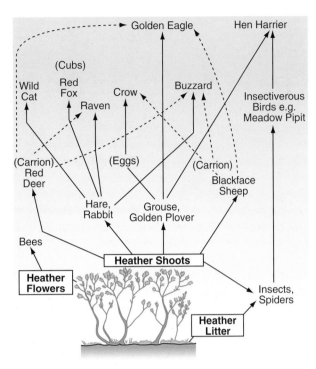

Fig 5.15 Heather moorland food web

(by stalking and live capture of deer, selected according to age, sex and condition) has allowed successful regeneration. Young pines and birch are now emerging through the moorland cover to resume their dominant status.

◆ **Return of the wolf?** This has been suggested by some conservationists. Allowing the wolf and other extinct species, e.g. the lynx, to prey on the excessive red deer population would be a natural way of re-establishing the total pine-and-birch system. The response to this idea is varied.

Re-creation of the Great Wood of Caledon, along with other regeneration schemes is an attempt to increase significantly the small area (less than 2% of Scotland's land surface) covered with native tree species (oak, Scots pine, birch, etc.). It is an attempt to create the **climax vegetation**, discussed below. Increasing woodland would mean re-establishing **biodiversity**, particularly in the uplands, once described as 'man-made deserts' by the famous ecologist Sir Frank Fraser Darling.

saplings are planted in a random pattern. Remote stands of non-native lodgepole pine are felled and left to rot, acting as a habitat for seeds of native trees. Deer fencing, however, is not without its problems: the RSPB removed some 40 m of fencing because they caused one-third of capercaillie deaths on its Abernethy estate.

◆ **Extensive deer culling** has been carried out, for example, by the RSPB at Abernethy and at Creag Meagaidh National Nature Reserve. Careful culling

5.3 Plant Succession, Climax Vegetation and Biomass

Plant communities change with the seasons and over the years. This chapter began with a description of how plant communities succeeded one another from the end of the Ice Age to 3000 BC. Even today,

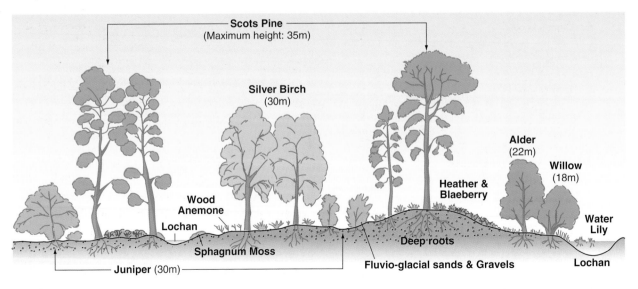

Fig 5.16 Pine-birch ecosystem

when a plant community is gaining a foothold, conditions do not often seem to be very favourable.

Take, for example, a plant-free site, for example, a pile of recently frost-shattered rock on a scree slope. There is no soil, seemingly no nutrients, and a lack of moisture because the bare rocks do not retain rainwater. Yet this seemingly uninviting environment is a **niche** for a **pioneer community** of algae and lichens. These simple plants spread, form a crust on the rocks and absorb rainwater and minerals for nutrition. Gradually, a thin soil starts to form because:

◆ the lichens help to weather the rock chemically, and

◆ as the plants die, small amounts of organic material (**humus**) build up.

Humus helps to retain moisture and provide nutrition for new plant growth. Usually mosses and grasses take advantage of these new niches. They are the next generation of colonisers and are more demanding. As they live and die, they further modify their environment by creating more humus and more shelter, with deeper soil for root anchorage, and thus they provide more nutrition for the next plant community (Figure 5.17). And so it continues until, eventually, shrubs, such as willows and then woodland (e.g. birch) develop. Over the many years involved, the physical conditions (soil, shade, light, etc.) will have changed so much that some early plants, unable to compete with newer, taller, more aggressive species, will have been completely replaced. Others survive, no longer as dominants but as part of a stratified plant community, such as an oakwood.

Fig 5.18 Hydrosere: The deepest water is occupied by yellow water-lilies and the shallow water by a second stage community of reeds. Willow trees are on the silted fringe

Fig 5.19 Halosere: This area of salt marsh is covered by salt water every high tide. The pioneer species is Salicornia. It slows down the ebb tide and encourages the deposition of mud

Such a sequence of plant communities inhabiting the same site is called a **plant succession** or **sere**. The one just described is called a **lithosere**. Several types of sere are shown in Figures 5.18–5.21. In each case there have been several seral stages on a particular site as it changes through time from the **pioneer stage** through a **building stage** to a final, mature or **climatic climax** stage. Figure 5.22 is a model diagram of such vegetation succession. It shows that overlapping changes have occurred in the number, height and density of plants as well as soil conditions and the **biomass** of the community (see below). Climatic climax is the ultimate stage because the vegetation no longer changes, apart from replacement of dead plants (and animals). Climax communities such as the oakwood or Scots pine-and-birch are said to be in a state of **equilibrium** or harmony with the climate and soils of the local environment and are relatively stable and self-sustaining ecosystems. The inputs of energy and nutrients are balanced by the

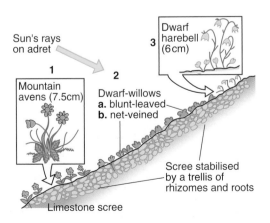

Fig 5.17 Plant colonisation on screen slopes (Swiss Alps)

Fig 5.20 Lithosere: Lichens and mosses are the initial colonising plants on an Icelandic lava flow

Fig 5.21 Psammosere: This shows an area of fixed dunes with a wide range of plant types, including lichens, compared to the shifting or yellow dunes

output. In reality, however, as we have seen, there can be **arresting factors**, usually due to human action such as controlled muirburn which encourages heather and prevents climatic climax (Figure 5.23). Where a plant community is prevented from fully developing, it is described as a **plagioclimax**. Not only is heather moorland an excellent example but, strictly speaking, so is much of the world's so called natural vegetation, for example, the Prairies of North America.

Throughout the world, in many different locations, a variety of inorganic surfaces are constantly being laid bare. This may be the result of (i) natural processes,

such as new lava flows, land-slides or a fresh exposure of sandy coastline, and (ii) human activity, for example, the creation of industrial waste heaps and abandoned building sites. Pages 151–159 give two examples of plant development (i) on sand dunes, and (ii) on derelict land, to illustrate in detail the principles of plant succession from pioneering phase to a mature, stable climax community.

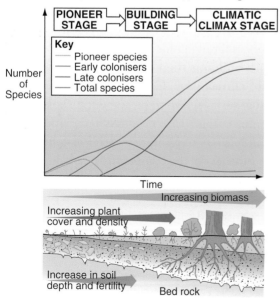

Fig 5.22 Stages of plant succession on a new site

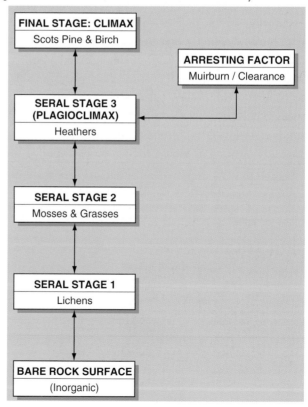

Fig 5.23 Heather moorland as a plagioclimax

To Take You Further

Biomass

Ecosystems vary enormously and so does the amount of living material. Biomass is the name given to the total weight of living matter in a given area, whether it occurs above or below the ground. Plant biomass, therefore, includes, leaves, branches, shoots, trunks, roots, fruits, etc. Animal biomass is small in comparison and this is reflected in the pyramid structure of trophic levels (see page 141).

Biomass present in any area can be measured by collecting, drying and weighing the vegetation from sample areas. The results obtained are influenced by (i) the types of plants and animals; (ii) the climate and soil; (iii) the time of year; (iv) the stage of plant succession. In the pioneer stage, biomass is minimal; by climax stage (e.g. a mature forest), it is at a maximum. As usual, (v) the influence of humans can be critical. Thoughtless clearing of vegetation is no more than plundering long accumulated reserves of biomass.

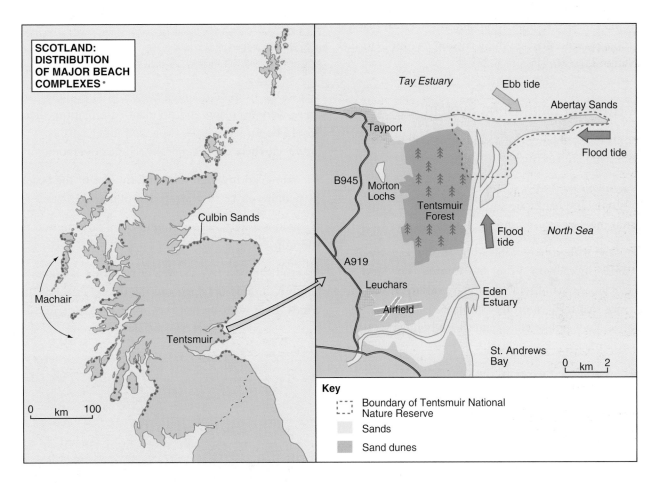

Fig 5.24 Tentsmuir: location map

5.4 Plant Succession on Coastal Sand Dunes

Tentsmuir, East Fife, is one of the most important and attractive coastal sand dune systems in Scotland. In winter, the offshore sandbanks are home to many waders – grey plover, oyster catcher and sanderling, as well as greylag geese, eider duck and seals. Immediately inland there are extensive pine plantations with many kilometres of tracks (see Figure 5.24). Very popular for recreation, including walking, mountain-biking, bird-watching and beachcombing, this area is mainly administered by (i) the Forestry Commission, and (ii) Scottish Natural Heritage, who manage the Tentsmuir Point National Nature Reserve (Figure 5.25).

Fig 5.25 Tentsmuir National Nature reserve. First designated as an NNR in 1954, the management plan aims to maintain its distinctive dune system and biodiversity while still allowing public access

A Growing Sand Dune System

For the development of a sand dune system the following are essential:

◆ a plentiful supply of sand. Figure 5.24 shows that offshore sand banks are sustained by coastal currents from the Tay estuary and the River Eden

◆ sand is then blown inland by strong, easterly winds, especially in spring and autumn. The sand particles are mainly transported by **saltation** which means that they vigorously bounce or hop along the surface (not unlike table tennis balls bouncing on a hard surface)

◆ an obstacle then traps the sand. Usually it is a plant or a piece of sea weed that is located on the high tide line.

At Tentsmuir the sand dune area is being extended eastwards. One piece of evidence for such growth is the almost total burial of wartime anti-tank blocks (see Figure 5.26). Built on the coastline as it was sixty years ago, they are now up to 800 m inland in some places.

Difficulties Facing Colonising Plants

In general, the sand dunes that form nearest the sea are small, with a few scattered plants, and are liable to shift. Further inland, the dunes are larger, with a greater variety of plants, and are fixed. Plants are the key to the formation and character of sand dunes. If there were no plants and no plant colonisation, there would be no sand dunes. For the pioneer plants, however, there are several obstacles to overcome:

◆ strong stormy high tides and the use of the beach by people can destroy plants and dunes

◆ sand is a sterile environment often lacking humus and nutrients

◆ sand can be very dry. Water retention is poor because of rapid drainage and the drying effect of the winds. Even wetting from sea spray is of little use

Fig 5.26 Because of the eastward growth of the sand dune complex, the anti-tank blocks, originally located on the shore in the 1940s, are now well inland

because of its salty nature. Pioneer species, therefore, have to be able to withstand consistently dry conditions. Such drought resistant plants are classified as **xerophytic**

◆ colonising plants along the foreshore, generally, have to adapt to alkaline conditions because of shelly fragments. (At Tentsmuir, however, because of the plentiful supply of offshore sand, the calcium carbonate content of the shore sand is relatively low).

Sequence of Plant Succession

In the face of such obstacles, plants adapt quite remarkably to the initially unstable and harsh environment of the foreshore. Figure 5.27 is a transect through a typical dune complex showing the stages of seral succession. Moving inland, acidity increases, salinity declines, water retention and biomass increase, and exposure to wind is reduced. Plant coverage and diversity increase until the climax vegetation is established.

◆ the plants which colonise the embryo, fore and mobile dunes have to be (i) xerophytic (drought resistant) and (ii) have particular growth features. **Sea rocket**, **sea lyme** and **couch grass** are often the first plants to invade the foreshore (see Figure 5.28). They are especially resistant to immersion in sea water by spring tides. Their foliage traps the sand, causing it to drift over them. As they grow vertically and laterally, they bind the sand and build up embryo and foredunes

◆ once the tops of dunes have grown well above the highest tide level, **marram grass** is able to colonise. It is the main pioneering plant and thrives in the drier mobile sand of the yellow dune. Figures 5.29 and 5.30 show some of its distinctive features. At first there are bare areas of sand and no other plants compete. Thanks to its **rhyzomes** (long creeping underground stems) spreading vertically and laterally, the marram very efficiently keeps pace with fresh deposits of sand, growing upwards at up to one metre a year. On such yellow, mobile dunes, the

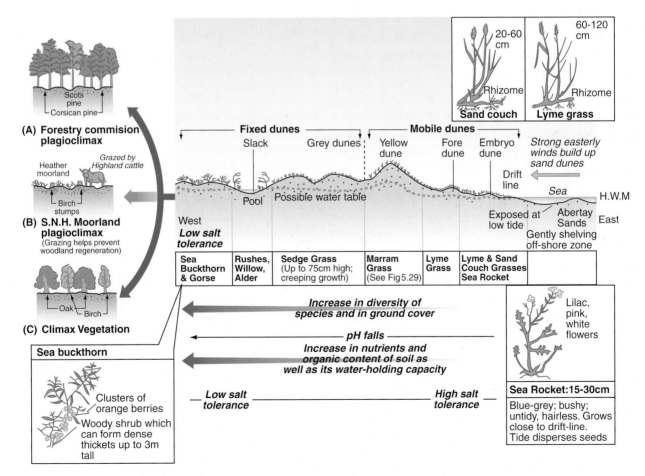

Fig 5.27 Tentsmuir: stages of seral succession

Fig 5.28 Tentsmuir species such as the sea-rocket (foreground) and sea-lyme (top left) are important pioneer plants which assist in the seaward accretion of the dune complex

Fig 5.29 Tentsmuir marram grass: sand collects around young tufts of marram grass, typical of the yellow dunes. Underground the long creeping rhizomes help to stabilize the dune

surface sand is still blown away and replaced by fresh sand (hence the adjective 'yellow') from the beach. The presence of these high dunes, however, reduces wind speed over the foredunes and allows them to grow shorewards

◆ on the landward, more sheltered side of the dunes, more plant species establish themselves. Mainly perennials, they increase the stability of the dunes. Vegetation cover is now complete. Such fixed dunes are called grey dunes because of (i) the

increased humus content of the soil and (ii) the surface growth of lichens (e.g. dog lichen). As the distance from the shore increases, sand no longer accumulates and marram grass cannot compete with the new colonising plants – **creeping fescue**, **sand sedge**, **buttercups**, **dandelions** and **bird's foot trefoil**.

◆ further inland, the older grey dunes support bushes such as **sea buckthorn** but more 'ordinary' plants are also present – **gorse**, **broom**, **heather**,

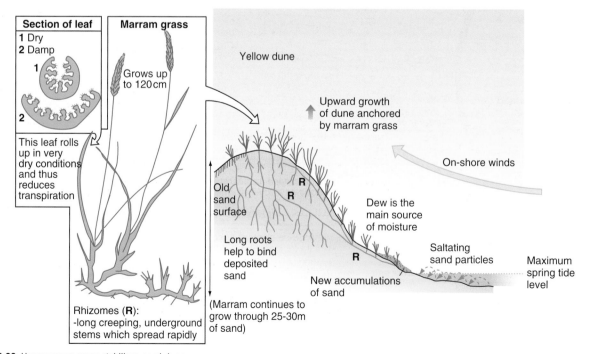

Fig 5.30 How marram grass stabilises sand dunes

The Biosphere

bramble and wild rose, not to mention rosebay willowherb. One factor is that the soil is now more acidic. Any lime from the shells has been leached (washed out) and the amount of humus is now greater.

◆ other 'ordinary' species occupy the dune slacks. These damp, low lying hollows have a higher water table, especially in winter. Figure 5.27 shows that hydrophytic (water tolerant) vegetation thrives: rushes, willows and alders are found here.

◆ finally, on this journey through history, the climax vegetation of woodland – Scots pine and birch giving way to oak – would be expected. Such an ideal sequence is rarely found. At Tentsmuir, the Scottish National Heritage management plan requires a plagioclimax of heathland. This has involved the clearing of woodland such as birch and the grazing of Highland cattle. In the area administered by the Forestry Commission, pine plantations date from 1924.

Such an ecological succession on sand dunes is known as a psammosere. Tentsmuir is a good example of the close interaction between plants and the sand dune environment. It is also a fragile environment, capable of alteration as a result of both natural causes and human activities.

5.5 Plant Succession on Derelict Land

In most large urban areas in the British Isles there is always a surprising amount of derelict land because of inner-city development, industrial decline and changing transport policies. There may be disused buildings, rubbly wasteground or temporary carparks where such buildings have been abolished, and abandoned railways, embankments, and cuttings. On the edge of towns, there are often rubbish tips, abandoned quarries and pits, and old spoil heaps where coal, oil-shale and (in Cornwall) china clay, were once extracted (see Figure 5.31). In Scotland alone, it is estimated that there are still 14 000 ha of derelict land, some badly contaminated.

Difficulties Facing Plants

As Figure 5.32 shows, pioneer plants colonising waste land usually have to adapt to very difficult habitats, although there may be exceptions.

◆ poor rooting conditions – loose spoil waste on bings and wide areas of concrete and tarmac do not make for an easy foothold

◆ drought is often a problem – water freely drains through piles of cement, stone, broken bricks and spoil waste. South facing walls and the steep, dark coloured slopes of a coal bing can become very warm when exposed to the sun

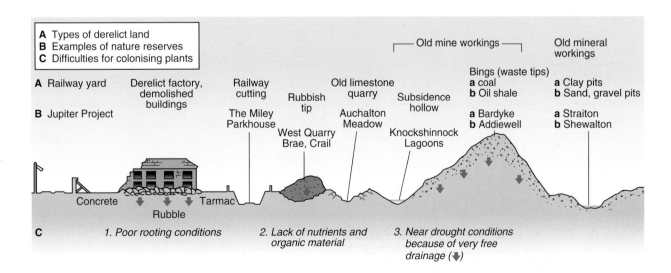

Fig 5.31 A classification of derelict land

155

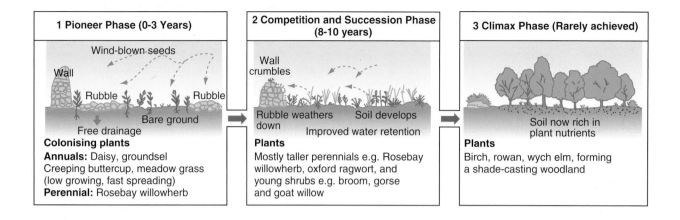

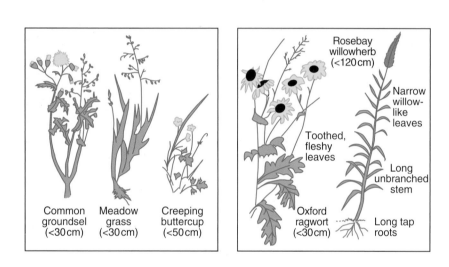

Fig 5.32 Plant succession on rubbly wasteground (based on observations in Glasgow)

◆ lack of nutrients – usually there is a lack of nitrogen, phosphorus and other minerals which are critical for plant growth. Also missing is organic material which helps to retain water and release nutrients.

Sequence of Plant Succession

Despite the difficulties which plants face when they colonise derelict land, they often spread rapidly. However, they rarely develop beyond the pioneer phase or indeed reach the climax stage. This is because such areas may only be temporary; bings are often levelled, while rubbly wasteground may be developed as a 'brownfield' site for new industry or offices. In some cases, the colonisation may be so successful that the areas become nature reserves.

Rubbly Wasteground

The sequence shown in Figure 5.32 reflects a gradual modification of the soil with an increase in organic material, improved nutrients and competition for light and root space.

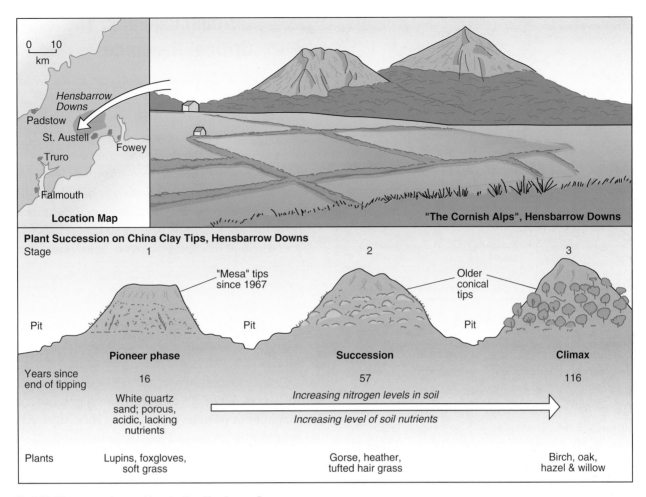

Fig 5.33 Plant succession on china clay tips, Hensbarrow Downs

China Clay Bings (Tips)

Plant successions have been observed in Cornwall where there are bings resulting from the mining of china clay (Figure 5.33). The diagrams show the age of the bings since the last tippings and sample plants from pioneer stage to climax. In spite of the difficulties facing plant pioneers, a critical factor was nitrogen cycling (Figure 5.34). Leguminous (nitrogen fixing) plants, such as gorse and tree lupin, were essential in the build-up of nitrogen in the ecosystem.

Bings and Conservation

Bings are now a vanishing feature of the landscape. They are removed because the spoil can be used for road metal and the infilling of derelict sites. Some conservationists argue that certain bings should be retained because:

◆ they are visible reminders of past (often dangerous) industrial activities and should be considered as monuments to our industrial history

◆ left undisturbed, some bings have an interesting plant history, ultimately developing a woodland climax (birch, willow and even oak). Sometimes

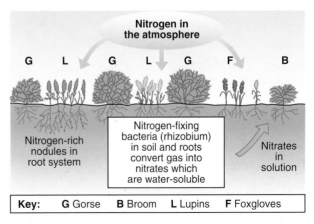

Fig 5.34 Fixing nitrogen in the soil

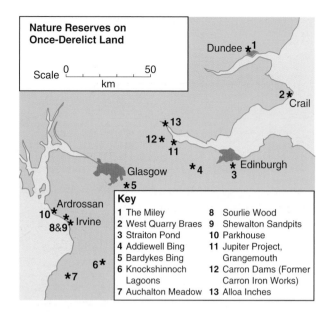

Fig 5.35 Nature parks on once-derelict land

they can be home to quite noteworthy species: Bardyke's Bing between Blantyre and Rutherglen is one of the few sites in the world where the orchid, Young's heleborine, is found (see Figure 5.35). In the case of Addiewell Bing, West Calder, after 50 years of colonisation, there are over 100 types of plants including heathers and orchids, as well as 28 species of birds and animals, including badgers and roe deer.

5.6 Introducing Soils: The Critical Resource

We build on it, we farm it, we play on it, we could not survive without it, and yet, we abuse it. Soil arguably, is our most important non-renewable resource. Home to countless living organisms, it forms a desperately thin layer lying on bedrock. Understanding the content, formation and characteristics of even just a few types of soil really requires a revision of the information in the earlier parts of this book. This is because soil is the central part of a system that links, as Figure 5.36 shows, the atmosphere, lithosphere, hydrosphere and the biosphere.

Soil Defined and Soil Content

Keep Figure 5.36 in mind when you (i) read the following definition of soil, and (ii) look at the pie chart (Figure 5.37). According to one definition, soil is a mixture of particles of weathered rock, decayed organic matter, water and gases in which living organisms are present. The pie chart shows that there are four main constituents of soil. It also shows a very approximate percentage composition of a typical topsoil. In reality, the amounts do vary, not least the ever changing amounts of water and air (gases).

◆ **Mineral matter:** this consists of minerals derived from parent material by physical and chemical weathering. By volume this is the most important component in soils, apart from peat soil. The parent materials are the fragments of rock called **regolith**

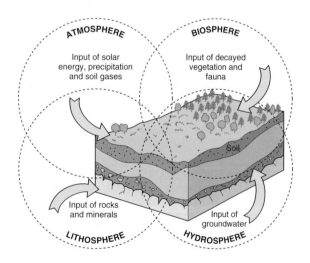

Fig 5.36 Soils and environmental inputs

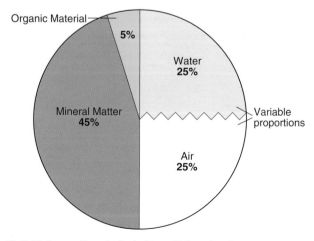

Fig 5.37 Composition of a 'typical topsoil' (by volume)

which can come from the underlying bedrock or (especially in Scotland) glacial deposits.

◆ **Organic material:** this is mainly derived from decaying roots, leaves, needles and remains of dead organisms. All such material is broken down by the action of countless micro-organisms and larger life forms such as worms and moles.

◆ **Air** and **Water** co-exist in ever changing volume and occupy the **voids** – the spaces in the soil. In a well drained soil, water forms a film around the mineral particles and allows space for the entry of air (gases). Soil air lacks light and has a higher percentage of carbon dioxide compared to 'atmospheric air'. Any well aerated soil can rapidly

change, however, when the soil is saturated with water.

Soil Formation and Soil Profiles

Figure 5.38 outlines three broad and very simplified stages showing how soils might develop. They can only give a brief idea of the many varied and interacting processes involved. This is because **pedology** (soil science) can be a demanding study. For **pedologists** (soil scientists), the fundamental unit for study is the soil profile which has developed through time (see Figure 5.39). A **soil profile** is a vertical section through the soil from surface vegetation to the bedrock. Information is usually obtained by digging a pit or by boring with a soil auger. By examining the (sometimes) distinctive layers or **horizons**, it is possible to work out the type of soil. Four horizons are picked out in the model profile (Figure 5.39). These differ in soil texture (the 'feel' of moist soil), colour and chemical composition.

1 The **Ao horizon**, when present, is the surface organic layer. Essentially, decaying vegetation, it can be subdivided into three layers:

L (litter) which may consist of leaves, pine needles, cones or dead heather shoots.

F (fermentation layer) where the organic material starts to decompose.

H (humus): the decomposed remnants of vegetation, animals and bacteria along with all their waste products. **Humification** (decomposition of the organic material) involves many organisms – worms, mites, fungi and bacteria, and produces an important source of nutrients for the soil below.

2 The **A horizon** proper is the main top layer and consists of a mixture of organic and inorganic material. It is here that the organic material is introduced from the Ao layer. Usually nutrient rich and fine textured, it is referred to as topsoil.

3 The **B horizon** is the subsoil, which contains less organic matter and is coarser in texture, reflecting the importance of weathering. Soluble soil material containing nutrients may be leached out of the A horizon into the B horizon. **Leaching** is the

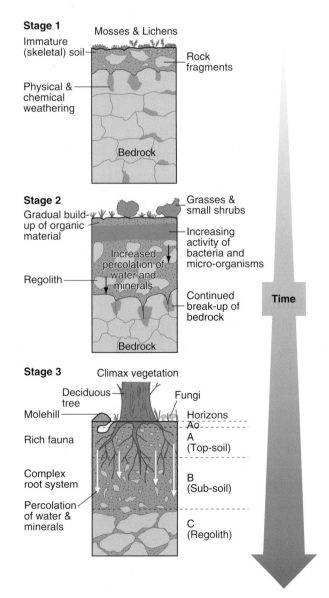

Fig 5.38 Simplified stages in soil formation

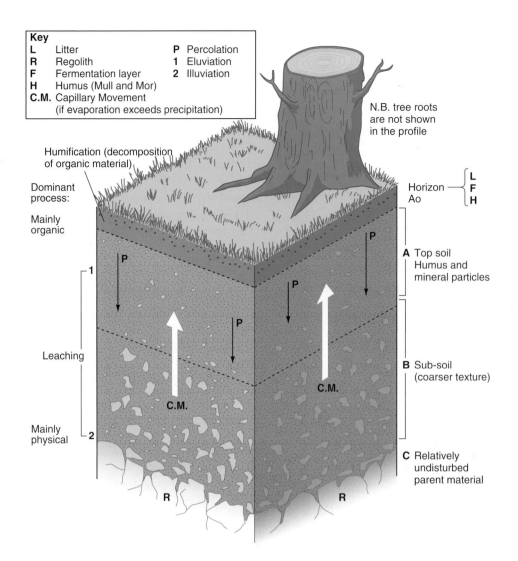

Key
L	Litter	**P**	Percolation
R	Regolith	**1**	Eluviation
F	Fermentation layer	**2**	Illuviation
H	Humus (Mull and Mor)		
C.M.	Capillary Movement		
	(if evaporation exceeds precipitation)		

N.B. tree roots are not shown in the profile

Humification (decomposition of organic material)

Dominant process:

Mainly organic

Leaching

Mainly physical

Horizon Ao — { L F H }

A Top soil
Humus and mineral particles

B Sub-soil (coarser texture)

C Relatively undisturbed parent material

Fig 5.39 A model soil profile

removal of soluble minerals and humus and involves (i) **eluviation** – washing out from the A horizon, and (ii) **illuviation** – washing in of the soluble material.

4 The **C horizon** is the zone of the **regolith** whose large particles sit upon the underlying bedrock. Physical and chemical weathering of parent material is a further source of nutrients.

Soil profiles vary in depth. In the humid tropics they can be up to 50 m deep, in Britain they average around 1.5 m.

Factors Influencing Soil Formation, with Especial Reference to Scotland

Profile photographs, representing four main types of soil found in Scotland, are shown in Figures 5.40A, B, C & D. If you examine them closely and read the accompanying notes, you see that they are quite distinctive in regard to texture, colour and chemical composition. Such differences reflect the varying influence of the following soil-forming factors. Pedologists, such as the American Hans Jenny, usually distinguish five main **interacting** factors. A sixth is now recognised because it is so important and this is the role of human activity.

1 Parent material: is especially significant in the early development of soil and its mineral content. It can vary from solid bedrock to a wide range of

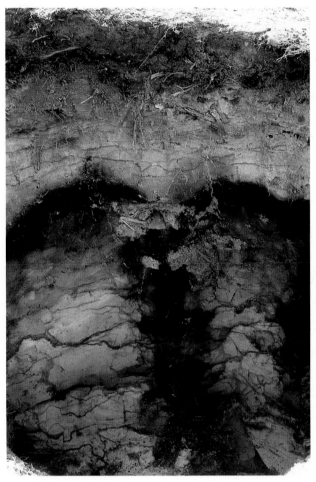

Fig 5.40A Podsol. The profile displays clearly defined horizons, including the acidic mor, the eluvial and illuvial horizons. Can you spot any others?

Fig 5.40B Brown Earth Soil. This profile shows a gradual lightening of colour with a downward decrease in organic content.

unconsolidated deposits including alluvium, river terrace gravels, wind blown sand and (most important of all in Scotland) glacial till. Parent material influences:

◆ **the rate of weathering:** over quite short distances, rocks can vary in their rate of weathering. Hard rocks, such as schist, gneiss and quartzite weather slowly, often leading to thin soils. Softer, sedimentary rocks, such as shale, weather more quickly.

◆ **chemical composition and soil colour:** minerals derived from quartzite and granite have a relatively high silica content and are acidic; soils on chalk and limestone are alkaline. Related to soil chemistry is colour: silica/quartz-rich acidic soils are light in colour; rocks such as basalt and gabbro have less silica but are rich in iron and are darker. Evidence of iron is seen in Strathmore's rich red soils, formed from glacial till derived from underlying Old Red sandstone.

◆ **soil texture:** (the 'feel' of moist soil) is influenced by the size of soil mineral particles (Figure 5.41). This is important because it helps to determine the permeability of the soil and the relative proportions of air (gases) and water. Particle size is classified as follows: Clay – less than 0.002 mm diameter; Silts – range from 0.002–0.05 mm; sand – 0.05–2.00 mm. Figure 5.41 shows the various classes of soil texture. Overall, sandy, porous parent material, rich in quartz and derived from granite, sandstone and schist, is easily leached and gives rise to podzols and brown earths (see page 166). Clay-like parent materials, such as basalt and shale, however, are not well aerated, are easily waterlogged and can result in gleys (see page 168).

2 Biotic factors involve the action of vegetation and a wide range of organisms, from bacteria to vertebrates (see Figure 5.42). They all interact, influenced by climate and the evolving nature of the soil itself, to produce the soil organic material or

Fig 5.40C Gley Soil Profile. This profile shows the grey-blue colour of the iron compounds in the soil

Fig 5.40D Raw Peat Soil Profile. A typical peat profile with very little inorganic material.

humus. This may lie beneath the L and F layers of the Ao horizon or it may be mixed through the whole of the A horizon. Various types are distinguished by pedologists and these include:

◆ **Mor humus**, which mainly develops beneath a coniferous forest or heather moorland and is usually associated with a cooler, wet climate and acidic parent material. The litter layer of pine needles, cones and/or heather shoots further encourages acidic soil reactions and discourages rapid breakdown of the plant material. Earthworms are not common so there is limited mixing of organic and mineral matter.

◆ **Mull humus** frequently develops beneath deciduous woodland whose leaves are rich in base materials as are the usually well-aerated soils. With a plentiful supply of litter and a rich soil fauna, there is no clearly defined humus layer, unlike the mor. Chemically almost neutral, mull is home to earthworms, which are very active decomposers.

Earthworms very capably ingest plant material and mix it with the mineral matter. Mull particularly combines with clay to form what is called a clay– humus complex. This helps the soil retain important plant nutrients which otherwise would be easily leached.

◆ **Moder** is an intermediate humus between mor and mull.

3   Climate is particularly important, especially seasonal and daily variations in temperature and precipitation. At low temperatures the rate of **pedogenesis** (soil formation) is slower, especially organic decomposition. Equally, warmer temperatures encourage decomposition and incorporation of organic material.

The input of water percolating through the soil horizons is not the same as precipitation but, provided precipitation exceeds evapotranspiration, **leaching** (removal of soluble minerals and humus) is a

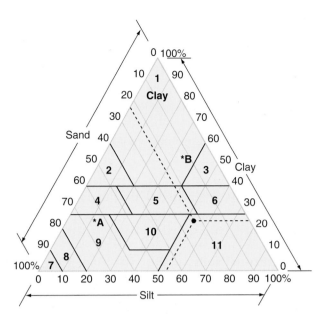

particularly important process not just vertically but also downslope (see Figure 5.43). Podsols and gleys are especially affected (see pages 165–7). On the other hand, if evapotranspiration exceeds precipitation, water and minerals in solution can be drawn upwards by the process of **capillary movement**.

4 Relief or topography of the land influences drainage and soil depth. Figure 5.43 shows the upper, well drained **shedding site** from which there is surface water runoff and throughflow. The lower **receiving site** gains water, organic and mineral matter. If the site is particularly badly drained, excess water accumulates and encourages the formation of gleys and peaty soils. Depending on the steepness of slopes, mass movement and even soil erosion can occur, thereby preventing the development of mature profiles.

Relief can also modify the effect of climate on pedogenesis. Shady north-facing slopes are colder and wetter compared to those with a south-facing aspect. This can slow up organic decomposition and encourage peat formation. Altitude is a related influence: with increased altitude, temperatures and

Key
- Soil sample containing: 23% sand; 53% silt; 24% clay
- *A&B other soil samples

2 Sandy clay	**3** Silty clay	**4** Sandy clay loam
5 Clay loam	**6** Silty clay loam	**7** Sand
8 Loamy sand	**9** Sandy loam	**10** Loam
11 Silt loam		

Fig 5.41 Soil texture diagram

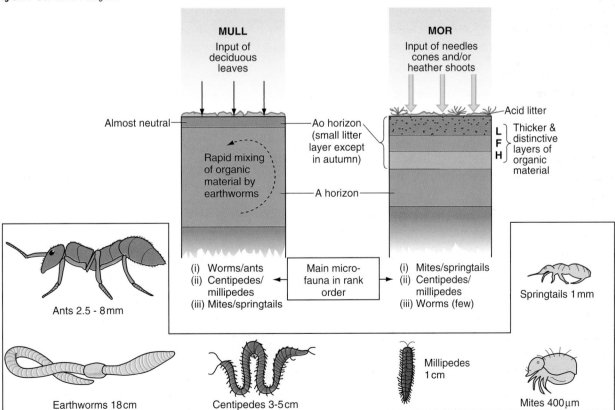

Fig 5.42 Mull and mor humus

the length of the growing season are reduced, and the precipitation increases.

5 Time is critical in the development of soils before they fully mature. When they are young, soils retain the features of the parent material. This is true of Scottish soils, which are relatively young as a result of the last Ice Age. Earlier soils, formed in warmer conditions, were effectively swept away by glaciers. Soils in Scotland, therefore, have mainly developed on glacial till in the last 10 000 years.

6 Human activity should not be underestimated. It started around 3000 BC, as prehistoric people began to fell the wildwoods and burn the heather plagioclimax, which accelerated naturally occurring soil erosion. More recently, blanket planting of coniferous plantations and the modification of soil chemistry by applying fertilisers, lime and pesticides have further altered soil characteristics.

Major Soil Types and Case Studies

Throughout the world, there is an enormous variety of types of soil. This is understandable, given the huge number of possible combinations of soil forming factors. As a way of better understanding the diversity of soils, pedologists come up with different soil groupings. These include: zonal, intrazonal and azonal.

◆ the **zonal soils** idea come from the work of Russian pedologists, including Vasily Dokuchaiev. In spite of different types of parent material, they observed that a small number of soil types broadly corresponded to the main biomes such as the temperate broadleaved deciduous woodlands and the boreal coniferous forests. Such soils have matured over a long enough period of time to allow the influence of climate and organisms to be particularly effective

◆ **intrazonal soils**, on the other hand, really reflect the influence of local factors not involving climate or organic material. Parent material can be particularly important: for example, the chalk and Jurassic limestone escarpments of south and east England develop lime-rich soil (known as calcareous brown earth). Another factor can be the local drainage conditions, which concentrate ground water. This can be seen in Gley soils (see below).

◆ **azonal soils** are essentially immature soils. There has not been enough time for all the horizons, especially the B horizon, to develop. They are found, for example, in recent glacial deposits, fresh river alluvium and volcanic soils.

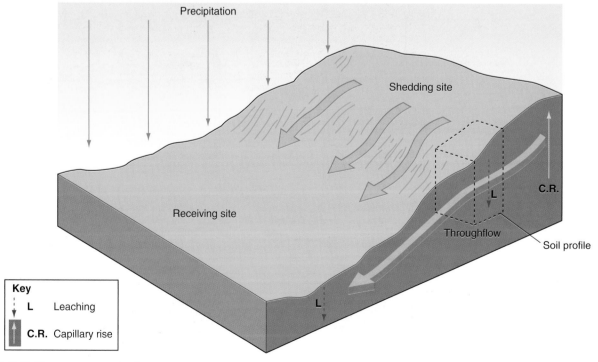

Fig 5.43 Water movement and soil

Case Studies of Selected Soils

The examples discussed match three of the profile photographs (Figure 5.40). Look at them as you work through this section.

Podsols usually have clearly defined horizons. This is mainly due to the process known as **podsolisation**. This involves pronounced leaching of material (iron, aluminium and humus compounds) from the surface layers to the lower layers.

They are found in the northern boreal coniferous forests with associated cold, wet climates in which precipitation exceeds evaporation. Podsols are also found in upland moorland areas. Figure 5.44 shows a typical profile. The Ao horizon forms from decaying plant litter which can be (i) the pine needles, cones and twigs from coniferous trees and/or (ii) dead heather shoots from upland moorland. Thanks to the cold climate, the organic material decays very slowly to form an acidic **mor humus**. Rain and melting snow combine with these organic acids and wash out (**eluviate**) the minerals from the A horizon. This produces an ash-coloured, bleached A horizon (*zola* is Russian for ash) mainly composed of insoluble silicates.

Lower down the profile, aluminium, iron, clay and humus are washed in (**illuviated**) and redeposited in the subsoil or B horizon. The presence of iron and aluminium help explain the reddish-brown colour of the horizon. If iron accumulates over a long enough period, a rust-coloured **iron pan** can form, often up to several centimetres thick. Iron pans can prevent: (i) the penetration of plant roots and (ii) the free drainage of the podsol resulting in waterlogging. The C horizon forms from a range of parent material (fluvioglacial sands or till) or may be derived from acidic parent rocks.

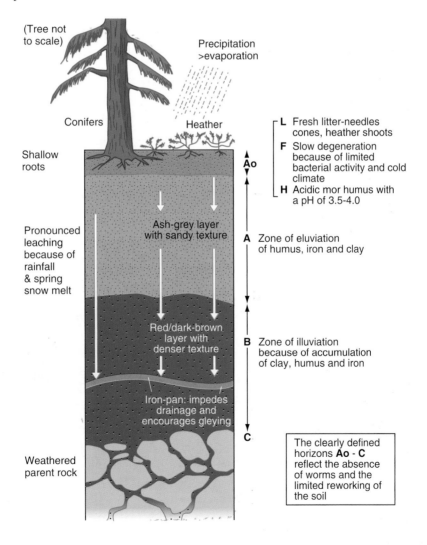

Fig 5.44 Podsol profile

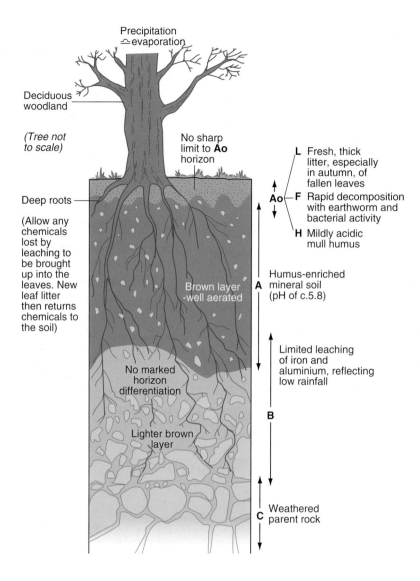

Precipitation
⇌ evaporation

Deciduous
woodland

*(Tree not
to scale)*

No sharp
limit to **Ao**
horizon

Deep roots

(Allow any
chemicals
lost by
leaching to
be brought
up into the
leaves. New
leaf litter
then returns
chemicals to
the soil)

Brown layer
-well aerated

No marked
horizon
differentiation

Lighter brown
layer

L Fresh, thick
litter, especially
in autumn, of
fallen leaves

Ao **F** Rapid decomposition
with earthworm and
bacterial activity

H Mildly acidic
mull humus

Humus-enriched
A mineral soil
(pH of c.5.8)

Limited leaching
of iron and
aluminium, reflecting
low rainfall

B

C Weathered
parent rock

Fig 5.45 Brown earth profile

Podsolisation, therefore, gives a soil profile with clear horizons. These are also encouraged by the lack of organisms, especially worms, with their characteristic ability to function like food mixers and obliterate the distinction between A and B horizons, as in the brown earth profile discussed below.

Podsols are not naturally fertile soils and cropping yields rapidly decline. Lime needs to be added to counteract the podsol's acidity and animal manure can boost the poor quality mor humus.

Brown Earth Soils (Figure 5.45) have developed beneath the temperate broadleaved deciduous forests of Europe, Russia and North America. The Ao horizon is rich in nutrients thanks to an input of decaying grasses and herbs and, in autumn, fleshy deciduous leaves. Compared to the podsol, this litter

decomposes relatively quickly because of the milder climate. The result is a characteristic **mull humus.** Less acidic than mor, the mull becomes well mixed with the soil minerals thanks to the activity of soil fauna, particularly earthworms, and soil bacteria.

The A horizon, therefore, is well aerated, with a loamy texture. It has a dark brown colour because humus replaces minerals as they are leached out. Leaching, however, is less pronounced because of a closer balance between evaporation and precipitation. The B horizon is not so distinct as in the podsol but usually is lighter in colour as humus becomes less abundant. The C horizon is derived from varied parent material which can range from limestone to schists. Figure 5.45 shows that tree roots can penetrate even the C horizon to extract minerals and ensure the efficient cycling of nutrients through the ecosystem.

Originally tree covered, these usually mildly acidic, brown earth soils, have been extensively exploited for agriculture since prehistory. Compared to areas of podsolic soils, the temperate areas with brown earths provide the best agricultural land and support a much higher population density (see pages 172–173).

Gley Soils (Figure 5.46) are intrazonal soils and are found in sites which are waterlogged, either permanently or temporarily. When soil is waterlogged for a long time, its pore spaces lose oxygen. Such a condition is called **anaerobic** and means that any decay of bacteria is slowed down. In addition, iron compounds in the soil are reduced chemically from their normally red-brown colour to a blue-grey colour (see Figure 5.46 for more chemical detail if you wish!). This process can be reversed because of any seasonal drying out of the soil. Gradually, as a result of seasonal changes, a gley profile develops a mottled appearance with orange-brown mottles set in a blue-grey matrix.

Figure 5.46 shows a sample profile which has developed under a badly drained grassy meadowland infested with rushes such as *Juncus effusus*. The Ao and A horizons are darker, reflecting the presence of organic matter. A large amount of organic material can accumulate because of the lack of bacterial activity necessary to create humus. Some red-brown mottling occurs along the root channels which allow oxygen to accumulate. The B horizon is predominantly blue-grey, indicating virtually continuous waterlogging. This has developed on a C horizon derived from an impermeable clay layer.

Gleying, therefore, is caused by the inability of soils to shed water quickly. Often found at the foot of slopes and in floodplains, gleys can support wetlands, permanent pasture and (with draining, ploughing and manuring) arable farming.

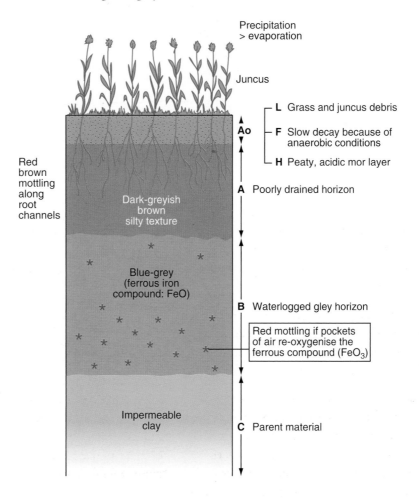

Fig 5.46 Gley profile

To Take You Further

1 Gleys are examples of hydromorphic soils, that is, they develop in areas of excessive water. Another exemplar is peat which, as Figure 5.47 shows, is one of the four main groups of soils found in Scotland. This map also shows that there are many combinations of the soil groups, differing in colour, texture and parent material, a reflection of the variety of soil forming factors. Changes in these factors can cause soils to alter. For example, podsolisation can alter brown earths into podsols and, in turn, the presence of an iron pan can help upland podsols develop into peaty gleyed podsols and, eventually, blanket peat.

2 Peat is an organic hydromorphic soil which has spread, since deglaciation, over large tracts of Scotland during colder and wetter periods. Under such conditions mineral nutrients were leached out of the soil. The resulting waterlogged environment encouraged plants, in particular, sphagnum moss, to accumulate at a faster rate (about one millimetre per year) than it can be broken down by soil organisms. Blanket peat is the main type and it accumulated in the upland areas, particularly in the north and west of Scotland. As the name suggests, it 'blanketed' the landscape and nowadays the original mineral soil can lie several metres below these extensive peat accumulations.

Peat and associated peaty soils (peaty gleys and peaty podsols) cover 50% of Scotland and support rough grazing, forestry and a variety of unusual and specialised plants, e.g. the midge-consuming sundew. The Flow country of Caithness and Sutherland is one of the largest expanses of blanket bog in the world, providing a habitat for a wide range of waders.

Catenas and Soil Development

On small scale maps, the soils of the British Isles are depicted in a generalised way with podsols in the north-west and brown earths in the south-east. As suggested in the last section, there are, in fact, large variations in types of soils, even in a comparatively small area. This can be illustrated in the soil sequences for a small part of the Eastern Grampians, shown in Figure 5.48.

Such a diagram is known as a **catena**. In a model catena, the sequence of soil profiles which develops downslope is related to the relief of the land. The changing slope angle affects the drainage conditions which then influence the amount of leaching and gleying. Ideally, the underlying parent material is the same and there is no marked change in climate. In reality, however, altitude influences the climate and there can be variations in parent material.

Figure 5.48 shows a (very) simplified catena from the rounded summit of Cairn O' Mount (435 m) to Glen Saugh (about 150 m) with four main soil types. Notice that the parent material changes south of the Highland Boundary Fault.

On the summit, blanket peat reaches a depth of 3–4 m and its presence reflects the cool, moist conditions and poor drainage. South-eastwards, these give way to podsols. Those profiles with pronounced iron pans prevent water from penetrating and this encourages gleyed conditions. Overall, soil nutrients are

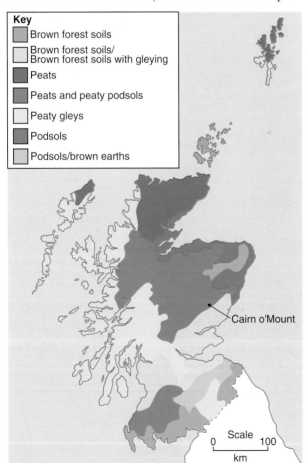

Key
- Brown forest soils
- Brown forest soils/ Brown forest soils with gleying
- Peats
- Peats and peaty podsols
- Peaty gleys
- Podsols
- Podsols/brown earths

Cairn o'Mount

Scale
0   100
km

Fig 5.47 Generalised soil map of Scotland

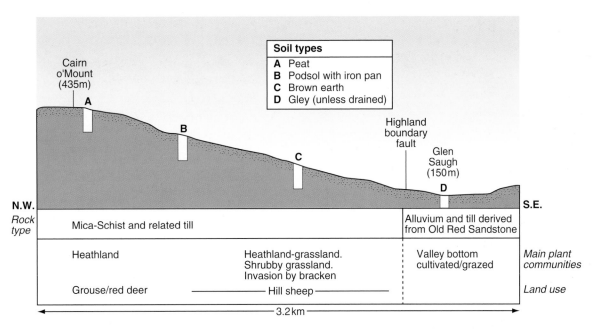

Fig 5.48 Very generalised soil catena in the Eastern Grampians

The Role of People in Soil Formation

transferred downslope and, where conditions are well drained, less acid brown earths can be found at about 250 m. Finally, unless the land is well drained, intrazonal gleys are found in Glen Saugh, which is the receiving site at the foot of the catena.

The Role of People in Soil Formation

It is now recognised by archaeologists and pedologists that soil has been affected by human activity, particularly as a result of farming activities from about 3000 BC onwards. Human impact can be seen in two ways.

1 Podsolisation has been encouraged in upland Britain. Excavation on the North York Moors has revealed 'fossil' brown earth soils preserved beneath Bronze Age burial chambers (barrows) yet the surrounding heathland landscape has podsols with iron pans. This reflects the history of land use. Initially, early farmers cleared small areas of deciduous forest, using a form of shifting cultivation. The introduction of grazing animals further prevented woodland regeneration and the resulting heather moorland (with mor humus) increased the soil's acidity compared to the original mull of the deciduous woodland. Podsolisation was further encouraged by the management of moorland for grouse and the twentieth century spread of

coniferous plantations on these uplands (Figure 5.49).

2 Related to podsolisation has been the waterlogging of soil. Figure 5.50 suggests various hydrological processes which can result in waterlogging, gleying and peat formation. Such a diagram reminds us of the ability of trees to recycle water, moderate the impact of a heavy storm and retain rainwater through the root system. Conditions are then ideal for the spread of sphagnum peat to spread and smother the ancient tree stumps.

Once woodland is cleared, therefore, the whole balance of the ecosystem is altered. As stressed throughout this chapter, all the components of an ecosystem interact. In geography, it is important to remember:

◆ the joint role of people and nature in our study of the biosphere

◆ the ability of people to totally transform extensive biomes, as illustrated next in the temperate deciduous woodlands.

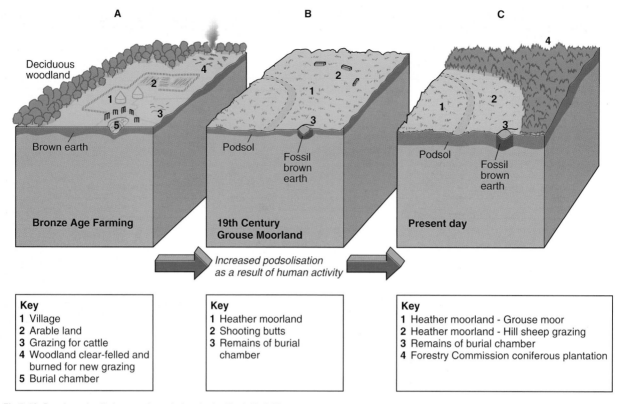

Key
1 Village
2 Arable land
3 Grazing for cattle
4 Woodland clear-felled and burned for new grazing
5 Burial chamber

Key
1 Heather moorland
2 Shooting butts
3 Remains of burial chamber

Key
1 Heather moorland - Grouse moor
2 Heather moorland - Hill sheep grazing
3 Remains of burial chamber
4 Forestry Commission coniferous plantation

Fig 5.49 People and soil changes through time in the North York Moors

5.7 Temperate and Coniferous Biomes: The Impact of People

Scotland lies at the interface of two major **biomes** – the temperate, broadleaved deciduous woodlands and the boreal (northern) coniferous forests. Figure 5.51 shows the one-time global distribution of these biomes and that deciduous woodlands once occupied most of Europe between 40° and 60° N. In eastern North America and eastern Asia, they ranged from 30° N to 50° N, while a smaller area occurred in Latin America on the lower, southern slopes of the Andes. Botanically, the formations differ between these areas, with Europe less rich in species since the end of the last Ice Age. Equally, the boreal coniferous forests have two main formations: the European and the North American, each with different dominants. In the European formation, for example, Norwegian spruce is common on damper soils, Scots pine occurs on drier, (often) fluvioglacial soils. Black spruce, white spruce and balsam fir are dominants in North America.

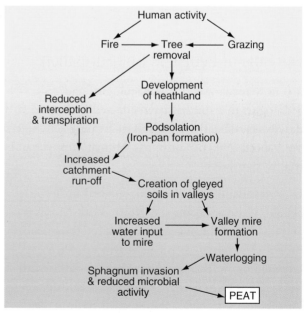

Fig 5.50 Human activity, hydrological processes and peat formation (very simplified)

Figure 5.52 is a transect through the European boreal and deciduous formations. It outlines the broad vegetation features and zonal soils. It also shows that there is no rigid demarcation line between the coniferous forests and the broadleaved deciduous trees to the south. Rather, there is a gradual change, a

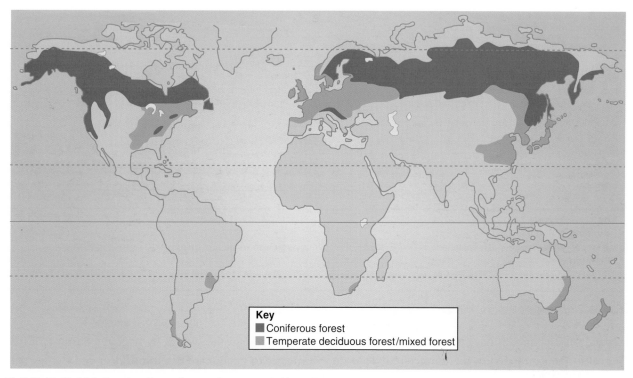

Fig 5.51 World distribution of coniferous forest and temperate deciduous forest

zone of transition called an **ecotone**. This is a mixture of coniferous and broadleaved forests reflecting a stalemate with regard to plant competition. The natural vegetation distribution shown on map and transect shows the climax situation. Throughout this chapter, however, the ecological role and impact of people has been stressed.

Since prehistory, people have recognised forests as a major resource. Forests have provided fuel for cooking, warmth and smelting metals, while timber provided shelter and construction materials for homes and ships. Recreation, from hunting to orienteering, can take place in woods and, when cleared, woodland provides land for arable farming and grazing.

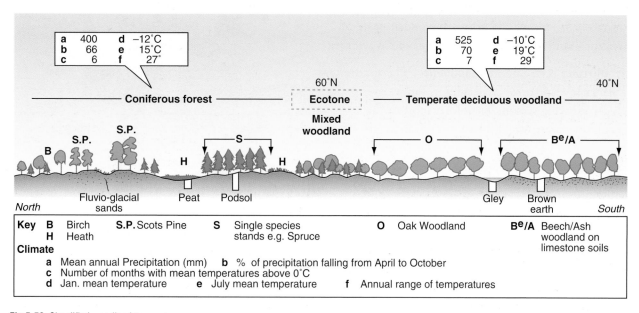

Fig 5.52 Simplified woodland transect

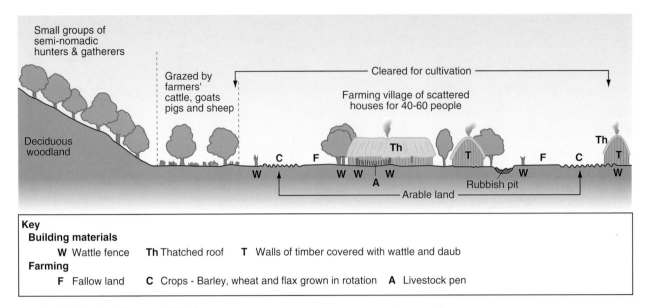

Key
Building materials
  W Wattle fence   **Th** Thatched roof   **T** Walls of timber covered with wattle and daub
Farming
  F Fallow land   **C** Crops - Barley, wheat and flax grown in rotation   **A** Livestock pen

Fig 5.53 Early farming in Europe c. 4500 BC

Overall, fewer inroads have been made into the boreal forests compared to the temperate deciduous forests. Several related reasons can be suggested:

◆ **location:** in relation to main centres of population, the boreal forests are relatively remote. Their exploitation for timber, especially with the increased demand for paper, pulp and other wood products such as chemicals, began in the areas of easiest access and cheaper transport. More remote areas of Siberia and northern Canada were left alone as extensive reserves. Today, with a changing attitude to forest resources and pressure from conservationists and ecotourist interests, these areas will not be cleared. For example, in northern Sweden some 55 areas of virgin forest have been designated as reserves

◆ **climate** is an important influence. The southerly, warmer lowland areas of deciduous forest have a longer growing season. Therefore, most of the deciduous forest was cleared for cultivation. In contrast, most of the coniferous biome does not favour agriculture. Although the longer hours of daylight in the higher latitudes of central Finland, for example, partly compensate for its shorter growing season, late frosts are common, and a prolonged snow cover and frozen ground encourages farmers here (and elsewhere in northern Scandinavia) to crop trees as part of their income

◆ **varying soil quality** has also been important from earliest times. Mesolithic hunters and gatherers

first modified the woods by selective burning of tree species to encourage grazing of deer and the growth of hazels. The greatest impact, however, came with the development of agriculture during Neolithic times (Figure 5.53). Spreading from the Middle East

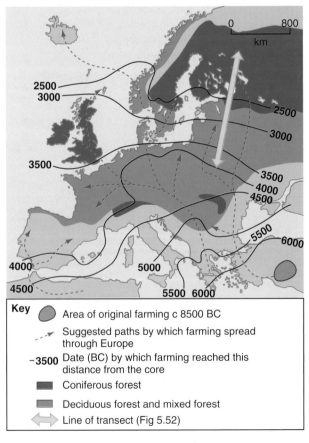

Key
  ⬭ Area of original farming c 8500 BC
  - - ▸ Suggested paths by which farming spread through Europe
  −3500 Date (BC) by which farming reached this distance from the core
  ▪ Coniferous forest
  ▪ Deciduous forest and mixed forest
  ⟷ Line of transect (Fig 5.52)

Fig 5.54 The spread of farming in Europe

and South-East Europe, the earliest farmers introduced wheat, barley, peas and lentils, which they grew on the fertile soils underlying the deciduous forest. The loamy texture of the nutrient rich brown earths and loess (a fine-grained wind-blown soil carried from newly deposited glacial materials) encouraged clearance, settlement and (see Figure 5.54) cultivation. On the other hand, the cooler climate and acidic, low nutrient podsols were less attractive to early settlers. Lime and animal manure were essential inputs for farmers. Another adaptation involved shifting cultivation or 'burnbeating'. Carried out in Scandinavia from

medieval times (until 1920 in Finland) it involved cutting down trees in small clearings, drying, burning the wood and then planting rye and barley in the fertile ash amongst the coniferous stumps. After a year or two, the clearing was abandoned for a fresh one.

These reasons, and the related differences in population density, effectively meant the virtual deforestation of the broadleaved deciduous forests. But even the boreal forests have been extensively modified and much is secondary woodland in the accessible areas.

ASSIGNMENTS

Scottish Wildwoods, Woodlands, Moorlands and Ecosystems
Read pages 136 to 148.

1 a) With the help of Figure 5.1, **outline** the broad changes in Scottish vegetation from the end of the last Ice Age to around 3000 BC.

b) Refer to Figure 5.2. **Briefly describe** the broad distribution of **Scotland's 'Wildwood'** around 3000 BC, mentioning actual species of vegetation. **Explain** the unwooded areas.

c) From Figure 5.2, **name** the likely surviving tree species found at:
(**i**) Beinn Eighe/Loch Maree, (**ii**) Wood of Cree, (**iii**) Ledmore Wood, (**iv**) Glen Nant.

2 a) What is meant by an **ecosystem**? **Give** some examples of ecosystems.

b) With the assistance of Figure 5.5, **draw** a summary diagram to illustrate the main features of a 'typical' **oakwood ecosystem**.

c) In an oakwood community, why do the various plants and animals have to occupy an **ecological niche**?

3 a) What is the ultimate source of energy in an ecosystem?

b) With the help of Figures 5.6 and 5.7, **explain** how the following play an important part in the functioning of an ecosystem such as an oakwood: **photosynthesis**, **food chain**, **trophic level**, **primary producers**, **consumers**, **decomposition** and **recycling of nutrients**.

c) Using Figure 5.8, **briefly describe** how an oakwood (part of the biosphere) 'interacts with the atmosphere, hydrosphere and lithosphere'.

d) Refer to Figure 5.9. In what ways have people 'managed' the oakwood system?

4 a) **Explain** why forest has given way to heather moorland over extensive areas of upland Scotland.

b) **Describe** the parts played by **grazing** and **muirburn** in the management of heather moorland.

c) With the aid of Figure 5.12, **outline** the **life cycle** of heather. What effect does muirburn have on the upland landscapes of Scotland?

d) Using Figure 5.15, **explain** the main features of a heather moorland **food web**. How does a food web differ from a food chain?

Plant Succession, Climax vegetation and Biomass
Read pages 148 and 158.

1 a) Study Figure 5.17. **Outline** briefly the ways in which different plants colonise, and then, succeed each other on a scree slope.

b) What is meant by **a plant succession** or **sere**? From Figures 5.18–21, briefly explain the terms: (**i**) lithosphere, (**ii**) hydrosere, (**iii**) halosphere, (**iv**) psammosere.

c) Study Figure 5.22. **Explain** how the various stages of **plant succession** lead to **climatic climax vegetation**.

d) **Explain** the term **plagioclimax**. In what ways is a heather moorland a good example of a plagioclimax?

2 Tentsmuir is an excellent example of a coastal sand dune system. It is gradually growing eastwards. The development of such a sand dune complex depends on successive stages of plant succession (**psammosere**) from pioneer stage to climax (see Figure 5.27).

a) **Briefly describe** the location of **Tentsmuir Point National Nature Reserve**. **Name** the two organisations mainly responsible for the ecological management of the area.

b) **Outline** the difficulties that face pioneering plants as they colonise coastal dunes such as Tentsmuir.

c) **Name** three plants which colonise the **foreshore** at Tentsmuir. What allows them to adapt to the foreshore environment? How do these plants encourage the development of **embryo** and **foredunes**?

d) With the aid of a sketch, **show** how **marram grass (i)** is adapted to the **'yellow' dune** environment, and **(ii)** encourages such a dune to stabilise (see Figure 5.29).

e) What are grey dunes and slacks? **Name** plants associated with these features.

f) **Briefly explain** why the natural climax vegetation at Tentsmuir takes the form of contrasting plagioclimaxes.

g) Referring to all the information on Figure 5.27, **describe and explain** the changes in plant types along the generalised transect from the coast inland.

h) Look at the plant survey on page 175. **Describe and explain** the changes in plant type with distance from the drift line.

3 a) Study Figure 5.31. **Outline** the main types of **derelict land** in Scotland.

b) What are the main difficulties facing pioneer plants colonising derelict land?

c) **Describe and explain** the sequence of plant succession either on an area of rubbly wasteground (see Figure 5.32) or on china clay bings (see Figures 5.33/34).

d) **Suggest** why plant communities on derelict land rarely reach the climax state.

e) Study Figure 5.35. Naming specific examples, **suggest** why certain bings should be retained rather than cleared.

Soil Formation and Soil Profiles
Read pages 159 and 165.

1 Refer to Figures 5.36 and 5.37. **Define** soil and **outline** the main components of a 'typical' top-soil. Why does the composition vary from time to time?

2 Carefully scrutinise Figures 5.38 and 5.39.

a) **Explain** the terms **pedology, soil profile, soil horizon, Ao horizon (including humus), A horizon, B horizon**, and **C horizon (regolith).**

b) **Draw** a large outline model soil profile. **Annotate** the diagram to show:
   (i) the key terms mentioned in a)
   (ii) soil forming processes including **humification, leaching, eluviation, illuviation, physical** and **chemical weathering** of bedrock.

The soil forming processes mentioned in (2) can be further understood by attempting questions 3 to 6.

3 a) **Give** examples of **parent material**. **Outline** the ways in which it influences soil formation.

b) What is the difference in particle size between sand, silt and clay?

c) Study Figure 5.41. The break down of a soil sample is shown. **Attempt** a similar break down by particle size for samples A and B, shown on the texture diagram.

4 a) Using Figure 5.42 to help you, describe and explain the differences between **Mull** and **Mor** types of humus.

b) Why might earthworms be compared to food mixers? How do they affect the differentiation of soil horizons?

5 a) How do variations in temperature influence soil formation?

b) **Explain** the following terms: **leaching, eluviation, illuviation** and **capillary movement**.

6 a) Study Figure 5.43. What are **shedding sites** and **receiving sites**?

b) Why is **time** a key influence on soil formation?

c) **Briefly, outline** ways in which people have influenced soil development.

Soil Types, Case Studies and Catenas
Read pages 165 and 170.

1 a) What is a **zonal soil**?

b) Look at Figures 5.51 and 5.52. State the predominant zonal soil to be found **(i)** in the coniferous forest **(ii)** in the temperate deciduous woodland.

c) In what ways do **azonal soils** differ from **intrazonal soils**?

2 **Zonal Soil Case Studies**. Study Figures 5.40 A and B, and Figures 5.44 and 5.45.

a) On a large piece of paper, side by side, **draw** outline soil profiles of a **podsol** and a **brown earth**. Each profile should be annotated as fully as possible to show the key features (colour, texture and composition) of the horizons and the main soil forming processes.

b) In what types of environment are podsols and brown earths found?

c) **Describe and explain** the effect of climate and vegetation on the development of the 'typical' soil profiles of podsols and brown earths.

d) Of these two soil types:
   (i) which has the most clearly differentiated horizons, and why?
   (ii) which is best suited to farming? **Justify** your answer.
   (iii) which is most likely to develop an iron pan? **Explain** your answer.

3 **Intrazonal Case Study**. Study Figures 5.40 C and 5.46.

    a) Under what conditions does **gleying** occur? When is soil described as **anaerobic**?

    b) On a large piece of paper, **draw** a sample gley soil profile. Your diagram should be annotated to show the main features of the horizons and soil forming processes.

    c) What effect would drainage have on **(i)** the horizons of a gley soil **(ii)** the agricultural potential of such a soil?

4 Study Figure 5.48. It would be useful to read the section 'To take you further' about peat formation.

    a) What is a soil **catena**?

    b) Briefly, **describe** the changes in the type of soil from Cairn O' Mount to Glen Saugh.

    c) Referring to changes in climate, slope and drainage, **explain** these changes in the type of soil from Cairn O' Mount to Glen Saugh.

5 Based on a soil survey, information about two soil profiles is provided below. **Draw** simple annotated profiles based on the sets of information. **Identify each** of the profiles, and **explain** your answers as fully as possible.

Temperate Deciduous and Coniferous Forest Biomes: The Impact of People
Read pages 171 and 174.

1 **a)** **Explain** what is meant by a biome (see page 140).

    b) Referring to Figure 5.51, **outline** the global distribution of the main formations of the coniferous forest and the temperate deciduous forest.

    c) With the help of Figure 5.52, **explain** the meaning of ecotone.

    d) **Describe** the main changes in vegetation in the transect diagram (Figure 5.52) in relation to changing climate conditions.

2 **a)** For what purposes have people exploited the coniferous and temperate forests?

    b) Study Figures 5.53 and 5.54. As fully as possible, **explain** why there has been less impact by people on the coniferous forests compared to the temperate deciduous forests.

Soil Survey Information

Location: Cairngorms
Altitude: 732 m
Slope: 4°
Vegetation: heather and reindeer mosses

0–5 cm: dark brown fibrous raw humus, pH 4.1, many plant features visible
5–13 cm: ash grey mineral horizon
13–23 cm: grey to medium brown sandy horizon, with quartz crystals and mica flecks. Some iron staining.
below 23 cm: parent material of grey schist boulders

Location: Lower slope of South Downs
Altitude: 67 m
Slope: 20°
Vegetation: oak woodland and grasses very deep roots

0–34 cm: limited plant litter, pH 6.8
3–18 cm: very dark brown loamy horizon
18–34 cm: very light brown, sandy texture
below 34 cm: chalky parent material

Psammosere Vegetation Survey at Kinshaldy Beach, Tentsmuir

Types of vegetation found at various distances from the drift line, east to west

12m: Sea Rocket, Saltwort
14m: Marram Grass, Saltwort, Orache
52m: Lyme Grass, Couch Grass
60m: Couch Grass
64m: Marram Grass, Lyme Grass
74m: Couch Grass
84m: Marram Grass
90m: Marram Grass

116m: Ragwort, Mouse-ear Hawkweed
117m: Young Scots Pine, Rosebury Willow-Herb, meadow Sweet, Elder, Stinkhorn, Marram Grass, Buttercup, Sea Holly, Rosebury Willow-Herb
150m: Couch Grass, Saltwort, Thyme
170m: Ladies Bedstraw
180m: Gorse, Guy Lichen, Scots Pine

Sample pH measurements at various distances from the drift line, east to west

14m: 7.6
74m: 7.6

130m: 6.2
175m: 5.7

Extra Assignments

1 With the aid of Figures 5.4 and 5.11, **explain** what processes are involved in pollen analysis and suggest its value in studying changes in vegetation.

2 a) Describe and explain the main characteristics of the vegetation and human exploitation of a 'typical' heather moorland.

   b) Explain the shrinking area of heather moorland and declining numbers of grouse.

3 a) What are the arguments for rehabilitating the Caledonian Woodland?

   b) Describe the various management methods that have been used in the regeneration of the Caledonian woodland. Is there a case for the return of the wolf?

4 a) With reference to Figure 5.49, **explain** the presence of a 'fossil' brown earth in an area of podsol soils.

   b) What is peat? What environmental factors encouraged its formation?

   c) Study Figure 5.50. **Explain** how human activity may initiate the formation of peat.

Key terms and concepts

- azonal soil (p. 164)
- biomass (p. 151)
- biome (p. 139)
- biosphere (p. 139)
- brown earth (p. 166)
- catena (p. 168)
- climatic climax vegetation (p. 149)
- coppicing (p. 139)
- dominants (p. 139)
- ecological niche (p. 140)
- ecosystem (p. 139)
- eluviation (p. 160)
- field layer (p. 140)
- fermentation layer (p. 159)
- food web (p. 141)
- gley (p. 167)
- ground layer (p. 141)
- humus (p. 159)
- illuviation (p. 160)
- intrazonal soil (p. 164)
- iron pan (p. 165)
- leaching (p. 159)
- litter layer (p. 159)
- moder humus (p. 162)
- mor humus (p. 162)
- muirburn (p. 142)
- mull humus (p. 162)
- nutrient cycle (p. 141)
- parent material (p. 160)
- peat (p. 168)
- plagioclimax (p. 150)
- photosynthesis (p. 140)
- pioneer community (p. 149)
- podsol (p. 165)
- pollen analysis (p. 139)
- psammosere (p. 155)
- receiving site (p. 163)

- regolithic (p. 160)
- shedding site (p. 163)
- shrub layer (p. 140)
- soil horizon (p. 159)
- soil profile (p. 159)
- soil texture (p. 161)
- trophic level (p. 140)
- zonal soil (p. 164)

Suggested Reading

Bridges, EM (1970) *World Soils* Cambridge University Press.

Broadley, E and Cunningham, R (1991) *Core Themes in Geography: Physical* Oliver & Boyd.

Darlington, A (1969) *Warne's Natural History Atlas of Great Britain*

Fitzpatrick, EA (1978) *An Introduction to Soil Science* Oliver & Boyd.

Galbraith, I (1990) *Ecosystems and People* Oxford University Press.

Gimmingham, CH (1975) *An Introduction to Heathland Ecology* Oliver & Boyd.

Pears, N (1985) *Basic Biogeography* (2nd edition) Longman.

Rackham, O (1990) *The History of the Countryside* JM Dent & Sons.

Riley, D and Young, A (1968) *World Vegetation* Cambridge University Press.

Thomson RD et al (1986) *Processes in Physical Geography* Longman.

Tivy, J and O'Hare, G (1981) *Human Impact on the Ecosystem* Oliver & Boyd.

Internet Sources

Biomes map:
www.snowcrest.net/freeman/geography/slides/biomes/index.html
British Society of Soil Science:
http://lurch.bangor.ac.uk/dj/bsss.html
Soil profile images:
www.mines.uidaho.edu/pses/teach_res/
Staffordshire Learning network:
www.sln.org.uk/geography
WWW Scotland:
www.wwf-uk.org/education/scotland/datasupport/
World Biomes:
www.uwsp.edu/acaddept/geog/faculty/ritter/geo101/biomes_toc.html

The Human
ENVIRONMENT

PROLOGUE TO HUMAN GEOGRAPHY

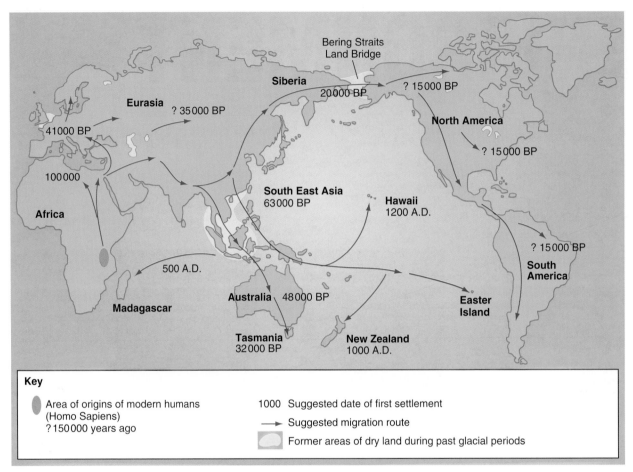

Fig 6.1 Migration of early people (after B. Fagan: 'World Prehistory')

In the earlier chapters of this book, the focus was on physical geography. However, reference was made to the role of human activities in changing the face of the earth, including: the contribution of fossil fuels and car emissions to global warming; the alteration of river channels in an attempt to curb the impact of flooding, a perfectly natural phenomenon; the initiation of mass movement by piling up unstable deposits of waste rock as at Aberfan; and by clearing woodland, regular muirburning and encouraging grazing, extensive heather moorlands were created and managed.

The second half of this textbook emphasises aspects of human geography, as a means of better understanding how people from the earliest of times have interacted with their natural environments. Chapter 7 covers population geography. A knowledge of the growth, spread and distribution of people across the earth is basic to any understanding of world history and contemporary problems. Figure 6.1 shows early migrations of people. Initially migrating from an African heartland, they gradually spread in small groups across the globe into the various biomes, adapting to different, and sometimes difficult, environmental conditions. Over thousands of years, the population of the world grew slowly but, as the generalised graph of population growth shows, there have been important surges in growth. These resulted from major revolutions in human history

The Human Environment

(Figure 6.2) associated with changes in agriculture, industry and settlement. Today, rapid global population growth, to over six billion people, means that the world faces many environmental challenges caused by past and present human actions – soil erosion, growing water shortages, loss of wildlife and plants, increased pollution, climatic change, and manifestly unequal shares of income, food, and shelter. Before food production, that is, agriculture, there was food collecting.

Indeed for 99% of human history, we were hunters, gatherers and fishers, living very successfully (although only for some 30 years on average) with roughly a 24-hour working week, and plenty of time for recreation and leisure. Athough hunter-gatherers are still found today, their numbers steadily shrank as people switched to farming. This transition is called the **first agricultural revolution** and involved people domesticating wild plants and animals. Dating from

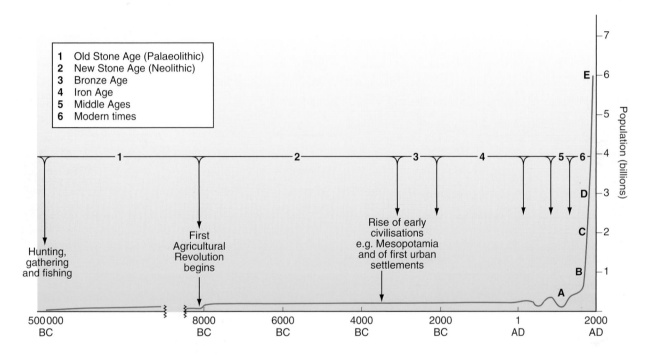

Significant Demographic Events since 1000 AD

A Black Death
First major outbreak of bubonic plague spreads from Central Asia to Europe and China. About one third of Europe's population died as a result

B Agricultural and Industrial Revolutions
These affected Europe and, later North America, by first increasing food production, and secondly, the rise of coalfield-based industry from 1750 (in Great Britain) onwards

C Improvements in Medical Care and Sanitation
Steady elimination of some diseases through inoculation e.g. smallpox and improvements in public health (water supply and sanitation) e.g. cholera

D 20th Century Agricultural Revolution
Development and diffusion of high-yielding varieties of wheat, rice and maize responsible for greatly increased food production to support rapidly increasing population

E The Population of the World Reached 6 billion in 1999

Fig 6.2 Global population growth

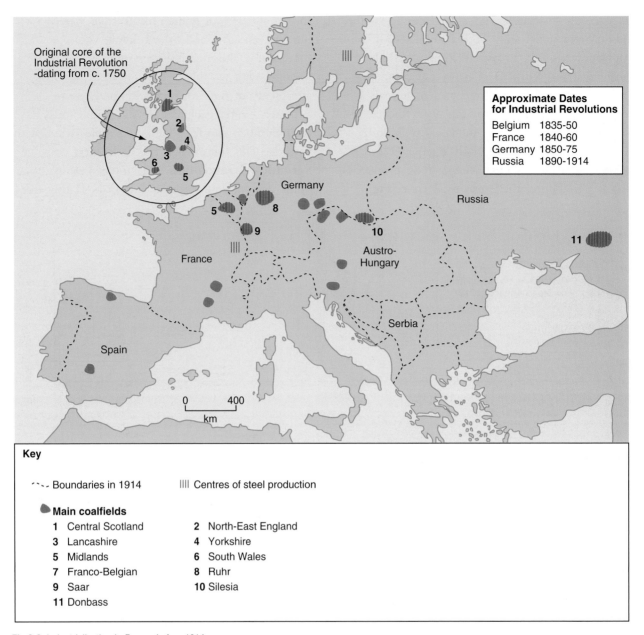

Fig 6.3 Industrialisation in Europe before 1914

roughly 8000 BC to 3000 BC, separate centres of food production developed independently, for example, in the Middle East, East Asia and Central America. In these areas, domesticated strains of wheat and rice and maize respectively, became staple crops and sources of food, with all the work involved in clearing fields, planting, weeding, harvesting and storing such crops. Whatever the causes of such a profound switch in human activity, the first agricultural revolution permitted a growth in world population to around 500 million by 1650. By that time, people, ideas, technology and disease were spreading from Western Europe, part

of the process of Europeanisation. This eventually led to the colonisation of large areas of Africa, Asia, the Americas and Australasia. In the following centuries, certain world agricultural systems, with their distinctive settlement patterns and population densities, developed, or were transformed, due to:

◆ new crops – groundnuts and sweet potatoes in the seventeenth century, and new varieties of rice in the twentieth century – were introduced to the tropical lands of East Asia with their long-established system of intensive peasant farming, allowing them to support an ever-growing population

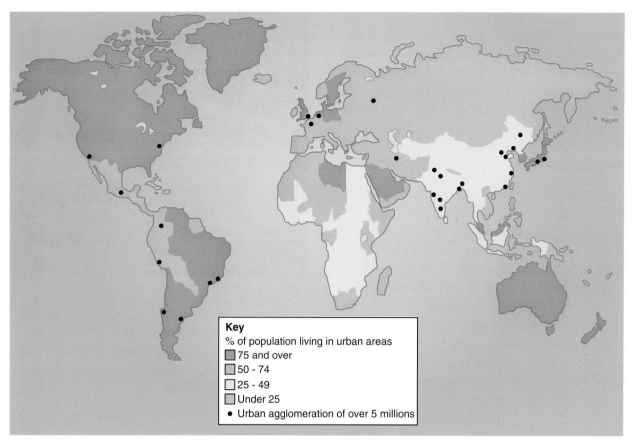

Fig 6.4 Global urbanisation

◆ new commercial plantations, new settlers and increased population pressures all combining to reduce the once widespread system of shifting cultivation in the humid tropics, from the sixteenth century onwards

◆ new lands, for example, the Great Plains of North America, were opened up in the last quarter of the nineteenth century, as new settlers and rail networks created a new geometric landscape associated with extensive commercial farming.

Many of these agricultural changes were designed to serve the growing industrial markets of the world. Industry is discussed in Chapter 8, beginning with the **Industrial Revolution**, another of the great transitions in world history. Down to early modern times c.1750, China and the Indian subcontinent were the main centres of manufacturing but then they were rapidly eclipsed by the rise of industrial Europe. New forms of industry developed from roughly the mid-eighteenth century and, increasingly, they were characterised by:

◆ the large scale production of goods (e.g. textiles) in factories

◆ the use of new machinery and technology

◆ a concentration of labour in factories instead of homes

◆ the use of coal and steampower

◆ the concentration of new factory-based industry in towns and cities.

The eighteenth century Industrial Revolution, which began in Britain, was based on the main coalfields and was centred on textiles and iron-making. Figure 6.3 shows the spread of industrialisation to continental Europe, initially to Belgium but later to France, Germany and Russia. By the 1870s, Britain was known as 'the Workshop of the World' but its industrial leadership was being overtaken by Germany and the USA. During the twentieth century, industrial growth further spread to the Pacific – Japan and California, and later to the so-called NICs (**Newly Industrialised Countries**) –

South Korea and Taiwan. Towards the end of the century, the NICs successfully competed with the older industrial countries, such as Britain, because they were able to adopt the latest technology, had cheaper labour costs and enjoyed state aid. By that time, industry in the European Union was a blend of (i) the latest high-tech innovations, and (ii) the struggling remnants of the earlier Industrial Revolution focused in the 'rust belts'.

Large-scale industrialisation was accompanied by massive population growth and increased rates of urbanisation. The final chapter mainly discusses the growth, development, townscape and problems of Edinburgh. It seems appropriate to conclude with such a case study since ours is now an urban world (Figure 6.4). While there is debate over what constitutes 'urban', 1999 saw not only six billion inhabitants but, for the first time in human history, half of them were estimated to live in cities. This fraction is forecast to rise to three-quarters by 2030. Cities are the main centres of growth and wealth in countries whose economies are increasingly dominated by the service industries. Their significance reflects the changing importance of the various sectors of employment. These changes are summarised in the **Clark–Fisher model** (Figure 6.5) which shows that:

◆ originally, most people were employed in the **primary sector** (Figure 6.6A). They exploited raw materials by mining, farming, fishing and quarrying. Such occupations were typical of the pre-industrial era, as characterised Scotland in the seventeenth century

◆ from the Industrial Revolution onwards,

Fig 6.6A Primary employment: West Virginia coalminer

increasing numbers of people obtained work in the **secondary sector** (Figure 6.6B). They were employed in manufacturing primary products, wool or cotton, as key raw materials for textile manufacturing

◆ gradually, the numbers employed in the secondary sector have been overtaken by those in the **tertiary** (Figure 6.6C). This sector involves the provision of services, and ranges from banking, insurance, tourism, education and transport to the provision of health services

◆ the model shows that ever fewer people remain in the primary and secondary sectors. This reflects increased mechanisation – the substitution of farm machinery for human power in the primary sector;

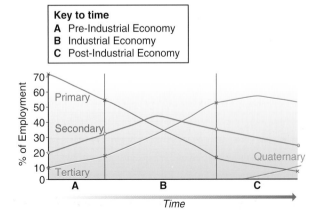

Key to time
A Pre-Industrial Economy
B Industrial Economy
C Post-Industrial Economy

Fig 6.5 The Clark-Fisher sector model

Fig 6.6B Secondary employment: Canadian steel worker

Fig 6.6C Tertiary employment: a travel agency – still relatively labour intensive

and automation and robotisation steadily replacing people in manufacturing, the secondary sector

◆ even the tertiary sector is not immune to change as computerisation replaces the workforce required (e.g. in banking and insurance). Some types of tertiary employment, however, are still relatively labour intensive – the health care services, retailing and tourism, partly a reflection of the need for 24-hour cover. Even supermarkets take on extra staff when they offer 'we never close' shopping

◆ finally, in the journey from a pre-industrial to a post-industrial economy, a **quaternary** sector of employment has recently emerged (Figure 6.6D). This reflects the switch to the knowledge-based economy and embraces a range of jobs from biotechnology to information technology.

Today, cities are key players in the modern, interlocking, global economy. The management of the ever-changing nature of cities is basically a matter for planning, essentially a form of applied human geography. In varying degrees, cities have to deal with an increasing range of problems: traffic congestion,

Fig 6.6D Quaternary Sector: genetic engineers in Califorinia

inadequate public transport and air pollution; social inequality, unemployment and urban decline in the inner city or peripheral housing areas; and fair access for all to clean water and sanitation. Yet cities are also the cultural, educational and political driving forces of an increasingly interdependent world where distance is becoming irrelevant. According to one influential architect, the challenge for an increasingly urbanised world is to have

'... *networks of cities across the world — sharing knowledge, technologies, services and recycled resources, and framing joint policies that both respect local cultures and implement common environmental objectives*'

Source: Richard Rogers *Cities for a Small Planet*

Key terms and concepts

◆ Clark-Fisher model (p. 185)

◆ first agricultural revolution (p. 182)

◆ industrial revolution (p. 184)

◆ newly industrialised countries (p. 185)

◆ primary sector (p. 185)

◆ quaternary sector (p. 186)

◆ secondary sector (p. 185)

◆ tertiary sector (p. 185)

After working through this chapter, you should be aware that:

◆ globally, there are marked differences in the rate of population growth in different countries and that the population structure of individual states varies with different levels of fertility, mortality and migration

◆ accurate demographic data assist government planning and there are various methods of gathering such information, for example, using a census. For a number of reasons, however, there are variations in the quality of the data and care must be exercised in their interpretation

◆ fertility and mortality rates have shrunk but at differing rates in the Economically More Developed Countries (EMDCs) and Economically Less Developed Countries (ELDCs), and there are various reasons for the changing rates

◆ age and sex pyramids provide a valuable picture of a country's population structure and show contrasts and similarities between EMDCs and ELDCs

◆ long-term changes in fertility and mortality can be shown in the different stages of the model of demographic transition

◆ countries such as Scotland, China and Italy have gone through various stages of demographic transition at different times and at different rates in response to different political, economic and social circumstances

◆ migration is an important feature of population change and can be classified, if not precisely, into varied types including refugee flows

◆ migration has many different causes, and its consequences are felt both in areas of inmigration and emigration, for instance, in China, in Western Europe and in the Mediterranean lands

You should also be able to:

◆ interpret population data in the form of maps, tables and graphs incuding population pyramids and flow diagrams

◆ construct and analyse sample population diagrams and maps

On 12th October 1999, the world's population reached six billion. Not for the first nor last time, many questions were raised about a range of issues in population geography resulting from yet another significant milestone in population history. These included:

◆ when will global population growth stabilise, given that, although birth rates are falling overall, the world's population is increasing by over 270 million every three years?

◆ how will the fastest rate of population increase affect the economic growth and, therefore, living standards in the so called 'developing' states or Economically Less Developed Countries (ELDCs)?

◆ by 2020, it is estimated that the so called 'developed' states or Economically More Developed Countries (EMDCs) will account for only 17% of global population. How will this affect the economic, technical and military balance in the world? Will the EMDCs still get far more than their fair share of the Earth's resources to sustain their high per-capita incomes?

◆ will the cityward migration continue in the ELDCs, will the 'flight from the cities' continue in the EMDCs, and what will be the consequences of such population movements?

◆ how will EMDCs, such as Italy, cope with a shrinking population? Is immigration the answer, and what will be the consequences? Is the 'greying' of population just a feature of the EMDCs?

Such questions highlight the importance of population geography, a branch of geography that impinges on so many aspects of physical and human geography.

7.1 World Population Trends

Population Contrasts: Growth and Decline

Population geography is closely related to **demography** or population studies. Demography's varied subject matter interests historians, economists, sociologists, statisticians, biologists and medical scientists. Geography's main demographic concerns are with: (i) variations in the rate of population growth between different places, and (ii) how rates have changed over time. Past influences are important: for example, the impact of changing birth rates from the period after World War Two (1939–45) still affects Britain's population structure today. Population geography emphasises, therefore, the key topics of **fertility**, **mortality** and **migration**, and how they have interacted spatially (from place to place) through time. Key demographic terms are outlined on pages 195–198 and should be referred to as you read.

Since the nineteenth century, and especially during the latter part of the twentieth century, global

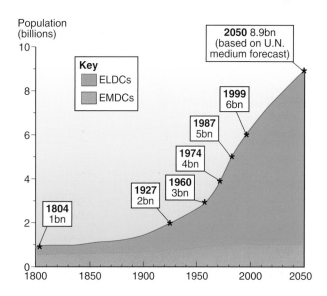

Fig 7.1 General population trends 1800–2050 AD (based on UN medium forecast)

population has grown rapidly, with the world becoming a much more crowded place. By October 1999, it reached six billion, a threefold increase from the late 1920s, and a sixfold increase from the early 1800s (Figure 7.1). This rise, however, was spread unevenly across the globe. Figures 7.2 and 7.3 show the different rates of population growth and that:

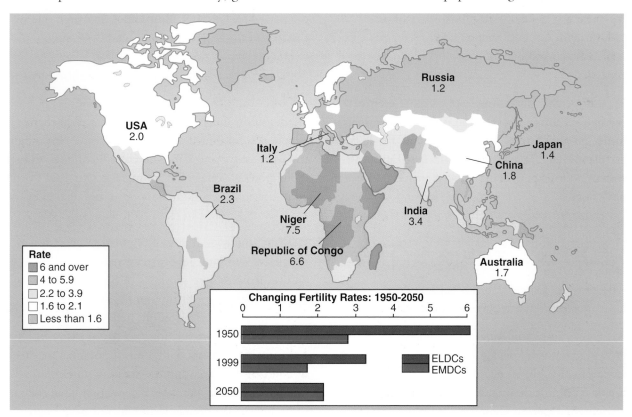

Fig 7.2 Global fertility patterns in 1999. (Fertility is the average number of children born to women during their lifetime.)

Area / Country	Population (mid 1999 Millions)	CBR	CDR	Infant Mortality Rate	Natural increase (Annual %)	Total fertility Rate	Doubling time at current rate	Population (est)		% of Pop'n aged		Life expectancy			% of Pop'n Urban	GNP (1997) per capita $
								2010	2025	<15	>65	Total	Males	Females		
World	5982	23	9	57	1.4	2.9	49	6883	8054	31	7	66	64	68	50	5170
EMDCs	1181	11	10	8	0.1	1.5	583	1216	1241	19	14	75	71	79	75	20350
ELDCs (excluding China)	3546	29	9	68	2.0	3.8	35	4272	5251	37	4	61	60	63	40	1510
Africa	771	39	14	88	2.5	5.4	28	979	1290	43	3	52	51	54	30	660
Europe (including Russia)	728	10	11	9	-0.1	1.4	-	731	718	18	14	73	69	78	73	13890
Brazil	168	21	6	41	1.5	2.3	45	190.9	217.9	32	5	67	63	70	78	4790
China	1254.1	16	7	31	1.0	1.8	73	1394.3	1561.4	26	7	71	69	73	30	860
China - Hong Kong	6.9	9	5	3.9	0.4	1.1	161	7.7	7.8	17	10	79	77	82	95	25200
India	986.6	28	9	72	1.9	3.4	37	1167.3	1414.3	36	4	60	60	61	28	370
Italy	57.7	9	10	5.5	0.0	1.2	-	57.5	54.8	15	17	78	75	81	90	20170
Japan	126.7	10	7	3.7	0.2	1.4	318	127.6	120.9	15	16	81	77	84	79	38160
Morocco	28.2	23	6	37	1.7	3.1	41	33.2	39.2	34	5	69	67	71	54	1260
Republic of Congo	50.5	48	16	106	3.2	6.6	22	70.3	105.7	48	3	49	47	51	29	110
Russia	146.5	9	14	17	-0.5	1.2	-	144.7	138.1	20	13	67	61	73	73	2680
Spain	39.4	9	9	5.5	0.0	1.2	1980	39.8	39.0	15	16	78	74	82	64	14490
United Kingdom	59.4	12	10	5.9	0.2	1.7	423	60.8	62.6	19	16	77	74	80	89	20870
U.S.A.	272.5	15	9	7.0	0.6	2.0	116	297.7	335.1	21	13	77	74	79	75	29080

Fig 7.3 Global population data (from Population Reference Bureau 1999)

◆ by far the greatest rate of increase is in the ELDCs with some 85 million additional people every year compared to only an extra 1.5 million in the EMDCs.

◆ Europe and Africa represent extreme continental examples. Europe's population is expected to shrink from 728 million in 1999 to 718 million by 2025. Germany, Italy, Russia and certain East European countries are experiencing quite dramatic declines in birth rates. European women are bearing on average 1.4 children, too low a figure to replace the population without immigration. On the other hand, Africa, with the highest birth rate of any continent, is projected to grow to 1290 million by 2025, reflecting the fact that, in sub-Saharan Africa, women have an average of six children each.

◆ in Asia and Latin America the population is growing relatively quickly but with marked variations. 1.3 billion people mean that China still tops the population league table in 2000. Although its one-child-per-family policy is not uniformly applied, population growth rates have slowed quite

dramatically. This will allow India, although one of the first of the ELDCs to adopt family planning, to overtake China's population by 2050. Japan's population, by contrast, is projected to shrink by 6 million by 2025.

◆ the USA, unlike the other EMDCs, is likely to experience continued population increase. Not only is it possible that American women will continue to bear 2 children on average but immigrants, attracted by the educational and economic opportunities offered by USA should ensure an extra 63 million by 2025.

Globally, the overall rate of population growth is slowing down but there is still a **momentum** which ensures that absolute numbers of births continue to increase even when birth rates decline. Effectively, therefore, there is the contrast between the continuing rapid population growth of many ELDCs, particularly in Africa, with a lot of potential mothers of child-bearing age, and the wealthy states of Western Europe where the key demographic issue is that of declining population, reflected in an ever growing bulge of elderly people.

Population Projections

What then of the future global demographic scene? Figure 7.4 shows three possible global projections, based on United Nation's forecasts. By 2150 it is estimated that world population will be between 3.6 and 27 billion! Yet, such a seemingly large discrepancy involves only small differences in childbearing levels:

◆ assuming women across the world bear an average 2.5 children, global population would ultimately reach 27 billion by 2150 according to the high projection

◆ the low projection, however, based on 1.6 children per woman would result in fertility rates rapidly dropping below replacement level. Ultimately, population would drop to 3.6 billion by 2150

◆ more likely is the medium projection of 2 children on average per woman with a resulting projection that population would stabilise at some 11 billion.

Whatever the ultimate outcome, the assumptions underlying such long term projections are many, complicated and speculative. Will health conditions necessarily improve in all countries? Will fertility rates continue to fall? Will future medical advances further prolong human life?

The difficulty with demographic projections is our inability to forecast the many political, economic, social and technological factors that affect fertility, mortality and migration. Imagine that it is 1950. Who, at that time, could have foreseen the demographic impact of the following influential events: the spread of AIDS, the impact of a one-child-per-family policy in China, the introduction and diffusion of the contraceptive pill, massive refugee movements, in Africa for instance, and the break up of the former USSR? The problem is that projections are a necessary part of demographic life because so many people and organisations need the best possible forecasts. These are based, of course, on current demographic data with all its associated limitations.

7.2 Demographic Head Counts

There are lies, damned lies and statistics

So said Benjamin Disraeli (1804–81), a heartfelt statement about the reliability and interpretation of statistics. How reliable are the statistics underpinning the statement that there were six billion people in 1999? In assessing demographic data, it is useful to look at how they are collected, their worth and their limitations.

Collecting the Demographic Data

There are several long established methods of acquiring population data, each with its limitations.

Censuses

Censuses have a long history and one early example is the Domesday Book. Initiated by William the Conqueror, published in 1086, it is a unique, though incomplete, survey of a newly conquered England. In 1755, the Rev. Alexander Webster, minister of Edinburgh's Tolbooth Kirk, helped compile Scotland's first unofficial census, according to which the country had a population of 1 265 380 persons. By 1801, the first British census was held, a practice since carried out every 10 years (decennially) apart from 1941. Census enumeration is now the responsibility of the Office for National Statistics in

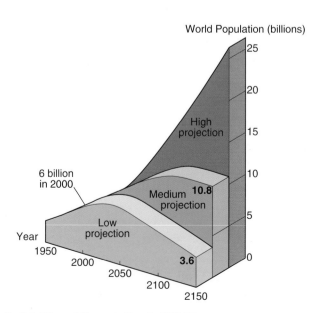

Fig 7.4 UN population projections to 2150 AD

England and Wales, and the General Register Office in Scotland. Essentially, a census is a statistical snapshot of private households and is conducted on a designated day (usually the third Sunday in April). Census data are gathered on the basis of small areas known as enumeration districts. All households in each district are visited by an enumerator who collects the completed census forms and ensures that all individuals are recorded. The 1991 census collected information that gave:

◆ details about individuals, including their age, sex, occupation, educational qualifications, birthplace, ethnic origins and (in Scotland) knowledge of Gaelic

◆ details about living conditions and life styles, including type of accommodation, housing tenure, health, amenities.

Once the results are collected, crosstabulated and published (and that, even with increased computerisation, may take several years) they are available for analysis by interested groups. Ideally, censuses should allow comparison of data over time, and between countries. Data collection methods, however, vary, reflecting differences in questions and timing of census taking. For example, the British census adopts a **de facto** approach. This records the actual place where an individual is living at the time of enumeration. Think how this method affects the results from a university and resort town such as St. Andrews. The **de jure** method counts people according to where they normally reside. This was first adopted in Roman Palestine:

> *In those days a decree was issued by the Emperor Augustus for a general registration throughout the Roman world. . . . For this purpose everyone made his way to his own town. (Luke Ch. 2)*

This method is employed in the USA. Doubtless, increased harmonisation of census taking will occur across the countries of the European Union to ensure uniformity.

Civil Registration

All so-called vital events – births, marriages and deaths (the so-called 'hatches, matches and despatches') – must by law be registered and are ultimately collated by the Registrar General of

Scotland. Such data permit more frequent updating than a decennial census and provide a useful means of monitoring population change at local, regional and national levels.

National, European Union and Global Surveys

A wide range of demographic and related social and economic information is also gathered, collated and disseminated via the Office for National Statistics (ONS), the Scottish Office, Eurostat, and various UN organisations, including UNESCO.

The Value and Uses of Demographic Data

Given, then, the variety of demographic sources and the sheer cost of data collection, why do organisations require so much accurate information? Historically, the earliest censuses were for purposes of taxation or army recruitment. Today, information from censuses and vital registration is used for varied demographic purposes and by different interest groups.

Maternity Care

Nursery Education

Primary and Secondary Education

Higher Education

Employment, Housing and Transport

Pension Provision

Geriatric Health Care & Sheltered Housing

Fig 7.5 The use of demographic data – 'From Cradle to Grave'

Every 10 years, the United States counts its citizens. In the early days of the Republic, this was not a particularly difficult task. Only free persons were counted; Indians, unless taxed, were excluded, the existing population was largely settled.

In the United States of 1990, the decennial census is a weightier burden: America is the most mobile nation in the world (estimates have it that the average American lives in over 10 places during a lifetime; it contains vast and shifting immigrant populations (some 5 million members of which are illegal); as much as one-tenth of the nation lives in official poverty, while the official 'homeless' number several million; and many urban centres are in a state of constant redevelopment or impenetrable anarchy.

Why, then is the census proving so troublesome? Being decennial, it has relatively few permanent staff; it has to go out and hire enumerators . . . (who) . . . cross social boundaries and, especially in urban areas, no longer deal with a stable, largely English speaking population. Within blocks of my house there are tenements with vast, floating populations of south-east Asians, Haitians and Central Americans, many of them illegal and hesistant to be counted. America is in a state of such rapid change, the question is can it be counted? I side with the pessimists: a margin of error of plus or minus 20%, especially where it counts.

'A Census not to be counted on' by Keith Bedsford.

Fig 7.6 Extract from from 'The Independent', 24 April 1990

Such information is used:
1) By the government, industry and commerce who need to know about changes in population size and age structure to assess the needs of people from the 'cradle to the grave' (Figure 7.5)

2) To forecast population trends and initiate population policies which may

◆ encourage births (a pro-natalist policy, as in pre-war Italy's active encouragement of large families)

◆ discourage births (an anti-natalist policy, as in China's one-child-per-family policy)

◆ promote migration (the transmigration policy in Indonesia, encouraging re-settlement from overcrowded Java to less densely-peopled Sumatra.

3) as a means of assessing the success of health care and family planning programmes.

Difficulties in Collecting Demographic Data

In the second half of the twentieth century, there has been a marked improvement in the quantity and quality of global demographic data. Thanks to UN encouragement and assistance, census taking and the use of sample surveys (usually 10% of the population), virtually all countries now produce demographic data. Generally, censuses are more reliable in EMDCs. Even there, however, countries experience difficulties. Figure 7.6 illustrates problems encountered in the USA during the 1990 census which 'lost' five million. Underenumeration also affected the United Kingdom's 1991 census: around one million people 'went misssing', probably to avoid the new poll tax. The following are some of the reasons for the distortion of census data:

Fig 7.7 It is not uncommon during census taking that a Jordanian male may respond that he has 'two children and three girls'

The figures given in "The Scotsman" today for the population of Nigeria do not make sense. The reported census figures include:

1963: Total population: 55 million (Northern Nigeria 29.8 million)

1973: Total population: 73 million (Northern Nigeria 51.4 million)

This implies an increase in population in Northern Nigeria of over 20 million in 10 years. If 15 million Northerners were female and 10 million were of child-bearing age, then each woman would have to have two children a year (assuming no net immigration, and that the female population of child-bearing age remained constant).

It was rumoured that the 1963 census figures for the North were exaggerated in order to secure increased political representation.

Fig 7.8 Extract from a letter in 'The Scotsman' 1 January 1974

◆ census taking is expensive and involves many people. In debt-ridden ELDCs, there are other priorities. The training of enumerators is critical but not cheap. It is also labour-intensive: over 1.7 million enumerators were required in India in their 1991 census

◆ although their numbers are declining, nomadic people, such as the Fulani of the Sahel, still cross boundaries and could either be recorded twice or not at all! Equally, illegal immigrants, for example, Mexicans in the sunbelt states of the USA or Moroccans in Italy, may be unrecorded

◆ underregistration can occur for social and religious reasons. The 1994 census in Jordan showed a marked majority of males in almost all groups. This is probably because females were omitted from the census. When asked about their family size, Jordanian males may respond 'two children and three girls' (Figure 7.7). In China, care must be taken with the interpretation of census data, especially the uneven sex ratio at birth. The apparent preponderance of boy babies is because many female babies are non-registered and/or widespread female infanticide is practised

◆ low levels of literacy mean that many people (and in many ELDCs, that frequently means females) find it difficult filling in forms. This is overcome by employing more enumerators which, therefore, increases the cost of census administration

◆ costs are also increased by the variety of languages in such countries as Nigeria and India. India's one billion people have 15 official languages and more than 1650 dialects. Language is a contentious issue and a badge of identity. Although Hindi is the main language in India, attempts to foist it on South India are resented, as are campaigns to do away with English, the former colonial language. Consequently, the 1991 census form was printed in 20 languages and multilingual enumerators were employed

◆ ethnic tensions and internal political rivalries may influence census accuracy. Nigeria, the most populous African state, eventually had to repudiate the results of its 1973 census. As the letter suggests (Figure 7.8), traditional rivalry over political representation between the north and south was reflected in vastly inflated population figures for certain northern states compared to the 1963 results

◆ rapid rural–urban migration makes it difficult to keep track of people. Daily, an estimated 300 rural families migrate to Mumbai (formerly Bombay), some to vanish statistically into overcrowded neighbourhoods like Dharavi, the largest slum in Asia

◆ poor communication links, difficult terrain and scattered settlements makes census taking more expensive. Sparsely populated countries like Botswana, therefore, pose an enumeration challenge for census administrators

◆ finally, human nature and fallibilty should not be forgotten. Worldwide, there can be: suspicion and distrust of official 'snoopers'; resentment at particular 'sensitive' census questions; census questions that are open to various interpretations; and it is all too easy, in spite of (or because of?) computerisation and ever-increasing number crunching, for mistakes and misprints to occur.

Quantitative data of a demographic nature are critical in the running of a modern state. Although there are limitations with the data, these are at least recognised. Also remember: (i) that the word 'statistics' derives from 'state' and (ii) ever changing issues of population dynamics involving fertility, mortality and migration, ultimately are issues dealt with by states themselves.

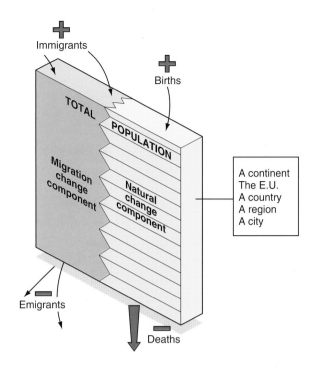

Fig 7.9 Components of population change

7.3 Population Dynamics: Components, Fertility and Mortality

Components of Population Change

We have seen that censuses are organised by and for individual states, of which there are over 230. Populations range in size from over a billion in China and India to the demographically miniscule, with around 10 000 in Vatican City and the Cocos Islands. Throughout history, the population of states has varied. While territorial expansion or loss is one possible explanation, another is the increase or decrease in their overall population. Population change may be highly dynamic, depending on interacting changes in fertility, mortality and migration.

Figure 7.9 shows that population change has two main components: natural change and migrational change.

1) **Natural change component:** The population of a state will experience **natural increase** if the number of births exceeds the number of deaths; if the number of deaths, however, is greater than births, then the population experiences **natural decrease**

2) **Migrational change component:** If there is a large inflow of immigrants to the country and it exceeds the number emigrating, then there will be a population increase. Alternatively, there can be a population loss if emigrants outnumber new arrivals.

A few observations need to be made about this model of population change. First, the components can also refer to a settlement, such as a rapidly growing city (e.g. Mumbai, formerly Bombay), and to a region, such as the Scottish Highlands and Islands, which has seen both decline and growth in the last 200 years. Equally, it can be used to explain the population dynamics of a group of states such as the European Union or a continent like North America. Second, the relative importance of the individual components has varied from place to place and at different times, as the components interact. The period from 1841–1861, for example, saw a steep decrease in Ireland's population from 8.5 million to 5.7 million. This was the result of (i) a marked natural decrease due to the potato famine of the mid-1840s, and (ii) wholesale emigration – 'a million died and a million fled'. Many Irish migrants became part of the massive trans-Atlantic flow to the USA from Europe which not only brought more immigrants but added to America's natural increase, since most of the immigrants were young adults.

A discussion of migration continues on page 211.

Declining Fertility

Basic Terminology

Fertility is the occurence of live births and is usually measured by the **birth rate**, or more fully, the **crude birth rate**. This is the number of live-births per 1000 people for a given year. As a measure, it is statistically 'crude' because it does not allow for the age and sex structure of the population. This is a drawback when it comes to comparing different countries. An alternative is the **general fertility rate** which expressses the number of live births per 1000 women aged 15–44 for a given year. Unlike the crude birth rate, it emphasises women of child bearing age rather than all age groups.

Global Trends

In the twenty-first century, fertility in many states is the critical component of population change, outstripping mortality and migration in importance. Unlike mortality, people at least have some choice about births and in most countries fertility is declining, though at varying speed. Some of the highest birth rates are in sub-Saharan Africa where countries such as Niger, Mali, Somalia, Angola and Chad experience birth rates of over 50; some of the lowest are in Europe with Italy, Russia, Slovenia and Spain, registering birth rates of 9. These European states have emphatically undergone a major 'reproductive' revolution. It is increasingly the case that many EMDCs have a fertility rate well below replacement levels. Yet, only a few hundred years ago, Scotland, England and France, in 1755/6, had high birth rates of 41, 33 and 41 respectively.

Factors Influencing Fertility

Fertility fluctuates in response to a wide range of overlapping political, economic, social, religious and demographic factors. The following factors have influenced the fertility of women at different times, at different rates and in different parts of the world:

◆ traditionally, children were a key part of the workforce (Figure 7.10), providing not only cheap labour on fields, down mines and in 'sweatshops', but also a form of family insurance for ageing parents

Fig 7.10 Child labourers laying bricks in blistering noon-day sun near Islamabad, Pakistan

in countries lacking formal social service provision for the elderly

◆ social pressures in certain parts of Africa, Asia and the Middle East encouraged a high rate of fathering as an indicator of virility and as a status symbol. For example, in parts of rural China, tradition favoured sons to work in the fields, to continue the family name and to provide for parents in their old age and after death

◆ children can be seen as an economic and social burden. Since the nineteenth century, this has been a growing response to compulsory schooling; factory legislation banning child labour; increased affluence with ever-rising material aspirations and consumer pressures; the changing status and employment of women; the decline of marriage and the 'conventional' family; and the increase in divorce and cohabitation. Some or all of these trends have particularly affected many EMDCs, and consequently delay and diminish fertility

◆ religion influences personal decisions concerning fertility. Islam, Roman Catholicism and Mormonism are examples of faiths traditionally associated with high fertility rates. It is, however, a complex issue. For example, many poorer Islamic countries also have high mortality rates which are a function of poverty and underdevelopment. High mortality rates encourage youthful marriages and an early start to childbearing. In general, as living standards improve (Figure 7.11) and urbanisation increases, women become better educated and enter the job market, and consequently fertility rates decline regardless of religion. This can be seen in the attitude to birth control in Roman Catholic areas. Traditionally, artificial birth control is against church teachings, yet Italy's birth rate has plummeted, reflecting the declining influence of the Vatican, widespread availability and acceptability of contraception, and increased acceptance of abortion. The regional exception is in the more rural and agricultural south, where there is a stronger adherence to the church's traditional teachings.

◆ improved educational opportunity, particularly for women, is a key influence, as was stressed at the 1994 Cairo Conference (see extract, Figure 7.12). Improved education provision for women drastically reduces fertility rates by altering aspirations for themselves and their families, and can initiate a virtuous circle of development (Figure 7.13)

◆ governments have attempted to influence fertility by adopting either **pro-natalist** policies or **anti-**

natalist policies. Pro-natalist policies were promoted before World War Two (1939–45) by states such as Italy, France and Germany, partly to compensate for the horrendous carnage of World War One (1914–18) but also for economic and military purposes. The most explicit pro-natalist policies of Nazi Germany involved: suppressing information about contraceptives; taxing unmarried couples; prosecutions for induced abortions; and encouraging large families through tax concessions and large family allowance payments. Examples of anti-natalist policies include those encouraged in certain Asian states. In Singapore, policies introduced in 1974 included: limiting income tax relief to the first three children; charges at hospitals increased for each birth; and limiting housing allocation for large families. More draconian policies have included: the one-child-per-family policy of China and, in India, a sterilisation programme which was so unpopular that it helped bring down the government in the 1977 elections

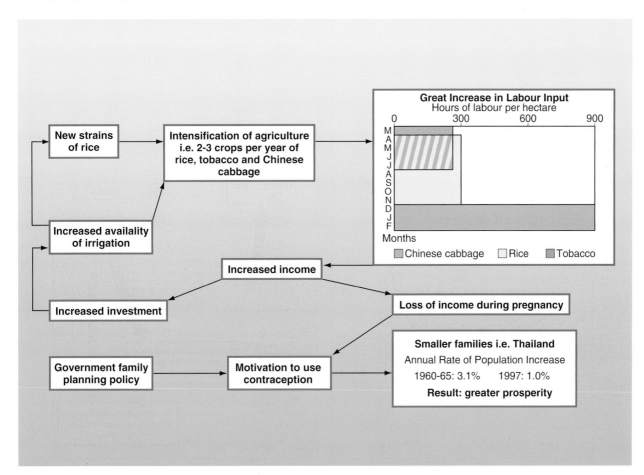

Fig 7.11 Case study: declining population growth – Chiang Ma Valley, Northern Thailand

Women hold key to population curb

The population explosion in the developing world is best controlled by education, giving young women an alternative to perpetual motherhood', the head of the United Nations Population Fund said yesterday.

Nafis Sadic, a gynaecologist and the first woman to lead a major UN programme, said the 'empowerment' of women was the key to controlling population growth . . . 'The fact is that women have been, and still are, disregarded and undervalued for everything they do apart from having children, preferably boys. Despite the advances of the past two generations, there is still a wide gender gap in both education and health care in most developing countries . . .' My personal view is that if you really looked after women's needs and women's health, everything else would take care of itself, because women take care of their families and give their highest priority to their children's health and future. Not allowing them to make decisions for themselves is really the main obstacle (to population control). 'Motherhood', Dr. Sadic said, 'should not be imposed as the only role in life.'

Continued . . .

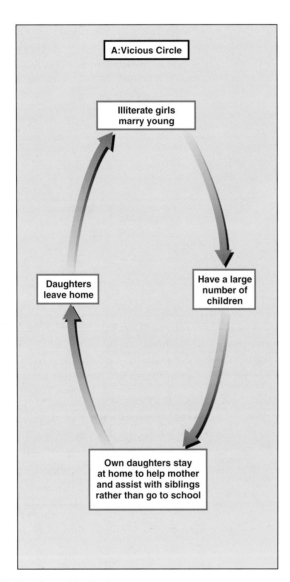

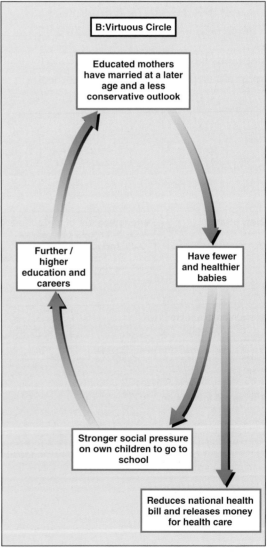

Fig 7.13 Education and family size

> *Educating women has been shown to decrease fertility rates dramatically . . . In Zimbabwe, for instance, women with no formal education have an average of seven children, women with some primary school education six children and women with secondary or higher education fewer than four children.*
>
> *A programme to tackle global population growth will be agreed by the world's governments at next month's population conference in Cairo. Dr. Sadik said current funding of $6 bn per year on family planning in developing countries needed to be increased to $17 bn in 2000. Unresolved issues included sections relating to abortion, opposed by the vatican, and contraceptives for teenagers, opposed by Islamic leaders who fear it would promote promiscuity.*

Fig 7.12 Extract from the Cairo Conference, 1994

◆ other demographic influences also influence fertility. For example, declining infant mortality rates reduce the necessity to have such large families by way of insurance for old age.

Overall, global fertility is declining but at varying rates. To quote the International Development Secretary:

> *The good news is that population growth has been in steady decline because women today have more choices, more access to family planning, access to education and greater freedom of choice. When people have the choice, they choose smaller families. Their children are better fed, their quality of life is improved, and this is carried on through the generations.*
>
> Source: Clare Short, October 1999.

Declining Mortality and an Ageing World

Basic Terminology

Crude death rate is the number of deaths per thousand people for a given year. As with the CBR, it does not take age structure into account. The following are useful measures:

◆ **Age Specific Death Rate**

$$= \frac{\text{Deaths in a specific age band}}{\text{Number in same age band}} \times \frac{1000}{1}$$

This allows for easier comparisons between states.

◆ **Infant Mortality Rate**, which is the number of deaths of infants under one year of age per thousand live-births in a given year. This is the most effective of the age-specific measures and is an excellent socio-economic indicator, which reflects the actual input

from health and social services. For the ELDCs, the average figure for the late 1990s was 62, for the EMDCs it was 8, but it was as high as 136 in Sierra Leone and 150 in Afghanistan, reflecting not just poor health and social provision but also the impact of civil war. Arguably, the best measure of mortality is the

◆ **life expectancy** (the average number of years from birth that a person can expect to live under prevailing mortality trends) with a global figure of 64 years for males and 68 years for females. Whether there is a limit to our potential maximum biological age is a moot point, but a figure of 130 years has been suggested. How many people will live to that age will depend on the continued effectiveness of public health measures and medical technology.

Global Trends

The twentieth century saw a marked decline in mortality, a consequent increase in life expectancy, and important changes in the main causes of death. For people in the EMDCs, declining mortality rates began to fall during the nineteenth century, continuing through the twentieth, though at a slower rate. Not only were death rates falling, but even larger falls in infant mortality were being experienced. Taking Scotland as an example, the crude death rate for 1870–72 was 22.3, falling to 15.2 by 1910–12, and 12.4 by the 1990s. Average life expectancy, therefore, rose throughout this century from 47 years in the early 1900s to 74.6 for men and 79.5 for women by 1996–97. Scotland, in common with many other countries, was experiencing, by the end of the century, the so-called 'demographic time bomb', ticking away beneath the health, welfare and pension provision for the elderly: a situation made worse by the fact that increased longevity is often accompanied by a decline in health.

In the ELDCs, the decline in death rates and increase in life expectancy came at a faster rate, particularly from 1950 onwards, and was to bring exponential rates of population growth to Africa, Asia and Latin America. Such a dramatic demographic impact is scarcely surprising. In the early 1900s the average life expectancy was only 25 years in many ELDCs. By the late 1990s, crude death rates had fallen dramatically, though not uniformly. Some countries such as Belize and Mexico had very low rates – 4 and 5, respectively; figures much lower than Sweden and Norway – 11 and 10, respectively, both prosperous countries with excellent health and social care provision. The difficulty lies with the use of crude death rates as a measure of mortality: the higher than expected death rate figures for EMDCs reflects their more elderly populations. They do not have such a 'juvenile' populations as the ELDCs, and consequently the increasingly high percentage of elderly people inflates the crude death rate figure, regardless of excellent health care provision. However, some of the highest death rates are recorded among ELDCs such as the sub-Saharan states: Ethiopia (21), Malawi (24), Niger (24) and Guinea-Bissau (21).

Factors Influencing Mortality Decline

Whatever measure is used, mortality rates have declined in response to the following overlapping causes:

Fig 7.14 New Delhi. Mohammed Nazamut, 10 months old, waits for polio immunization drops as part of a WHO sponsored scheme

Fig 7.15 Demonstrators with inflated condoms demand cheaper AIDS drugs during a protest in South Africa. More than three million people are HIV positive in South Africa

◆ **public health measures** have been critical, with the elimination of water and food-borne diseases such as typhoid, dysentery and cholera by improving water supply, sanitation and housing. Overall, such measures have been successful since their introduction to the rapidly urbanising states of Europe and North America in the nineteenth century. Sheer population pressure and lack of finance hinder their universal diffusion, particularly to the poorest ELDCs. Much, however, depends on the social and economic priorities of individual states – Sri Lanka and Cuba have seen public health care measures as a key priority

◆ **medical advances**, from the introduction of vaccinations against smallpox in the eighteenth century, to the use of antibiotics in curing once lethal infections such as the bubonic plague, have helped reduce death rates, sometimes dramatically, especially among children. The incidence of common diseases such as measles, smallpox, respiratory TB, polio and malaria has been reduced, particularly since the foundation of the WHO (World Health Organisation) in 1950 and its diffusion of various immunisation/eradication campaigns and publication of international health regulations (Figure 7.14)

◆ **improved diet and nutrition,** with the average intake of protein and calories having increased in most countries. Improved agricultural efficiency, stemming from the nineteenth century opening up of new areas of cultivation (e.g. the Great Plains), and the twentieth century Green Revolution, as in

India, has helped ensure that, on average, food production has grown faster than population. A combination of these changes with medical, educational and environmental advances, further explains the global reduction in mortality, particularly the lowering of infant mortality.

Generally, the success of these factors is shown in considerably increased life expectancy. Many states have undergone an 'epidemiological transition' as infectious diseases are replaced by those of ageing as the main cause of death. It is most likely that mortality decline will continue in most of Asia and Latin America, but there are still many **obstacles**:

◆ there is the ever prevalent problem of **new epidemics**. For one thing, the healthcare revolution now has to contend with mutations in the stocks of viruses and bacteria. There are new microbial strains of diphtheria and cholera; and malaria, despite the success of earlier eradication programmes, is undergoing a revival as mosquitoes have become resistant to the conventional pesticides. For another, deaths from sexually transmitted diseases are rising. There is the human immunodeficiency virus (HIV) which may result in AIDS (Acquired Immunodeficiency Syndrome) (Figure 7.15). Its impact is global but sub-Saharan Africa is most affected by the AIDS pandemic, adding to its overall high death rates. When this is considered along with poverty, civil war and environmental problems, it explains that some parts of that continent will not experience conspicuous ageing for many years

◆ not that Africa is the only problem area – parts of Asia are also affected by AIDS, and its incidence is growing. So are the number of deaths caused by some of the current main killers in the EMDCs. As living standards improve among the more prosperous ELDCs, smoking, alcohol and traffic accidents affect mortality rates. Equally, any breakdown in health care provision, increased unemployment and a decline in living standards can quickly reduce average life expectancy, as experienced in the former Soviet Union.

Conclusion

The massive global demographic explosion, experienced particularly last century, has been mainly determined by the changing relationship between fertility and mortality. The experiences of EMDCs and ELDCs have been different as their birth and death rates have declined but at varying rates. Such a historical shift of fertility and mortality rates is referred to as the **demographic transition** (see page 203).

7.4 Population Structure: Characteristics and Consequences

One of the most interesting and useful of demographic diagrams is the population pyramid. It displays the age and sex structure of (usually) a country or any other administrative area: a region, a local council or an urban area. A pyramid is drawn to show the numbers of males and females in five year age bands (year cohorts) building up from babies and infants (0–4 years) to the very elderly (over 80). Typically, pyramids show the percentage of males and females in each age group, making for easier comparison between countries. Some pyramids, however, are constructed using absolute numbers instead. Conventionally, the bars for males are drawn to the left of the centre line, and females to the right.

Figure 7.16 shows a variety of pyramids for different places and times. They reveal some of the characteristics and value of such diagrams.

1) For a start, most of them are not 'pyramids'. They are neither three dimensional nor do they have a particularly wide base. Those that are more 'pyramid'-like are typical of the ELDCs. For example, the pyramid for the Democratic Republic of Congo has: a wide base as a result of that country's high CBR; a narrow apex showing a relatively low average life expectancy; and a distinct upwards decrease from one age cohort to the next, reflecting a relatively high CDR. Overall, with such a high percentage of the population aged between 0 and 14, it has a high degree of juvenility.

At the other extreme is Japan, representing the EMDCs, and whose 'pyramid' is more coffin-shaped and displays quite different features, including: a much narrower base, the product of a low and declining CBR; a blunted apex and straighter sides

1: Sample EMDC
Japan 1992

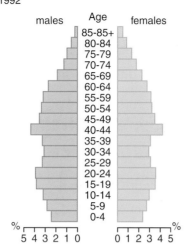

2: Sample ELDC
Republic of Congo 1994

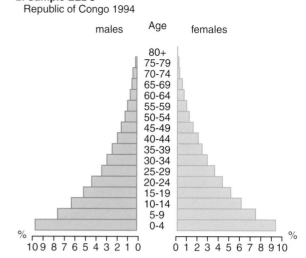

A–C: Pyramids Showing the Impact of War, Migration and Population Policies

A Paraguay 1871

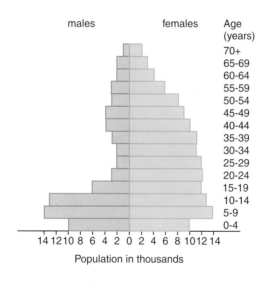

Population in thousands

B Kuwait 1990

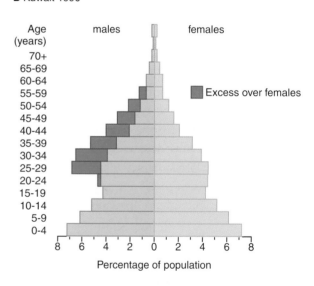

Excess over females

Percentage of population

C Central part of Shanghai 1980

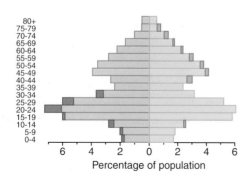

Percentage of population

A Shows the drastic impact of the war of the Triple Alliance (1865–70) when Paraguay fought against Brazil, Uruguay and Argentina

B In comparison with (**A**), the pyramid for Kuwait shows an excess of males over females. This is the result of male immigrants being 'pulled' to work in the oil and construction industries

C Shows the impact of inward migration to Shanghai by young adults in the 1970s, and the effect of one-child-per-family policy

Sources: 1 & 2: The Oxford School Atlas. 1997
             A, B & C: The Geography of the Worlds Major Regions. J.P. Cole 1996

Fig 7.16 Sample age and sex pyramids

which gradually taper upwards as a result of low CDRs and a long life expectancy. In fact, 'silvery' Japan leads the world: in 1999, life expectancy was 77 for males, 83 for females.

One feature shared by these two otherwise contrasting pyramids is the longer life expectancy of females. In nearly all countries, rich and poor, women outlive men.

2) Population pyramids provide a useful visual method of showing the relative proportion of the population who are either **dependents** or **economically active:**

◆ dependents are not economically active. They include non-working, **young dependents** aged 0–14 and 'elderly' (over 65) **old dependents**, usually of retirement age.

◆ the **economically active** or working population are aged 15–64.

Such socio-economic knowledge is useful for predicting and planning the education and health services, and assessing the taxation base which may be tapped to support such facilities.

3) Because the young and old dependents are economically inactive, they are dependent on the working adults. A useful measure is the **dependency ratio** which is calculated as:

$$\frac{\text{\% young dependents} + \text{\% old dependents}}{\text{\% economically active}}$$

If the dependent group is a relatively large percentage of the population, the dependency ratio is high; however, a large proportion of economically active people means a low dependency ratio.

4) Any discussion of dependency ratios must raise certain issues:

◆ what are the best age thresholds to act as boundaries between the three groups? In many ELDCs, poverty means that children start work either in the formal or informal sector at very young ages. Also, many people over 65, in the absence of pensions, continue to work. Also, what is 'old'? Take a woman aged 50, living in a rural district in Tanzania, for example, who has worked beyond her physical capacity, has experienced ten pregnancies

and has been undernourished since childhood. Will she not be 'old' by 50?

◆ Is a young dependent more of a burden than an old dependent? Much will depend on the quality and quantity of education and health provision. In the EMDCs, for example, geriatric health provision costs are rising, not just because of increased longevity, but because of increasingly sophisticated but expensive medical advances.

5) Population pyramids reflect long-term trends in birth and death rates and demonstrate differences between EMDCs and ELDCs. They also show the impact of events such as war, natural disasters, population policies, epidemics and migration flows, resulting in distinctive notches and bulges in pyramid sides (Figure 7.15). Perhaps their main value is that they encourage questions about the consequences of the population structure by those involved in planning for the future in both EMDCs and ELDCs.

Consequences of Age Structure

In EMDCs, the main overall demographic issue is that of ageing and, for some states, population stagnation, and even decline. Demographers maintain that women should be having 2.1 children on average to sustain population replacement level. In 1999, however, sample figures for the United Kingdom were 1.7, for Japan 1.4 and Italy 1.2. Consequently, for such countries, there are several related implications:

◆ how will so many elderly dependents be supported by a decreasing number of economically active people? Any political agenda must consider issues such as: the provision of adequate public pension schemes and the encouragement of more private pension provision; ensuring access to good health and care services (e.g. geriatric wards, 'meals-on-wheels' services and day-care centres), as well as new, but often expensive, medical technology; the cost of sheltered housing and residential accommodation. Also, there will be greater demands for public transport

◆ with proportionately fewer young dependents, issues of falling numbers, resulting in the closure of schools and maternity wards, have to be set against the costs of high quality education and health provision

◆ if there are fewer people in the workforce, perhaps the following changes are inevitable. Retirement ages must rise – in the UK the age of retirement for case of women will rise from 60 to 65 from 2010 onwards. Greater efforts may be made to ensure that the unemployed and the work shy are employed. Female participation rates in the workforce will continue to increase. More than ever, retraining will be a necessary part of one or more careers to ensure that people keep up with the demands of the 'knowledge-based' economy.

◆ dependency ratio can also be reduced by encouraging more immigrants and easing access for asylum seekers. International labour migration has a long history, particularly from the Industrial Revolution onwards. Post-war north-western Europe enlarged its labour pool from countries such as Spain, Portugal and Turkey or former colonies such as Algeria and the West Indies. In the twenty-first century, such an obvious measure unfortunately raises issues of ethnic tension and migration restrictions.

In ELDCs, particularly in the poorer states of sub-Saharan Africa, the very different age structure raises the following issues:

◆ youthful populations and consequent high dependency ratios means that states, such as Kenya and Tanzania, have to spend disproportionately large sums on education and child healthcare. Given the significance of education, particularly of girls, as a key factor in the reduction of fertility, such countries

cannot afford not to invest in adequate school provision

◆ such a young population also involves **population momentum**. This means that, as the younger age cohorts move into the reproductive age cohorts (15–49), they are potential parents. They are also a future workforce with demands for services, housing, infrastructure provision, and jobs. Unfortunately, a large percentage of the population of ELDCs is already unemployed or underemployed. Even with a dramatic reduction in the CBR, absolute totals will continue to rise.

◆ population momentum also carries through into an ever-increasing number of old dependents. According to a recent Earthscan report, the number of old dependents in the ELDCs will more than double to reach 850 million by 2025. Certain countries, such as Cuba and Sri Lanka, will have higher proportions of over-65s than the USA in the late 1990s. This report also emphasises that, in spite of the undoubted difficulties facing many elderly people, 'instead of being a problem, they will be part of the solution ... Healthy older persons are a resource for their families, their communities and the economy'. (Figure 7.17)

Conclusion

The varied and ever-changing shapes of population pyramids reflect many features of a country's demographic history. Long-term changes in structure can be seen in the five model pyramids (Figure 7.18) which broadly correspond to the different stages of the model of demographic transition now considered. As ever in geography, however, each country has its own distinctive demographic features.

7.5 Demographic Transition: Theory and Reality

Demographers, by studying population data and trends over several hundred years, have shown that shifts in mortality and fertility rates can be shown by a generalised diagram called the **model of demographic transition** (Figure 7.18). First proposed in the 1940s, it was based on a study of

Fig 7.17 Elderly yet active: this man is carrying a hefty load of bananas in Karnataka, India

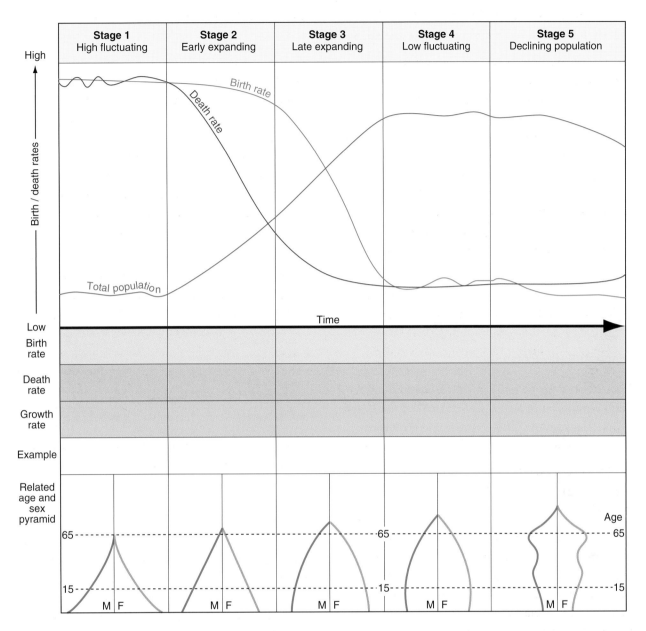

Fig 7.18 Model of demographic transition model and related population pyramids

North-West Europe, whose countries, such as Sweden, have a long history of enumeration. The model shows how CBRs, CDRs and population alter over time by responding to changing economic and social circumstances. The original model suggested that the broad demographic experiences of countries could be divided over time into four stages; however, because of falling CDRs in countries such as Italy and Hungary, some demographers maintain that a fifth stage can be distinguished. General features of each stage are outlined below, along with references to the experiences of Scotland and sample ELDCs.

Stage 1: High Fluctuating

Both birth rates and death rates are high, life expectancy is low as is the average age of the population. Population growth is slow, often reduced by periodic wars, famine and epidemics (collectively called 'Malthusian Checks' – see below) but, in due course, population levels are usually restored, assisted by the high birth rate. One major exception was the fourteenth century plague known as the Black Death, responsible for killing off up to a third of the population in parts of Europe. Examples of the High Fluctuating Stage can be seen in:

◆ Scotland before 1760 – birth rates in a pre-industrial Scotland were high, reflecting the absence of family planning and birth control, high infant mortality rates and the need for children to help on the land. Parish registers show, that in common with other European states, mortality crises were frequent from the sixteenth to the early eighteenth centuries. War, disease e.g. outbreaks of bubonic plague in 1597–99 and 1600–9, and famine (e.g. in the 1690s, which killed off one person in seven), all helped to check population growth

◆ today, the few remaining traditional tribal societies in isolated parts of the Amazon basin and New Guinea, with limited contact with so-called 'western' civilization, typify stage one. So did many ELDCs until the 1950s, when they moved into the next stage.

Stage 2: Early Expanding

Population starts to grow at an exponential rate. Such rapid growth results from a marked fall in the CDR. Because the CBR remains high, there are high rates of natural increase as the gap between the two rates widens. Life expectancy increases, infant mortality rates fall and the proportion of the population under 15 increases. The causes of this dramatic transition vary in timing and impact:

◆ in Scotland from the 1760s to 1830s, mortality rates fell because of: improved nutrition and water supplies; increased resistance to disease; rising wages and the ability to buy, wash and change more garments; and the introduction of vaccination with the resulting decline in smallpox. However, very rapid urbanisation (one of the fastest growth rates in Europe at that time) inflicted great strains on housing, water supply and sanitation, resulting in increased deaths from cholera, typhoid, etc. Consequently, it was not until later in the nineteenth century, during Stage 3, that the full impact of health measures, improved water supplies and medical advances was felt (Figure 7.19). CBRs continued at a high level but social influences, such as marrying at a later age, influenced fertility.

◆ in many ELDCs, the increased provision of better medical facilities, improved diet, sanitation, personal hygiene and disease eradication programmes such as

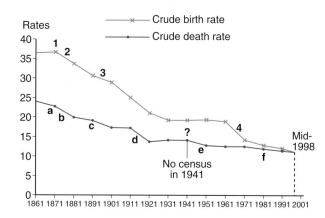

Factors influencing Mortality

a Public Health Act
b Last major smallpox epidemic
c Main causes of death: T.B. bronchitis, pneumonia, measles, and whooping cough
d 74000 deaths in World War I
e 34000 deaths in World War II
f Main causes of death: heart disease, cancer and strokes

Factors influencing Fertility

1 1872 Education (Scotland) Act
2 Bradlaugh trial encouraged the diffusion of contraceptives
3 Overall improvement in living and increased demand for consumer goods
4 Increased consumerism, more working women

Fig 7.19 Graph of changing birth and death rates in Scotland 1861–1998

the WHO anti-malaria campaign, caused a rapid drop in CDRs and infant mortality rates. CBRs continue at a high rate, reflecting a basically, traditional rural society with low per capita incomes, and where social and religious factors discourage family planning and birth control. While some ELDCs (the Republic of Congo, for example) were still at Stage 2 in the 1990s, many countries, where fertility has declined, had moved into the next stage.

Stage 3: Late Expanding

Population continues to grow but at a progressivly slower rate. The fall in the CDR results from further improvements in and spread of health care and sanitation provision, coupled with further medical advances. Much more dramatic is the impact of the decline in the CBR. This is in response to several factors:

◆ Figure 7.19 illustrates some of these for Scotland. By the 1870s, Scotland was one of the most urbanised, industrialised and wealthy countries in the world. The full impact of a sustained decline in

CDRs can be seen, but from this time onwards CBRs fell markedly in response to: the adoption of family planning, assisted from the 1880s by the steady diffusion of contraception (partly encouraged by the publicity surrounding the trial of Charles Bessant and Annie Bradlaugh, propagandists for contraceptives); compulsory schooling after the 1872 Education (Scotland) Act and the increased costs of keeping children. Mortality rates declined as more children (and mothers) survived childbirth; and a wish to partake of an ever increasing range of consumer goods available after the 1870s as illustrated in the following extract:

[The late nineteenth century] was the era when mangles and gas cookers, bicycles and pianos entered working-class and lower-middle-class homes, and when those who had a little surplus cash and not too many familial encumbrances could begin to enjoy occasional trips by train or to the music-hall and even a week at the seaside.

Source: M. Anderson and D.J. Morse in *People and Society in Scotland 1830–1914*

Two additional influences which lowered fertility were the loss of life during World War One, and the impact of mass unemployment and high emigration rates of the depression in the 1930s. This stage of the model approximately dates from 1870 to 1950 in Scotland and is reflected in the fall from an average of six children per family to 2.5 during that period

◆ as late as the 1970s certain North African countries, for example, Algeria, Morocco and Tunisia, were still at Stage 1, yet by the 1990s they had reached Stage 3. Why was the transition so rapid? Essentially, within the space of a generation, women were having fewer children in response to increased urbanisation, better educational opportunities resulting in not only higher female literacy rates but also more participation in higher education, and improved health care and declining infant mortality rates (Figure 7.20). Another important influence has been that of migration. Migrants from North Africa have settled in France, Belgium and the Netherlands. The money (remittances) sent home encourages more consumerism, and when migants return, either permanently or just for a holiday, their aspirations and values are less pro-natalist.

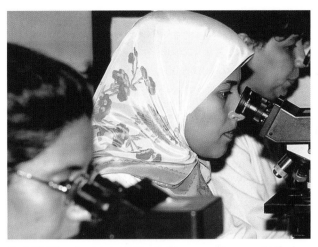

Fig 7.20 Educational opportunities for Moroccan girls are gradually improving

Stage 4: Low Stationary

By this stage, CBRs have declined and now equal the CDRs. Unlike Stage 1, it is the CBR which fluctuates more than the CDR. This is because it responds to changing economic circumstances. CDR does rise slightly, however, because of the ageing of the population.

◆ Scotland entered Stage 4 at roughly 1950. Figure 7.19 shows that CBRs had dropped and by the 1990s there were fewer children than at any period since the 1840s. There were surges (not picked up by the graph because of the time interval), notably the 'baby booms' of 1947 and 1964, but by the late 1990s the CBR was 11. The growing affluence of post-war Scotland, increased consumerism, more working women with new aspirations, an increase in couples starting families in their thirties rather than their twenties (if they chose to have any family at all), and migration of 20–40 year olds from Scotland meant that the average number of births to women born in or after 1978 fell to 1.8 by the 1990s, that is, below replacement level. Over the same period, the population continued to age. However, the causes of death had changed: cancer, heart disease, strokes and road accidents are essentially diseases of affluence. Given the growing convergence of CBR and CDR, Scotland seems to have completed the demographic transition but with a projected decline of population from 5.1 million to 4.8 million by 2030, perhaps there is some merit in the inclusion of Stage 5 in the model.

◆ one of the criticisms of the model is its Eurocentric character. Is it realistic (if not arrogant, given the differing histories and environments of the ELDCs) to assume that the countries of Latin America, Africa and Asia, will follow the same demographic paths as Europe? Certainly, the pace and timing of transition will be different. Also, regional variations should be borne in mind: urban areas/rural areas and rich regions/poor regions can be at differing stages of transition. There is always the need to take each country as a unique case.

Stage 5: Declining Population

A fifth stage has been suggested that is appropriate to Italy, Germany and Hungary, countries whose fertiliy rates, particularly among the 15–29 age group, are among the lowest in the world. At the Cairo Conference, Italy's report predicted a fall in its population from 57 million to 19 million over the next century. The cause was familiar: increased affluence as the country shifted from an agricultural to an increasingly urbanised and industrialised society. This came at a later time compared to other Western European countries but its impact was more sudden. In addition, as one demographer emphasised, Italians lavish more 'protective and obsessive care' on their

offspring. Italy is a country where a one-child-per-family policy has evolved, in spite of the devout Roman Catholic traditions still seen in the more agrarian south with a higher fertility rate. In contrast, China's one-child policy has developed under quite different political, economic and social circumstances.

China's Changing Population Policies (1949–Present)

By the end of the twentieth century, China, with a population of some 1.3 billion, topped the world demographic league table. According to some demographers, the world's population would have reached six billion in 1996 but for China's policy on family size. By restricting couples to one child in the 1980s, China has pursued one of the most severe demographic engineering policies ever.

In 1949, the communists, under Mao Zedong, gained control of mainland China. This resulted in strong government control over many aspects of everyday life, including population policies. Broadly speaking, for the next twenty years China generally pursued pro-natalist policies; from the 1970s, there was a complete about turn and anti-natalist policies were adopted.

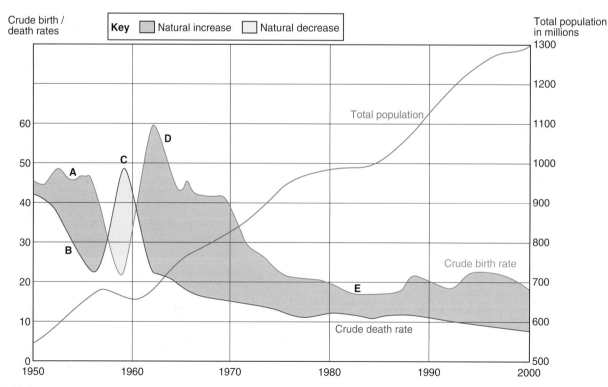

Fig 7.21 Graph of population change in China 1950–2000

N.B. The effect on the population of:

A The famine following the failure of the 'Great Leap Forward' policy
B The increase in birth rate during the Cultural Revolution
C The introduction of the 'One-Child-Per-Family' policy
D The relaxation of the 'One-Child' policy
E The personal decisions to limit the number of children as affluence increased especially in urban areas
   (Fertility rate in 1996: 1.42 for China, ranging from 1.78 in Hainan to 0.84 in Beijing)

Fig 7.22 Population pyramid for China, 1996. (The percentage of the total population, male or female, is given for the age groups up to and including 40–44, e.g. male 7.5%)

Pro-Natalist Policies

Figure 7.21 shows that in 1949 both birth and death rates were high. Improved health care measures in the early 1950s reduced death rates. High birth rates were encouraged by the government. According to Mao:

Of all things in the world, people are the most precious. Under the leadership of the Communist Party, as long as there are people, every kind of miracle can be performed.

Overpopulation was not a problem, and if China was to survive a nuclear war, the more people the better. Strict controls were placed on abortions, sterilisation was forbidden and family benefit was paid for every child.

One catastrophic demographic disaster resulted in some 20 million deaths in 1959–61. A combination of unfavourable weather and a massive neglect of agriculture during a fanatical drive to industrialise ('The Great Leap Forward') triggered famine conditions. China's population went into decline. One consequence was a rapid decline in fertility and

some 30–35 million births were lost. This is reflected in the 1996 population pyramid (Figure 7.22).

Pro-Natalist policies were revived during the early years of the Cultural Revolution. During this period of social disorder, the crude birth rate rose again, peaking at 45 per thousand, with a consequent baby boom.

Anti-Natalist Policies

After the death of Mao in 1976, there was a relaxation of economic controls and an increased emphasis on reducing and eventually stopping population growth. Demographer Liu Zheng argued that by 2080 the target population should be 700 million to ensure a balance between population and resources, particularly water resources.

Initially, a two-child policy was adopted, with its slogan of 'later, longer and fewer'. In 1979, however, the Chinese Communist Party launched its **one-child-family** policy. Initially, the policies were draconian:

◆ women who became pregnant for a second or subsequent time had enforced abortions

◆ compulsory sterilisation policies were adopted for persistent 'offenders'

◆ massive contraception campaigns were waged

◆ early marriages were discouraged

◆ food and clothing rations were withheld

◆ public denunciation and visits from unpaid, often elderly, neighbourhood watch members (the 'granny police') exerted social pressure on married couples

Assessment of the One-Child Policy

This policy and associated penalties have been applied throughout China except among the non-Han minority populations, for example, the Tibetans. It is in the cities that the policy has been most successful because (i) it is easier to enforce, and (ii) educational standards and levels of economic development are higher. Overall, the government claims that China's population is some 300 million lower than it would have been without the policy. It is arguable, however, that the decline in birth rate would have taken place anyway as a result of rapid economic development, particularly in the eastern coastal regions.

In rural areas, there has been more resistance to its implementation. People are more conservative and the traditional desire for sons and large families to

Fig 7.23 'Little Emperor' – spoilt, pampered only child in China

work on the farm is still strong. Some 70% of China's population is rural. Recognising the difficulties of implementing the policy the government relaxed its enforcement in rural areas to allow a second child, particularly if the first was a girl.

Problems of the One-Child Policy

Although fertility has declined, there are still more than 20 million babies born each year. China's population is expected to peak at 1.6 billion by 2050. By then, however, some 25% of the population will be over 60, with implications for the scarcely developed pensions system.

Another issue is that of the 'Little Emperors' or 'Little Empresses' (Figure 7.23). China now has around 65 million only-children, who are characterised as spoilt, pampered and selfish. This is because predominantly urban parents focus money and attention on the education and comfort of these children. Since the 1980s, China has run parenting classes for adults in the cities on how not to spoil such children and how not to over pressurise them to perform well at school. Interestingly, under the family planning rules, only-children who marry other only-children will be permitted to have two children.

Perhaps the biggest long term problem is the sex imbalance between boys and girls. When nature is left to herself more boys are born than girls. Strict family limitation, the help of ultrasound technology and selective abortion is resulting in an unnaturally high proportion of males to females: 115 boys are born for every 100 girls. In the long term, major social problems may result. A related issue, highlighted by the television documentary *The Dying Rooms*, is that many girls are being abandoned because parents want a son. The documentary demonstrated that children, especially girls and handicapped boys, are left to die from neglect in a number of orphanages.

To Take You Further

The Reverend Thomas Malthus (1766–1834), clergyman and economist, published his 'Essay on the Principle of Population' in 1798, revised in 1803.

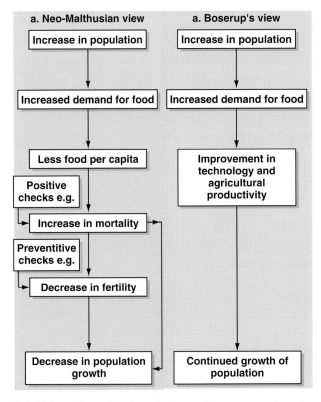

a. Neo-Malthusian view

Increase in population
↓
Increased demand for food
↓
Less food per capita
↓
Positive checks e.g.
→ Increase in mortality
Preventitive checks e.g.
→ Decrease in fertility
↓
Decrease in population growth

a. Boserup's view

Increase in population
↓
Increased demand for food
↓
Improvement in technology and agricultural productivity
↓
Continued growth of population

Fig 7.24 Population and food supply: two possible sequences of events

Writing at a time of population growth, he maintained that, unless controlled, population increases geometrically (1,2,4,8,16,32, etc.) but food supply increases arithmetically (1,2,3,4,5, etc.). Consequently, population would far outstrip any increase in food supply until eventually disaster would strike in the form of famine, disease, or even war.

According to Malthus, population increase could be kept under control by two types of 'checks':

◆ **positive checks**: those which 'shorten the natural duration of human life'. These included lowered resistance to disease, war, famine and infanticide.

◆ **preventative checks**: marrying later in life and only when a man is capable of supporting a family. Malthus urged 'moral restraint' on the part of young men who should stop 'pursuing the dictate of nature in an early attachment to one woman'.

Malthus was writing at a time when Britain was moving into Stage 2 of the model of demographic transition, and population was increasing. Understandably, he was unable to foresee several demographic developments. These included the dramatic increases in productivity in agriculture,

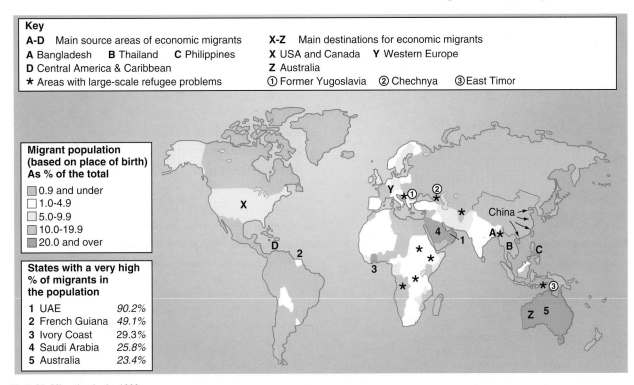

Key

A-D Main source areas of economic migrants
A Bangladesh **B** Thailand **C** Philippines
D Central America & Caribbean
★ Areas with large-scale refugee problems

X-Z Main destinations for economic migrants
X USA and Canada **Y** Western Europe
Z Australia
① Former Yugoslavia ② Chechnya ③ East Timor

Migrant population (based on place of birth) As % of the total
☐ 0.9 and under
☐ 1.0-4.9
☐ 5.0-9.9
☐ 10.0-19.9
☐ 20.0 and over

States with a very high % of migrants in the population

1	UAE	90.2%
2	French Guiana	49.1%
3	Ivory Coast	29.3%
4	Saudi Arabia	25.8%
5	Australia	23.4%

Fig 7.25 Migration in the 1990s

industry and transport during the nineteenth and twentieth centuries, coupled with the import of cheap food from, for example, the Great Plains. Nonetheless, despite criticisms of his ideas, various 'doom and gloom' forecasts in the 1960s and 1970s echoed those of Malthus, and advocates of such pessimistic views are referred to as neo-Malthusian. Others, in their discussion of the balance between population and food supply, adopt a more optimistic view, stemming from the work of Danish economist Esther Boserup. Essentially, she argues that population growth stimulates increased food production (Figure 7.24). Boserup's views do raise questions: What happens when food supply cannot be increased any further? What are the environmental consequences of continued intensification of food production?

7.6 Population Dynamics: Migration

Migration and Global History

Earlier in this chapter, migration was mentioned as one of the main components influencing the overall size of the population of a particular area. Migration is a significant topic, not least because it has been critically important throughout world history. For example, it has involved:

◆ the initial peopling of the globe as groups of hunter-gatherers spread from Africa from the earliest of times (see Figure 6.1)

◆ massive movements of 44 million people in the nineteenth century from Europe to North America and Australasia; and the shipping of African slaves between the sixteenth and nineteenth centuries to the Americas and the Caribbean

◆ the movement of people for religious reasons, with the founding of new states. Examples are the transfer of millions of Hindus and Muslims when newly independent India and Pakistan emerged in 1947, and the arrival of thousands of persecuted Jews in the recently founded state of Israel in 1948

◆ the involuntary uprooting of millions of refugees during the 1990s in various parts of the world –

Kosovo, Chechnya, Rwanda and Burundi. (Figure 7.25)

Migration, therefore, reflects many of the great events of world history. However, for the majority of people, migration takes place for more mundane reasons. This does not mean that it is a straightforward process. Migration is a highly personal matter, often involving breaks with the ties of family, community and country.

Migration: General Considerations

Meaning

Migration is a complicated topic, involving people taking a wide variety of factors into consideration. Essentially, it is a form of **mobility** which embraces all types of movement by people. One short-term type of mobility is called **circulation**, a name that includes daily commuting to work, holiday-making, recreational trips and shopping excursions. Unlike migration, however, there is no question of permanently moving residence. **Migration**, therefore, can be defined as the movement of people from one place to another, involving a permanent change of residence.

Motivation

Something of the reasons why people move can be seen from Figure 7.26. The diagram suggests that the main reasons for migration can be grouped into: 'Pull', at the place of destination, and 'Push' at the place of origin. 'Pull' factors are those that attract people to another location; 'push' are those that encourage them to leave. There are many examples of these factors, some of which are shown in Figure 7.26 and, broadly speaking, they can be grouped into four categories – political, economic, social and environmental. What is always an interesting question is: which carries most weight, push or pull?

Figure 7.26 also shows that a related, but important factor concerns the role of **networks**. These provide the essential information and the actual means of migrating. Migration networks usually involve friends and family, who provide important information about the proposed destination and current job and

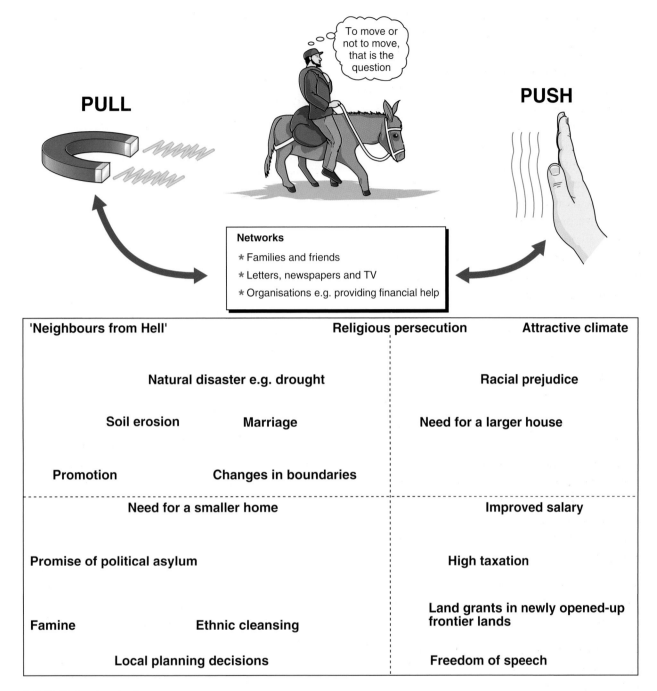

Fig 7.26 Motives for migration

housing opportunities, and give financial help. Alternatively, they can include employers, governments (an example is the assisted passages to Australia, to encourage 'the £10 pom' in the 1950s), migrant organisations, and even smuggling groups transporting illegal immigrants. Networks, therefore, link the pull and push factors, ensuring that potential migration becomes reality.

Types

As ever, any classification of migration is fraught with difficulties. Figure 7.27 is one attempt and takes into account:

◆ **the difference between migration within a country and international migration**: On the face of it, international migration would seem to be a far more daunting affair, often involving long distances. Differences in language, customs and life styles usually mean that a great deal of adjustment is involved, as in the great nineteenth century trans-Atlantic migrations. It seems a reasonable divide. Yet, internal differences in a country can also be a challenge: at one end of the scale, the socio-economic, linguistic and religious differences between the Pacific coast of China and its interior provinces such as Xinjiang; at the other end of the scale, migration between French speaking Wallonia and Flemish speaking Flanders, posing a problem of cultural distance

◆ **voluntary versus forced migration:** Movements of slaves and refugees are examples of **forced migrations**, as are movements resulting from natural disasters such as drought, religious and political persecution; **voluntary (or free) migrations** are usually the result of economic and social factors, for example, choosing to retire to a coastal town, or move abroad to a better paid job and what is perceived to be a potentially more satisfying lifestyle

◆ **destination**: migrants may move from an urban to a rural area or from rural to urban. In nineteenth century USA, it might have involved moving to a new frontier area, such as the Great Plains; in

twentieth century Brazil or Indonesia, similar frontiers are being developed in the rainforests.

◆ **duration**: migrations vary in length. Some migrants never return, others return within a few months or after several years. Examples of temporary migrations include students undertaking study in the EU (e.g. the Erasmus scheme), or the nineteenth century agricultural labourers, nicknamed the 'swallows', who travelled from Italy to Argentina for seasonal work. Return migrants are a particularly important category and include those returning from France or Germany in the 1970s to southern Italy, Greece and Turkey.

Selectivity

This answers the question, Who migrates? Usually, refugee movements involve whole communities. Voluntary migration, however, is normally selective, as can be seen with regard to:

◆ **age**: often the migrants are young adults of working age, between 20 and 35. Alternatively, they may be of retirement age, opting to migrate to rural or coastal areas – the so-called 'Costa Geriatrica' of south-west England, Spain's Costa del Sol or Italy's 'Chiantishire'.

◆ **sex**: according to circumstances, males or females

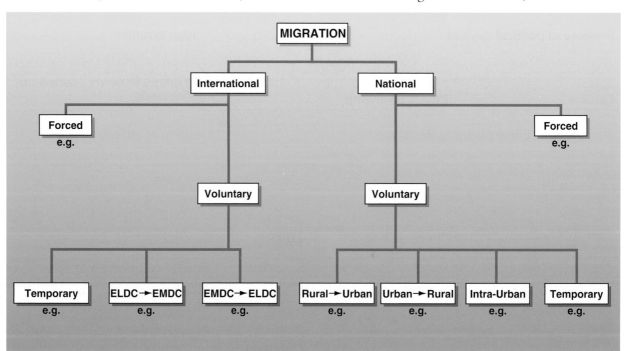

Fig 7.27 Types of migration

may move, usually in search of jobs. One example of seasonal, selective migration was the 'gutting quines'. Traditionally, fisher women ('quines') of the north-east of Scotland followed the fishing fleet, which followed the migrating shoals of herring from Stornoway to Yarmouth. Once the catch was landed, the mainly female workforce gutted and packed the fish in salt barrels.

◆ **education**: those who have spent longer in higher education and are better able to take advantage of the knowledge economy, are more likely to migrate between core areas and cities, which are often the headquarters of major transnational companies.

Some of these general principles can now be illustrated from Scotland (1800–1945), China (1949–present) and the Mediterranean.

Scotland and Migration (1800–1945)

Sample Extracts

Extract A

I left Donegal in 1908. They all come to Scotland at that time and I came over first in May and then I used to go back then come back to the harvest again. They carried it under their oxter their hooks, for to cut the corn. I mind that as well. I come on the – did ye ever hear about the Gola boat – a boat that used tae leave from where we stayed just and we got the boat there and she left us in Glasgow. We only paid four shillins.

Source: Mrs Coll from *Odyssey* edited by Billy Kay

Extract B

William Chisholm in the Parish of Farr charges that he was removed from his farm without the legal warning of removal; that in June 1814, being from his home, his house was burned about his family's ears; that Margaret Mckay, his wife's mother, an infirm old woman was confined to bed in his house when set on fire; that it was at the peril of his wife's own life that she rescued her said aged mother from the flames.

Source: *Eviction in the Parish of Farr, 1814* from *History of the Destitution in Sutherlandshire* Donald Macleod

Extract C

For the old ploughman of the farmtouns looking over the land in the 1930s one thing was becoming abundantly clear: machines were to be the chariots of agricultural advance. Their own day and that of their Clydesdale pairs was over . . . In time, and in despair, they moved away these farmtoun men, to jobs in the town – as milkmen, carters and cleansing department drivers. And there were other forces that drove men away . . . There was hardly a small town within a bothy-lad's bicycle reach that did not have its cinema. Farmer's boy and kitchen maid alike were susceptible to the glorious life it depicted.

Source: *The Ballad and the Plough* by David Kerr Campbell

Extract D

Scotland is gradually being emptied of its population, its spirit, its wealth, industry, art, intellect and innate character. If a country exports its most enterprising spirits and best minds year after year, for fifty or a hundred or two hundred years, some result will inevitably follow.

Source: *Scottish Journey* by Edwin Muir

Extract E

There were Yiddish posters on the hoardings, Hebrew lettering on the shops, Jewish faces, Jewish butchers, Jewish bakers with Jewish bread, and Jewish grocers with barrels of herring in the doorway . . . One heard Yiddish in the streets, more so in fact than English, and one encountered figures who would not have been out of place in Barovke.

Source: *Coming Home* by Chaim Beraint

Scots have long been settling abroad, and people have long been coming to settle in Scotland (Figure 7.28). During the sixteenth and seventeenth centuries, thousands of Scots migrated to Scandinavia, Poland, Ireland and England as merchants, peddlars, farmers and mercenaries. In fact, it has been suggested that migration almost became an accepted way of life. During the nineteenth century, Scotland lost some two million people between the 1820s and World War One, sharing, with Ireland and Norway, some of the highest rates of migration per head of population in Europe, supporting the assertion that Scotland's main 'export' has been people.

Fig 7.28A April 6th is now 'Tartan Day' in the USA

Fig 7.28B Andrew Carnegie (1835–1919) was born into a weaver's family in Dunfermline and made his fortune as an industrialist in Pittsburgh.

The motivations for migrating at that time were varied. Extract B shows that William Chisholm had no choice. Like many others in the Parish of Farr, he was evicted from his farm to make land available for extensive sheep grazing. As a result of such clearances and the later potato famine, there was a mass exodus from the western Highlands and Islands; many migrated to the growing industrial cities such as Glasgow and Dundee, others joined the massive trans-Atlantic stream of migrants to North America. Most of the Scottish emigrants, however, came from the towns rather than the rural areas and frequently were skilled or semi-skilled, reflecting Scotland's advanced economic situation in the late nineteenth century. Overall, it has been estimated that there are (an estimated) 25 million people of Scots extraction in Canada, New Zealand, Australia, England and elsewhere. During the the nineteenth century, Scotland was also a nation of immigration.

Mrs Coll (Extract A) was just one of the many thousands of Irish who came to Scotland. Eventually she settled in the Gorbals district of Glasgow but, at first, she only came over for seasonal farmwork. The Irish, who formed Scotland's largest immigrant group, concentrated in the industrialising cities, especially Glasgow and Dundee, and rapidly growing towns such as Coatbridge and Airdrie, as well as in the mining communities of Ayrshire, Lanarkshire and the Lothians. Many fled the destitution that followed the Potato Famine of 1846–47 and formed what was initially a largely unskilled and semi-skilled labour force that played an important role in Scotland's economic boom in the nineteenth century.

Another much smaller but distinctive group of immigrants were the Jews, who were forced by anti-semitic pogroms (organised massacre) to flee Russia and Eastern Europe. By the 1920s some 10 000 lived in the Gorbals district of Glasgow, maintaining their identity with their own ('Kosher') shops, schools and synagogues (Extract E). Over the years, as they became more prosperous and the Gorbals was redeveloped, many moved to the more affluent district of Giffnock. Italians were another group but, unlike the Jews, were more dispersed. 'Pushed' by the poverty of southern Italy, they opened up hundreds of small, family run fish and chip shops and ice cream parlours throughout Scotland from the 1890s onwards. Among many others immigrants, Lithuanians came to mine coal in Lanarkshire and Polish soldiers stayed on after World War Two rather than return to a newly established Communist state.

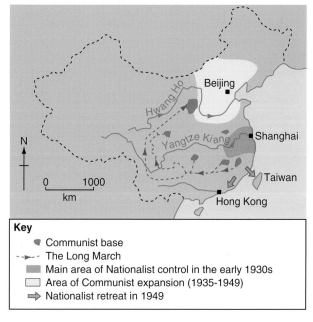

Fig 7.29 Sample forced migrations in China (1934–1949)

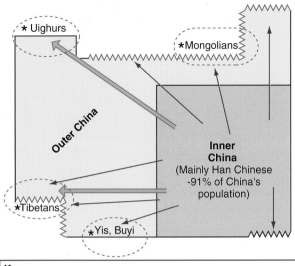

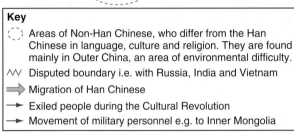

Key

( ) Areas of Non-Han Chinese, who differ from the Han Chinese in language, culture and religion. They are found mainly in Outer China, an area of environmental difficulty.

∿∿ Disputed boundary i.e. with Russia, India and Vietnam

⇒ Migration of Han Chinese

→ Exiled people during the Cultural Revolution

→ Movement of military personnel e.g. to Inner Mongolia

Fig 7.30 Sample migration flows for China (1949–1978)

China and Migration (1934–Present)

Since the establishment in 1949 of Communist control in China, there has been a wide variety of migration flows. In fact, the Communists might not have come to power but for a significant migration flow in 1934, when they were forced on the 'Long March'. This was an attempt to evade the Nationalist forces then threatening the Communist strongholds in the south. By 1935, after enduring many hardships, they regrouped in Yan'an. In 1949, however, with the Communist success, the tables were turned and the Nationalists escaped to Hong Kong or Taiwan, where they established their own government (Figure 7.29).

Since 1949, there have been varied migration flows, responding to changing social, economic and, above all else, political circumstances.

Migration Flows 1949–78

During these years, migration flows included the following examples:

◆ millions moved to the cities, including peasants

seeking work in the industrialising cities, and former city dwellers returning after the civil war. The size of this flow was made worse by the failure of the 'Great Leap Forward' (1958–60) when the resulting catastrophic famine caused millions to flood into the cities. Between 1949 and 1960, China's urban population grew from 11% to 19%.

◆ political and strategic factors influenced two related flows towards Outer China. The majority Han Chinese were encouraged to move to areas of minority discontent, notably Xinjiang and Tibet, in an attempt to alter the ethnic balance. Tibet, itself, lost thousands of refugees, who fled to India after an unsuccessful uprising in 1959–60. During these years, large numbers of military personnel were moved into other equally sensitive border provinces because of boundary disputes (Figure 7.30)

◆ a particularly distinctive cause of urban–rural migration came as a result of the Cultural Revolution (1966–69). During this anarchic period, schools and colleges were closed and many politicians, teachers, students and scientists were either imprisoned or sent to work in the countryside. Jung Chang remembers:

. . . in 1969 my parents, my sister, my brother and I were expelled from Chengdu one after another and sent to distant parts of the Sichuan Wilderness. We were among millions of urban dwellers to be exiled to the countryside . . . According to Mao's rhetoric, we were sent to the countryside to be reformed . . . On 27 January 1969, my school set off for Ningan. Each pupil was allowed to take one suitcase and a bedroll. We were loaded into trucks, about three dozen of us in each . . . The column of trucks bumped up and down country roads for three days before we reached the border of Xichang . . . On the fifth day, the trucks unloaded us at a granary at the top of a mountain . . . A couple of dozen peasants were there to help us with our bedrolls and suitcases. Their faces were blank and inscrutable, and their speech was unintelligible to me . . . At dusk we reached the lightless village . . . We had a few days . . . to get all our necessities like water, keresone and firewood organised; after that we would have to start working in the fields.

Source: *Wild Swans* by Jung Chang

Rural–Urban Migration (1979–present) – 'The Great Escape'

Migration Patterns

With the death of Mao Zedong in 1976 and the development of a more market-orientated economy, there has been a massive flow of migrants from rural to urban areas. Figure 7.31 shows some of the key factors influencing this huge migration. The diagram is based on Lee's model of migration. This develops the simple 'push'/'pull' model by showing that both the place of origin and the place of destination have a wide range of features, some positive, some negative and some neutral. Another feature of the model is that of intervening obstacles. In China, from 1958 until the early 1980s, strict control over migration was maintained by the state through a compulsory system of household registration (hukou). This gave citizens access to basic goods and services but these could only be claimed at the place of registration, thereby acting as a critical obstacle to migration. Even a visit to a another part of China required official permission.

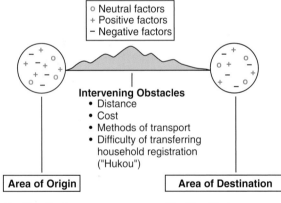

Fig 7.31 Lee's model of migration applied to rural-urban flow in China (1979 to present day)

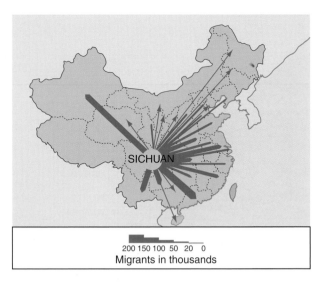

Fig 7.32 Inter-provincial migration from Sichuan (1985–1990)

Figure 7.32 shows some of the most frequent migration flows from Sichuan Province (with its mainly poor, agricultural population), emphasising the attraction of the rapidly growing cities of the eastern and southern provinces. Most of the flow, therefore, is to the areas of higher income (Figure 7.33). With their better economic opportunities and foreign investment, these include areas such as the SEZs (Special Economic Zones), the Pearl River Delta, Beijing and Shanghai. Migration also continues to provinces such as Xinjiang, reflecting Government spending and encouragement.

However, not all rural–urban migration is interprovincial. There is also intraprovincial movement as shown in the following extract about Chonquing, the main industrial city in Sichuan. Located on the Chang Jiang (Yangtse River), it lies west of the Three Gorges Project, and by the 1960s, was China's third industrial centre. Today, its state-owned smokestack industries are declining and a new class of urban unemployed is rapidly growing. Once looked after by the state, with free schooling, medicine and housing (the so-called 'iron rice bowl'), they are now competing for jobs with ever increasing numbers of migrants from the countryside and, potentially, thousands of people relocated by the vast Three Gorges reservoir.

Many migrants can now be found on building sites, street markets or working as bamboo pole carriers on the steep steps above Chonquing's docks. This is where Qin Guoneng has chosen to ply his trade. He looks older than

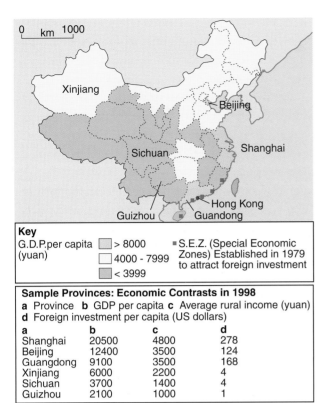

Key

G.D.P.per capita (yuan)		
	> 8000	■ S.E.Z. (Special Economic Zones) Established in 1979 to attract foreign investment
	4000 - 7999	
	< 3999	

Sample Provinces: Economic Contrasts in 1998
a Province **b** GDP per capita **c** Average rural income (yuan)
d Foreign investment per capita (US dollars)

a	b	c	d
Shanghai	20500	4800	278
Beijing	12400	3500	124
Guangdong	9100	3500	168
Xinjiang	6000	2200	4
Sichuan	3700	1400	4
Guizhou	2100	1000	1

Fig 7.33 Regional inequalities in China

his 46 years, though he says he can still carry three large rice sacks without any trouble. He came to Chongqung three years ago and now works as a bangbangjun (carrying-pole porter) . . . His home is in a small village north of Chongqing, a five-hour, or 15 yuan, bus ride away. He is married with two children and an elderly mother living at home . . . His family now raises a few pigs and chickens and they grow rice, corn, rape, beans and several kinds of vegetables . . . Qin's family can produce enough food on their plot, but they need cash for fertilisers, coal and school fees. Qin is keen to see his children – especially his son – educated, and this can cost him as much as 500 yuan a year. Eventually he would like to build a new, two-storey house, but with his present earnings he expects this to take him about ten years. In the meantime he seems satisfied with his migrant life, sleeping fifteen to a room in a squalid back-street hostel. Few workers, used to the relative comforts of the iron rice bowl, would tolerate such hardship.

Source: *Chongqing: A city at the crossroads* by Stephen Hallett in *China Review* 1998

Consequences of Migration

Taking China as a whole, there are an estimated 120 million people (almost 1 in 10 of the population)

living in a different province from that in which they were born. The consequences of temporary and, increasingly, permanent migratory flows are profound for both areas of origin and destination.

1) Areas of origin (losing population). These vary with location, and from family to family. There can be:

◆ a loss of usually younger, skilled people, with a slightly higher proportion of males, often with higher educational standards and potential initiative

◆ an age imbalance because of a residual population with a disproportionately larger number of elderly people left to tend the fields, leading to a potential decline in productivity

◆ yet the flow of remittances and return migration can boost local consumption and investment, and encourage modernisation by transferring scientific and technological information to remote communities

2) Areas of destination (expanding population). The impact of migration is a mixed blessing.

◆ housing the large number of temporary (but effectively permanent) migrants is a problem, as construction rates have not kept pace with growth. Makeshift, squatter type housing can now be seen in the marginal areas of cities (Figure 7.34). Overcrowding and a lack of basic amenities in such shanty towns pose a problem for public health care.

Fig 7.34 Overcrowded squatter housing in Beijing. Such areas are gradually being redeveloped

◆ additional infrastructure burdens are also placed on city authorities by increased overcrowding on the already strained public transport systems. A 1989 report by the authorities in Guangzou suggested that the city's 914 000 floating population had utilized an extra 286 000 tonnes of water per day, consumed 2 353 600 kWh of electricity, and produced 394 000 tonnes of waste water and 573 tonnes of solid wastes

◆ city administrators find it difficult to impose the one-child policies. On the other hand, immigration and the relaxed natural birth rate are slowing down the ageing process in cities such as Shanghai

◆ it is also suggested that migrants are responsible for an increase in crime. A further issue is that there is little integration by migrant communities in host cities because of cultural and linguistic differences

◆ the migrants provide a valuable contribution to the urban labour force, fitting in to a varied range of jobs – construction work, retailing, maids, pedicab drivers, serving at food stands, etc., and are willing to accept lower wages.

The Mediterranean Lands and International Migration: 1950s–Present

The Mediterranean has long been an area of migration flows of varied types, ranging from: the colonies of Ancient Greeks expanding westwards to Sicily and eastwards to Asia; the colonisation by France of Algeria in the nineteenth century; and massive flows of Portuguese and Spaniards to Latin America, Italians to the USA and Greeks to Australia.

Migration Patterns 1950s–70s

During the period from the 1950s to the 1970s, there was a 'Mediterraneanisation' of many of the cities of north-west Europe. Mediterranean states acted as a critical labour reserve. Immigrants from Italy, Spain, Portugal, Turkey, Greece and the former Yugoslavia, helped rebuild the war-damaged economies of states such as the former West Germany, France, Belgium and the Netherlands. They accepted a variety of jobs, in factories, at construction sites, and in low-grade service sector work, usually accepting lower wages, longer hours and poorer conditions than local people.

Gradually, these various immigrant groups formed distinctive ethnic communities in districts such as the inner cities (e.g. the Kreuzberg district of Berlin), peripheral housing estates (e.g. Bijlmermeer, Amsterdam) and industrial towns (e.g. Pforzheim, near Stuttgart). The tendency of 'guestworkers' ('gastarbeiters' in Germany) to form distinctive communities increased, as wives and children joined their husbands. By the early 1970s, however, migration flows had started to change. In the former West Germany, for example, restrictions were placed on the recruitment of foreign labour as a result of economic recession. Incentives in the form of grants were offered to Turks to return home but few took advantage of them. Migrants, who did return to the Mediterranean, were driven by a combination of unemployment, the need to care for their parents, nostalgia and a desire to have their children reared in their 'native' culture.

Migration Flows 1980s–Present

By the 1980s, however, the situation was changing rapidly. Countries such as Italy and Spain were experiencing a migration turnaround. From being net exporters of people, they were were becoming net importers, drawing on migrants from (i) other EU countries and (ii) southern Mediterranean countries such as Morocco and Tunisia. Many of the latter group are clandestine or illegal workers. These new migratory flows from the Mediterranean have been for the following reasons:

◆ high unemployment rates and lower income levels are key 'push' factors in north African countries

◆ many north African countries are experiencing high rates of population growth (though declining in states such as Morocco and Tunisia)

◆ there has been a growing demand for labour in Italy and Spain in times of economic prosperity, as in the late 1980s, when Spain reaped the benefits of entry to the EU. However, even when unemployment rates are high, as in the early twenty-first century, there is still a demand. Migrant workers are willing to undertake tasks that are very low paid or socially undesirable, and that are rejected by unemployed Spaniards.

This is a trend that is likely to continue, given the projected population decline of Italy and Spain. This reflects the remarkably low levels of fertility and an increasing ageing of the population in these countries.

Moroccans now form the largest immigrant groups in both Spain (Figure 7.35) and Italy. Although many are initially illegal (often smuggled in by well organised gangs taking advantage of the long, not easily policed coastlines of Italy and Spain), there are occasional 'regularisation' schemes granting official residence permits. This means that Moroccans can legally pursue a wide range of activities including street hawking, harvesting, working in small factories and hotel work for males, while females work as nannies and domestic servants.

Consequences of Migration

As in China, the consequences vary from place to place.

1) **Areas of origin:** In rural areas of Morocco and Tunisia,

◆ there is an outflow mainly of males, often unmarried, and generally aged between 20 and 35 years of age, leaving a residual elderly population and a lack of innovative leadership

◆ remittances and return migration are helping with economic development and encouraging a decline in fertility rates

◆ return migrants often invest in new housing and may take up a job in the service sector, for instance, a taxi business or a shop

2) **Areas of Destination**

◆ immigrants provide an important workforce, willing to undertake dirty, unskilled jobs, helping to overcome labour shortages, and, by contributing through taxes, help support the growing number of elderly dependents in the host country

◆ immigrants usually occupy poorer quality housing and are prepared to tolerate overcrowding

◆ immigration has become a major domestic issue as right-wing parties encourage racial tension. This situation can be made worse by ethnic segregation, fears of high birth rates among migrants, language

Fig 7.35 Moroccan immigrant workers face Spanish riot police following stabbing incidents in El Ejido, southern Spain

differences and a perception that immigrants contribute to the rising crime rates.

For the future, the Mediterranean is seen, by some, as Europe's Rio Grande, forming a demographic frontline against a growing wave of illegal immigration from North Africa. What is impossible to predict is the unexpected flow of refugees from politically volatile areas, as in the flow of Albanian asylum seekers to Italy in 1993 and the flight of Kosovan refugees to Albania in 1999.

7.7 Geography of Uprooted People: Refugees

The twentieth century rightly has been described as the century of the 'Uprooted Man'. Rwanda, Burundi, Angola, Afghanistan, Bosnia and Herzegovina, Kosovo, East Timor and Chechnya were just some of the locations from which there were major refugee movements in the 1980s and 1990s.

Numbers

As with all demographic data, it is difficult to provide an accurate figure for the global number of refugees. According to the UNHCR (United Nations High Commissioner for Refugees), figures ranged from 17 million in 1991, 26 million by 1996, dropping to 22 million by 1997. Not only are the actual outflow figures difficult to record but many return home as quickly as conditions permit. For example, an estimated 850 000 ethnic Albanian Kosovans fled from the Kosovo region of Yugoslavia in April 1999 but this was matched by such a dramatic repatriation that, by October, UNHCR had closed down most of the remaining refugee camps in neighbouring Macedonia and Albania. However, in countries such as Jordan, Palestinian refugees have been residents since the formation of the state of Israel in 1948.

Definition

Accurate figures also depend on a clear definition of 'refugee': according to the UNHCR, a refugee is legally defined as a person who has 'a well-founded fear of being persecuted for reasons of race, religion, nationality, membership of a particular social group or political opinion', and lives outside his/her own country. If not, he/she is defined as 'internally displaced'. Overall, the total number of people displaced either inside or outside their own country was estimated to reach at least 50 million by the end of the twentieth century.

Special Migrants

Refugees are a special type of migrant because:

◆ they are the ultimate example of forced migration and, consequently, they have experienced persecution, human rights violations, loss of relatives and friends, dispossession of property and mental trauma

◆ they are a marked case of non-selective migration. Generally speaking, all age groups, regardless of ability or sex, may be forced to migrate. However, there can be exceptions. During the ethnic cleansing of Kosovo in 1999, there were examples of ethnic Albanian women and children being driven out while their menfolk were systematically killed.

Distribution

In the immediate post-1945 period, refugees were mainly found in Europe, driven by the impact of World War Two and the imposition of the 'Iron Curtain', and in the Middle East, following the outflow of Palestinian refugees from Israel to neighbouring countries, especially Jordan. By the end of the twentieth century, the main focus of refugees centred on sub-Saharan Africa, Asia, the former Yugoslavia, east Timor and Chechnya. The key 'push' factors, as always, were political threat, ethnic tension, persecution and war. Such is the strength of these pressures that not all refugees opt for repatriation. Some prefer to stay in a host state, possibly as a stepping stone to achieving asylum status in the EMDCs (Economically More Developed Countries).

Problems

Refugee flows can result in a variety of problems including:

◆ resentment on the part of the host population. This situation often is exacerbated by the sheer scale of the problem and the poverty of the host country, with its own housing difficulties. Traditional hostility and differences of language and religion can also make matters worse

◆ environmental pressures on local water supplies and fuel resources, as in the case of refugees from Afghanistan in Pakistan in the 1980s.

CASE STUDY:
former Yugoslavia

During the period from 1945 to the 1980s, Yugoslavia survived as a federal state, with its six republics, five nations, four languages, three religions, two alphabets and one leader – President Tito. He was the key figure and, after his death and the decline of Communist power, demands for independence were made by republics such as Slovenia, Croatia and Macedonia. In addition, demands were made by Serbian nationalists for an enlarged Serbia to include parts of Slovenia, Croatia and Bosnia-Herzegovina, all of which had minority groups of Serbs. Although Macedonia (with a Serb population of 2%) gained independence peacefully in 1991, matters increasingly deteriorated, particularly in the ethnically mixed Bosnia and Herzegovina. The bloody conflict resulted in large scale population movements both within the former Yugoslavia and outwith, ensuring that the term 'ethnic cleansing' became commonplace (Figure 7.38).

Figure 7.36 shows the flow of refugees following the break up of Yugoslavia in 1991–3. Several reasons can explain the variations in the number of refugees and include the following:

◆ Germany was the main recipient of refugees because of (i) the existence of a large number of Yugoslav 'guestworkers' and (ii) a tradition of generous public support for asylum seekers by providing clothing, housing schooling, food, medical care and a social allowance
◆ Switzerland and Austria also have 'guestworkers' and are geographically close

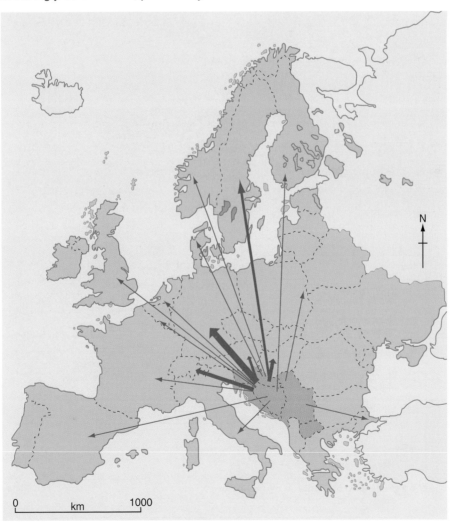

Fig 7.36 Map of flow of refugees from former Yugoslavia (1991–1993)

◆ countries further away seem less affected but Sweden does have a tradition of generosity to refugees and also hosts Yugoslav workers.

Further refugee flows began when violence broke out in Kosovo, a republic in southern Serbia, in 1998. Matters intensified the following year when ethnic Albanians were subjected to a programme of ethnic cleansing by the Serbs. Whole villages were burned and the Albanians were driven out, a situation made worse by a NATO campaign of air strikes against Serbia (Figure 7.37). Although the vast majority of refugees were to return relatively quickly, repatriation did not bring an end to the suffering in Kosovo. Many Serbs, in turn, opted to flee to Serbia because of reprisals, both real and perceived.

Fig 7.37 Kosovan refugees are shown. The picture brings out the non-selective nature of enforced migration

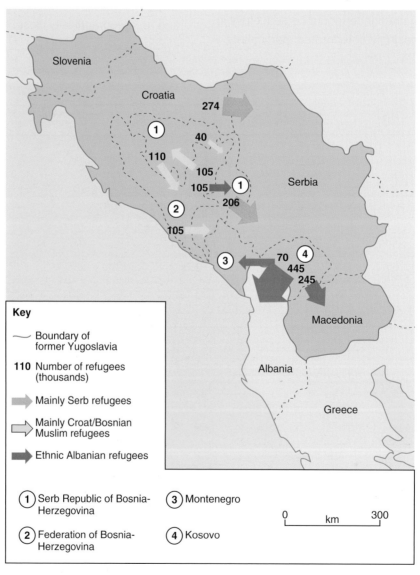

Key

— Boundary of former Yugoslavia

110 Number of refugees (thousands)

➡ Mainly Serb refugees

➡ Mainly Croat/Bosnian Muslim refugees

➡ Ethnic Albanian refugees

① Serb Republic of Bosnia-Herzegovina

② Federation of Bosnia-Herzegovina

③ Montenegro

④ Kosovo

0   km   300

Fig 7.38 Map of refugee movements in former Yugoslavia (1991–1999)

Too often, raw statistics convey little of the plight of refugees. Too often, as in the case of former Yugoslavia, it is easy to emphasise refugee issues close at hand and forget those further away. The following extract is a reminder of the suffering of real, live, innocent people.

Nothing prepares you for it. Nothing prepares you for the sight and smell of death. At Manigi refugee camp near Goma in eastern Zaire (now Republic of the Congo), the bus pulled up in the morning. The journey of less than 5 km to the camp took nearly two hours. On the way, the streets were lined with people – thousands and thousands of men and women, boys and girls and children. Many lay still and silent in the morning sun. Dead. Hundreds of them were wrapped in cheap, straw mats, tied at the bottom, and lying where they fell and died hours earlier.

These are the first days of a tragedy beyond belief, the result of a bloody civil war in Rwanda that spilt over into horrific tribal massacres . . . Now there are an estimated

1.2 million Hutus seeking refuge, mostly within a 32 km radius of Goma . . . As we proceeded, we passed more bodies lined along the side of the road . . . It reminded one of Auschwitz but in colour – the lovely rainbow clothes of the dead still draped on warm corpses . . . And so on to Hell. The crying, the hopelessness, the helplessness sink into your bones at Munigi. On the mudslopes and hills on either side of the road thousands sit, stare and weep amid the ashes and their belongings. In the centre where a few doctors distribute rehydration salts to only a fraction who need it there are at least 500 bodies on straw mats and rags of old dresses and shirts.

Twagiramungu, a moderate Hutu said 'This war is a landmark in the life of every Rwandan. We cannot have such another apocalyptic situation such as this one . . . If you look at what has happened here you really wonder if it was done by human beings. The savagery is beyond what the mind can imagine.'

Source: *Rwanda: Hell on Earth* by Michael Hand in
Scotland on Sunday 24/7/99

ASSIGNMENTS

Population Trends, Demographic Head Counts and Population Dynamics (fertility/mortality)
Read pages 187–200

1 Using Figure 7.1, briefly **describe** the main global population trends from 1800 to 2050.

2 What are the meanings of the following demographic terms?
(i) demography (ii) crude birth rate (iii) crude death rate (iv) infant mortality rate.

3 With the help of Figures 7.2 and 7.3, **outline** the different rates at which population is growing, bringing out the differences between Europe, sub-Saharan Africa, Asia and the USA.

4 **a)** Briefly **describe** three methods of collecting demographic data.
 b) With the aid of Figure 7.5, **explain** the value of accurate demographic data.
 c) Referring to named countries, **outline** the difficulties in obtaining accurate census data in the ELDCs (Economically Less Developed Countries).

5 **a)** Using Figure 7.9, give the four main **components of population change** in a country (or city, etc.).
 b) What is meant by the following demographic terms?
 (i) general fertility rate (ii) life expectancy (iii) natural increase (iv) natural decrease.
 c) What is the advantage of using **general fertility rate** instead of **crude birth rate** as a demographic measure when comparing the population growth potential of countries?

6 **a)** Using a spider diagram (or 'mind map' diagram), **summarise** the main factors influencing **fertility rates**. Use a large piece of paper to accommodate all the information.
 b) With the help of Figures 7.12 and 7.13, **explain** how improved education for women alters fertility levels.

7 **a)** Using a spider diagram (or a 'mind map' diagram), summarise the main factors influencing **mortality levels**. A large piece of paper is helpful.
 b) It is believed that mortality rates will continue to

decline. What obstacles might prevent this happening?

Population Structure and Demographic Transition
Read pages 200–211

1   Two model **age and sex 'pyramids'**, one typical of an ELDC, and one typical of an EMDC are shown below (Figure 7.39). Add the following labels to demonstrate the main differences between the diagrams:
very wide base; high fertility rates; narrow pyramid apex; narrow base; low life expectancy/high mortality rates; high longevity/low mortality rates; low/declining fertility rates; sharp upward decrease in age cohorts; age cohorts of almost even size; high degree of juvenility.

2   a) **Explain** the terms **dependent population** and **economically active population**.
   b) Write out the formula for measuring a country's **dependency ratio**.
   c) Using the data in Figure 7.3, complete Table 1. Comment on your results.
   d) What are the economic and social consequences arising for:
      (i)   a country with a population structure typical of an EMDC?
      (ii) a country with a population structure typical of an ELDC?
   e) Which do you think pose the greatest challenge to health and social service provision – an ageing population or a 'baby boom'? Give reasons for your answer.

3   a) **Explain**, in simple terms, what is shown by the **model of demographic transition**.
   b) Make a large copy of the model of demographic transition (Figure 7.18).
      (i)   Complete the first two lines using the words **high**, **low**, or **declining**
      (ii) Complete line three using **rapid increase**, **stable** or **declining**
      (iii) Complete line four by adding **Republic of Congo**, **Italy**, **Morocco**, **Scotland**

1950–present and **remote parts of New Guinea**.
Also add **Scotland before 1760, Scotland 2010, Scotland 1760–1870** and **Scotland 1870–1950**.

   c) **Describe and explain** the population changes between Stages 1 and 2, and between Stages 3 and 4 of the **model of demographic transition**.

4   In the period 1949 to the present, Communist China had undergone remarkable population changes in response to changing demographic policies.
   a) **Explain** the changing demographic policy terms **pro-natalist** and **anti-natalist**.
   b) Study Figure 7.21 and match the letters A–E with the following:
      Great Leap Forward resulted in a severe famine and natural decrease; high birth rate as a result of pro-natalist policies; dropping death rate because of improved health care rates; birthrate decline because of one-child-per-family policy; very high birthrates during the anarchy of the cultural revolution.
   c) Name the Chinese government policy that allowed China speedily to pass through Stage four of the demographic transition model. What methods have been used to implement this policy?
   d) How successful has the Chinese government been in implementing the **one-child policy**, and what have been its consequences?

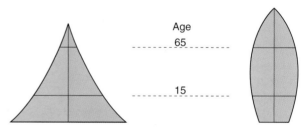

Fig 7.39 Two population pyramids

Table 1

Country	% of Population Aged		Dependency Ratio
	<15	>65	
United Kingdom			
Italy			
Japan			
Morocco			
Rep. of Congo			

5 Figure 7.18 shows five model age and sex pyramids which match each stage of the model of demographic transition.

   a) Briefly **describe and explain** the size and shape of each 'pyramid', mentioning how it links with the appropriate stage of the model.

   b) Look at Figure 7.22 and **explain** how it reflects the changing demographic policies of the Chinese Government.

Population Dynamics (Migration) and Geography of Uprooted People
Read pages 211–224

1 **a)** Briefly **explain** the difference between the terms **migration** and **circulation**.

   b) Study Figure 7.26 on the Motives for Migration. Try to group the various motives into four groups: political, economic, social and environmental in a copy of Table 2 below.

   c) Giving examples, **explain** the importance of **networks** in migration.

2 **a)** Make a large copy of Figure 7.27, **Types of Migration**. Match the following examples with the diagram: African slaves shipped to the West Indies; a British student undertaking VSO (Voluntary Service Overseas) in Malawi; ethnic Albanian refugees fleeing Kosoyo; transhumance from the Camargue to the French Alps; retiring from Edinburgh to Skye; a Turkish family moving to former West Germany in 1960 and permanently settling there; moving house from a small inner city flat to a large suburban bungalow; migrating on a long term contract (later extended) with a multinational company from Britain to Hong Kong; moving from the Scottish Borders to London to teach.

   b) From the five extracts illustrating Scottish migration, pick out phrases which show **voluntary** and **forced** migration; **permanent** and **temporary** migration; and the **selective** nature of migration.

3 Since 1949 China's population has been on the move, demonstrating different types of migration flow.

   a) With the help of Figures 7.29 and 7.30, give examples of different types of migration flow from 1949 to 1978.

   b) With reference to Figures 7.32 and 7.33, **describe and explain** the pattern of **migration flow** from Sichuan Province (1985–90).

   c) What are the **consequences** of such rural–urban migration for:
      (i) the **areas of origin**, and
      (ii) the **areas of destination**?

4 The countries of the Mediterranean have long been a source of migrants.

   a) **Outline** briefly the **migration flows** from Mediterranean countries to north-west Europe during the period from the 1950s to 1970s.

   b) Why did some migrant workers **either** choose to remain (e.g. in Germany) **or** return to their country of origin (e.g. Turkey).

5 **a)** Increasing numbers of Moroccans have been migrating to Spain. Make a large copy of the diagram on page 227, drawing simple outlines of Morocco and Spain in the country 'box'. Complete the data sections using Figure 7.3 to assist you.

   b) Name the jobs available to Moroccans. Suggest likely **'push', 'pull'** and **'network'** factors likely to encourage a move to Spain.

   c) **Outline** the consequences of such a migration flow for a rural area in Morocco and a Spanish city such as Madrid.

   d) Using the data provided, draw:
      (i) a statistical diagram to show the origin of immigrants to Madrid
      (ii) an age and sex pyramid for Moroccan immigrants in Madrid.
      Comment on what each of the diagrams shows.

6 **a)** Figure 7.25 shows Migration in the 1990s. Identify the areas with large-scale refugee problems.

   b) What is meant by a **refugee**? Suggest why it is difficult to provide an accurage global number for refugees.

   c) In what ways do refugees differ from other types of migrants and what problems can result from a large influx of refugees?

   d) Figure 7.36 shows the flow of refugees from the former Yugoslavia. **Describe and explain** the flow of refugees.

Table 2

	Political	Economic	Social	Environmental
'Push'				
'Pull'				

MODEL OF LABOUR MIGRANT FLOW

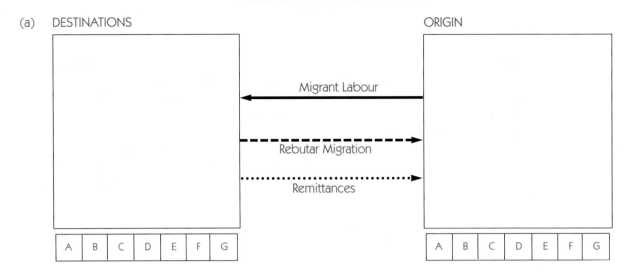

(a)   DESTINATIONS   ORIGIN

Migrant Labour

Rebutar Migration

Remittances

A   B   C   D   E   F   G   A   B   C   D   E   F   G

A: Population (millions)   B: CB Rate   C: CD Rate   D: National Increase (annual %)
E: Fertility Rate   F: Doubling Time of Population   G: GNP per capita ($)

(d)

Origin of Immigrants to Madrid

COUNTRIES	% OF TOTAL IMMIGRANTS
USA	13.63
FRANCE	8.07
ARGENTINA	7.79
GERMANY	7.49
PORTUGAL	6.05
UK	5.80
CUBA	5.56
MOROCCO	5.28
ITALY	4.55
OTHERS	35.78

Age & Sex data for Moroccans in Madrid

Age	Male %	Female %
65+	1.6	2.1
60–64	1.0	1.1
54–59	1.1	1.5
50–54	3.0	1.5
45–49	3.3	1.9
40–44	4.8	2.9
35–39	7.1	4.6
30–34	7.6	5.2
25–29	5.1	6.4
20–24	5.1	4.5
15–19	4.1	3.7
10–14	4.1	4.7
5–9	3.4	3.4
0–4	3.2	2.0

Extra Assignments

1 With the help of Figures 6.1 and 6.2, **describe** the broad patterns of peopling and population growth from the earliest of times.

2 With reference to Figure 7.11, **examine** the relationship between fertility decline and economic development.

3 Does the model of demographic transition need a fifth stage? Justify your answer.

4 Who was Thomas Malthus? Outline his demographic ideas, and assess their relevance today. Are his ideas more pessimistic than those of Boserup concerning food supply and population?

5 The reasons for migration are very complex. Discuss this statement with regard to Lee's model of migration applied to China (Figure 7.31) and/or another country of your choice.

Key terms and concepts

- age specific death rate (p. 198)
- anti-natalist (p. 196)
- census (p. 190)
- circulation (p. 211)
- crude birth rate (p. 195)
- crude death rate (p. 198)
- demographic transition (p. 203)
- demography (p. 188)
- dependency ratio (p. 202)
- economically active (p. 202)
- fertility (p. 195)
- forced migrations (p. 213)
- general fertility rate (p. 195)
- infant mortality rate (p. 198)
- life expectancy (p. 198)
- migration (p. 211)
- mobility(p. 211)
- mortality (p. 198)
- natural decrease (p. 194)
- natural increase (p. 194)
- old dependents (p. 202)
- one-child policy (p. 208)
- population momentum (p. 189)
- positive checks (p. 210)
- preventative checks (p. 210)
- pro-natalist (p. 196)
- refugees (p. 220)
- registration (p. 191)
- voluntary migrations (p. 213)
- young dependents (p. 202)

Suggested reading

Kenefick, B (1998) **Demography** in Cooke et al *Modern Scottish History* Vol 2, Tuckwell Press.

Broadley, E and Cunningham, R (1991) *Core Themes in Geography: Human* Oliver & Boyd.

Dwyer, D (1994) *China: The Next Decades* Longman.

Champion, T (1998) **Demography** in Unwin, T *A European Geography* Longman.

Jackson, S (1998) *Britain's Population* Routledge.

Geddes, J and Muir, K (1987) *Aspects of Social Geography* Edward Arnold.

Gray, M (1990) *Scots on the Move* Economic & Social History Society of Scotland.

Hornby, WF and Jones, M (1993) *An Introduction to Population Geography* Cambridge University Press.

Guiness, P and Nagle, G *Advanced Geography: Concepts and Cases* Hodder & Stoughton.

Money, D (1996) *China in Change* Hodder & Stoughton.

Ogden, P (1984) *Migration and Geographical Change* Cambridge University Press.

Population Reference Bureau – Various reports and data sheets (issued annually) distributed in the UK by Population Concern, Studio 325, Highgate Studios, 53–79 Highgate Road, London NW5 1TL.

Randel, J et al (1999) *The Ageing and Development Report* Earthscan and Help Age International.

Robertson, AJ (1999) *The Population of Jordan: Census 1994* SAGT Journal No 28.

Tranter, N (1998) **Demography** in Cooke, A *Modern Scottish History* Vol 1, Tuckwell Press.

Waugh, D (1994) *Geography: An Integrated Approach* Nelson.

Internet sources

Cairo Conference: www.mbnet.mb.ca/linkages/cairo.html

Population Reference Bureau: www.prb.org

International Organisation for Migration: www.iom.ch/

Japan Information Network: http://jin.cic.or.jp/stat/

CIA World Factbook: www.portal.research.bell-labs.com/cgiwald/dbaccess/411

Sources of European Population Statistics: www.nidi.nl/links/nidi6js.html

8 RURAL GEOGRAPHY

After working through this chapter, you should be aware that agriculture:

◆ is part of a complex system

◆ operates at different levels of intensity in different parts of the world, and for different purposes

◆ is affected by the interaction of the hydrosphere, the atmosphere and the biosphere, as well as by human and economic factors

◆ has changed, and is changing, as a result of developing technologies, in part related to the Green Revolution of the latter part of the twentieth century

◆ has, according to the type, a distinctive human landscape with its own pattern of settlement and communications

◆ has, again according to type, a characteristic population density, structure and distribution

You should also be able to analyse:

◆ land use data and crop yields in map and diagrammatic form

◆ the results of farm surveys

◆ and annotate field sketches and photographs of rural landscapes.

This section will be studied by considering three different types of agriculture in widely contrasting environments:

1) extensive arable farming – grain growing, on the Great Plains of the USA

2) shifting cultivation in the Amazon Basin

3) intensive peasant farming in Malaysia (see Figure 8.1).

Although the world's population is becoming increasingly urbanised (now 50% of the total population), rural areas are still of major importance because:

◆ in some parts of the world, Monsoon Asia for example, they are very densely populated, and food may be sometimes in short supply

◆ in other parts of the world which are thinly populated, for example the Great Plains, a very large proportion of the world's food supply is produced.

There are many different types of rural landscape, depending on the population density, the history of settlement and the level of technology.

8.1 Types of Agriculture: Some Definitions

A simple classification is based on whether the outputs are crops (**arable farming**), livestock (**pastoral farming**), or both (**mixed farming**). Arable farming is found in the most favoured areas (Figure 8.2) in terms of relief, soil and climate, while livestock farming is usually a response to difficult conditions (Figure 8.3) in terms of height, slope and rainfall (either very high or very low).

Overlapping classifications take into account the purpose of the farming carried on, its spatial distribution and its scale of operations, that is:

◆ whether it is for **subsistence** (to feed the local people without having a surplus), or for **commercial** purposes – to produce a small surplus for sale at the local market, or a larger surplus, often for distant markets. Subsistence farmers may lack capital and modern technology, but their methods often involve a high degree of skill and ingenuity. Commercial farmers make much use of all kinds of technology (mechanical, chemical and biological), need much capital and are dependent on efficient transport links

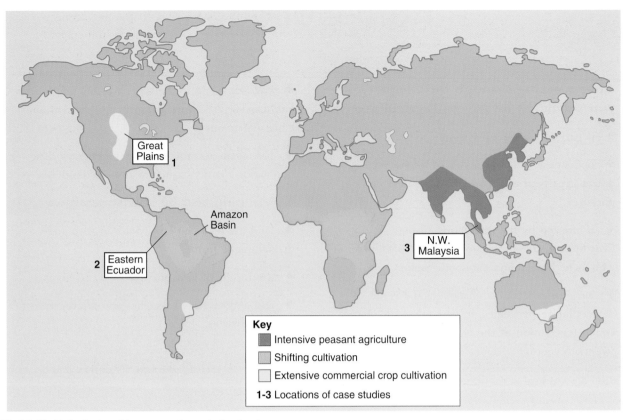

Fig 8.1 World distribution of three contrasting agricultural systems

Key

Intensive peasant agriculture

Shifting cultivation

Extensive commercial crop cultivation

1-3 Locations of case studies

◆ whether it is **shifting** (usually subsistence farming) or **sedentary**. Shifting cultivation can only take place where the population density is low, and there is thus limited demand for food. There must be plenty of land available, although of limited fertility, to allow the people to move both their cultivated plots and settlements as necessary. Most types of farming in the world are sedentary, that is, with both the farmland and permanent settlement in fixed locations, but there may also be seasonal settlements if **transhumance** takes place (Figure 8.4)

◆ whether it is **intensive** or **extensive**. Intensive arable farming is usually on a small scale. A limited area of land is farmed, involving a high input of

Fig 8.2 The landscape of arable farming in East Lothian, one of the most productive farming areas in the UK

Fig 8.3 Posso Farm, a 1600 ha hill-sheep farm in the Manor Valley

Fig 8.4 Transhumance: a flock of sheep being moved to summer pasture in the Vercors, a limestone region in the French Alps

labour and skill, but often with a low input of capital. In contrast, much capital, and high inputs of all the agricultural technologies are needed to farm very large grain farms extensively. In livestock farming, the fattening of cattle on feedlots and battery poultry units are examples of intensive farming, whereas a hill sheep farm or a cattle ranch exemplify extensive farming.

Today, a growing number of farms, usually intensive, commercial, and often mixed, are also classified as **organic** farms if they have ceased making use of chemical technology – fertilisers, pesticides and herbicides.

8.2 Farming as a System

All types of farming can be seen as systems, with inputs and outputs. The location of different types of agriculture at all scales (local, regional, national, continental, global), is the result of the interaction between:

◆ the different components of the physical environment – the biosphere, the lithosphere, the atmosphere and the hydrosphere

◆ the physical environment, economic/political factors and cultural (human) factors

Figure 8.5 suggests that commercial farming is a very complex system in which the physical environment

plays a significant part. However, as a business, commercial farming is very much influenced by external factors, including changing technologies, and to a large degree is controlled by external financial and political factors. Farmers use their experience and understanding of all these factors in coming to a decision about their plans for the use of the farmland. Where neighbouring farmers reach similar decisions, a particular pattern of land use emerges in that area, although there are always exceptions to any pattern. The relative importance of the different types of inputs varies according to the level of development. Such variations in inputs are responsible for the different types and patterns of farming throughout the world. A systems model can be applied, at different levels of complexity, to all types of farming, no matter the scale or location.

8.3 Extensive Commercial Farming

Extensive farming can be defined as a system of farming which is carried out on very large holdings, with a very high reliance on technologies, and where the relatively low yields are compensated for by the very large areas under cultivation. The human input is minimal, although decisions taken by the farmer, or the corporation, are of great importance. Extensive farming is often found in areas where land is relatively cheap, population density is low, and the climate is marginal. It could be argued that the growing of cereals in South East England is now increasingly carried on in an extensive way on farms of 1000 ha or more, where hedges have been removed to create fields of 40 ha growing heavily-subsidised crops of cereals and oil-seed rape. However, the land is costly and the population density is relatively high, so this system is best examined through a study of extensive farming on the Great Plains of North America.

Farming on the Great Plains

Figure 8.6 shows the Great Plains, which are drained by the right-bank tributaries of the Mississippi between the Rocky Mountains in the west and the Central Lowlands in the east. In latitude, this area

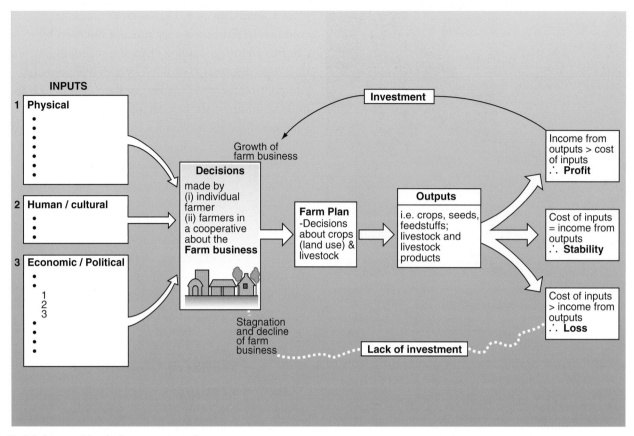

Fig 8.5 Commercial agriculture – a systems diagram

extends from northern Mexico to the three Prairie Provinces of Canada and, as it does so, the growing season/frost-free period diminishes to less than 100 days. It rises westwards in two steps, separated by an escarpment, and its undulating surface is occasionally broken by residual hills of crystalline rock, including the Black Hills of Dakota and, in the west, outliers of the Rocky Mountains. It is an area of extreme climate, marked especially by harsh winters and low, erratic rainfall. Longitude 100° W, which more or less coincides with the 500 mm isohyet, is sometimes taken as the dividing line between the lower plains where rainfall is more reliable, and the semi-arid area to the west. It also marks the divide between the eastern tall-grass prairie and the short-grass prairie to the west (see Figure 8.7). This prairie grassland was the vegetation encountered by the first ranchers and homesteaders. Woodland is confined to river banks (mainly oaks and cottonwood) and hill slopes.

Figure 8.8 shows the part of the Great Plains within North Dakota and eastern Montana, including the location of Williston, climate details of which are shown on Figure 8.9. The human landscape and land use have changed rapidly in the last 150 years, from a hunting and gathering economy with semi-nomadic Indian tribes to today's extensive commercial farming. The pacification of the Sioux Indians and the westward extension of the railways formed the prelude to the large-scale settlement of the Great Plains, in this case by farmers from the Mississippi–Ohio Lowlands and the eastern states, as well as by European immigrants.

My father and two uncles came from Norway, from mountains and fiords, to make their living in this flat, empty land, the Red River Valley and the bordering drift prairie. West of the big, rich valley, the flat bottom left by the glacial Lake Agassiz, we entered the rolling plains of the drift prairie, broken randomly by ravines and sandy ridges, erratic signatures of the advancing, then retreating ice-sheets.

Source: *Listening to the Land: Star Tribune*, Minneapolis 19/12/93

The geometric appearance of the present-day landscape (Figure 8.10) was created by surveyors before the land was settled.

The new arrivals found themselves in country that defeated the best efforts of the eye to get it in sharp focus.

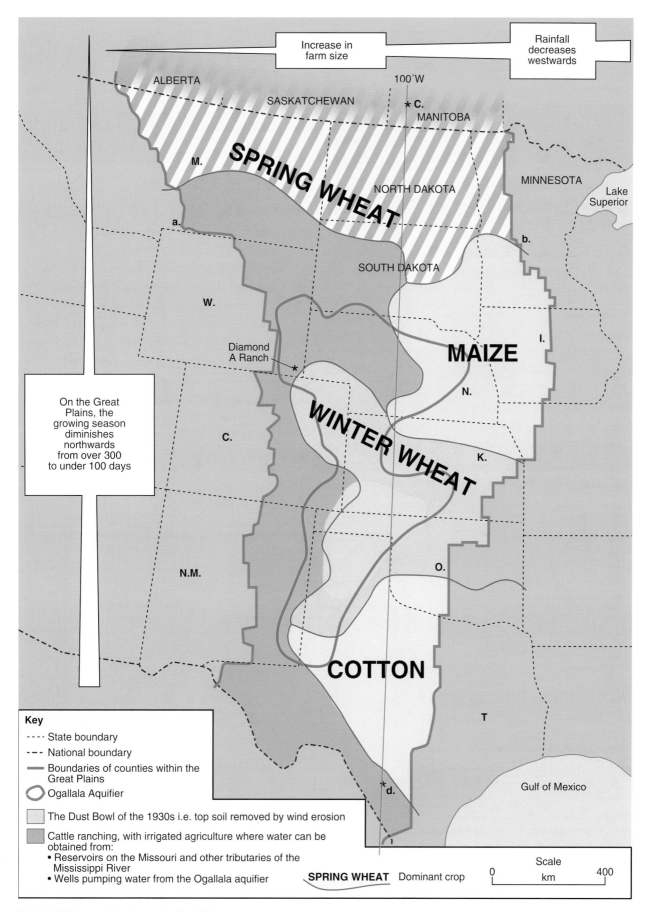

Increase in farm size

Rainfall decreases westwards

100°W

ALBERTA

SASKATCHEWAN

★ C.

MANITOBA

M.

SPRING WHEAT

NORTH DAKOTA

MINNESOTA

Lake Superior

a.

SOUTH DAKOTA

b.

W.

MAIZE

I.

Diamond A Ranch

★

N.

On the Great Plains, the growing season diminishes northwards from over 300 to under 100 days

WINTER WHEAT

C.

K.

N.M.

O.

COTTON

T

Gulf of Mexico

Key

---- State boundary

-·-· National boundary

—— Boundaries of counties within the Great Plains

⬭ Ogallala Aquifer

☐ The Dust Bowl of the 1930s i.e. top soil removed by wind erosion

☐ Cattle ranching, with irrigated agriculture where water can be obtained from:
• Reservoirs on the Missouri and other tributaries of the Mississippi River
• Wells pumping water from the Ogallala aquifer

SPRING WHEAT Dominant crop

★ d.

Scale

0   km   400

Fig 8.6 Main types of farming on the Great Plains

Fig 8.7 Short grass prairie used for extensive cattle grazing in eastern Colorado

It went on interminably in every direction ... It was not quite raw land, but nor was it a landscape. The northern plains had long been grooved by dainty-footed buffalo, then lightly patterned by winding Indian trails. Ranchers, driving cattle from Texas to Montana, left ribbons of trampled ground as broad as superhighways. The army, under generals like Custer, Miles and Terry, built compass-straight military roads that marched up hill and down dale, disdainful of contours. The railroad companies ran tracks along the creek and river bottoms. Yet all these routes added up to no more than a few hairline scratches on the prairie. You would need to know what to look for in order to notice the really important landscaping work ... survey teams from the federal office turning it into a grid ... of townships, each sub-divided into 36 sections, with each section pegged out into quarters.

Source: Jonathan Raban: *Bad Land*. Picador. 1996

Figure 8.11A shows how this was done in the Dakota Territory before the influx of immigrants just over 120 years ago. Although the Great Plains had been categorised a desert by the early explorers and then taken from the Indians to create vast open-range cattle ranches, the later nineteenth century settlers believed that the climate had changed for the better, and that ploughing the Plains would literally increase the rainfall to the amount required to grow wheat. A major incentive was that settlers were given 64 ha of free land, providing they built a house on their quarter-section and cultivated their holding. The land at this time was much more densely populated than it had ever been before, or has been since, and the farming was intensive rather than extensive.

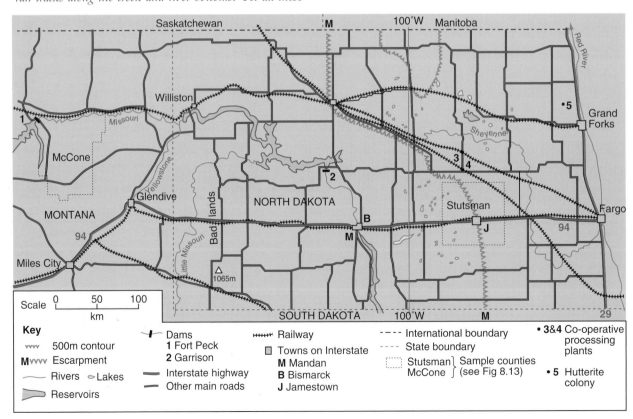

Fig 8.8 Landscape of extensive commercial farming: Northern Great Plains, USA

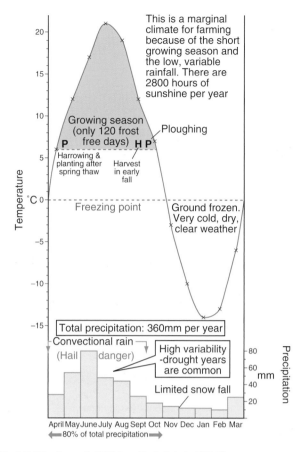

This is a marginal climate for farming because of the short growing season and the low, variable rainfall. There are 2800 hours of sunshine per year

Growing season (only 120 frost free days)

Ploughing

P H P

Harrowing & planting after spring thaw

Harvest in early fall

Freezing point

Ground frozen. Very cold, dry, clear weather

Total precipitation: 360mm per year

Convectional rain

(Hail danger)

High variability -drought years are common

Limited snow fall

April May June July Aug Sept Oct Nov Dec Jan Feb Mar

◄─80% of total precipitation─►

Fig 8.9 Climate graph: Williston, North Dakota (48° N)

Settlers drained the aquifers as if they would magically replenish. They tore up the prairie which had taken thousands of years to develop, and wondered why it blew away in the dry years. Unlike communities back east where settlement rested on a three-legged stool of resources – water, timber and land – on the Plains there was only the land.

Source: *Star Tribune* 19/12/93

Early twentieth century

Many of the first quarter-section homesteads did not survive because of recurring, prolonged droughts. Surviving farmers were allowed to increase their holdings to a full section (256 ha). Improved agricultural technologies allowed them to cope, initially, with the cycles of drought years and wet years which followed. These innovations included:

◆ the new steel plough, disc harrow, reaper, binder, etc.

◆ pumps and windmills

◆ barbed wire for fences

◆ new strains of fast-growing spring wheat, imported from the Steppes of Russia.

In some of the more arid areas, south-east Wyoming for instance, many sections were bought up by more prosperous individuals or companies, and reformed as extensive cattle ranches – the Diamond A Ranch once consisted of 130 homesteads, which now form one holding of 84 sq km. Cereal farming in the more humid areas was also affected by external economic pressures and changing demand for their products. A major problem in the southern part of the Great Plains in the 1920s and 1930s was accelerated erosion of the soil by wind, which culminated in the Dust Bowl (see Figure 8.6). Rural depopulation then increased as a result, and farms became larger, that is, more extensive. Irrigation of the semi-arid western areas of the Great Plains began. Where this was not possible, dry farming was used – land was left fallow and wheat grown only every second year, which is still done today on the Waller Farm in the semi-arid McCone County, Montana (see Figure 8.13).

Late twentieth century

As the rate of rural depopulation continued to increase, farms became extremely large (Figure 8.11B); the dependence on agricultural technologies expanded, and output soared as new strains of wheat were developed, and more land was brought into production. Large surpluses of cereals resulted in federal government payments to farmers to cut back on production. Irrigated farming expanded very rapidly, resulting in the equally rapid depletion of the Ogallala Aquifer (see Figure 8.6). Elsewhere on the

Fig 8.10 The geometric landscape of spring wheat cultivation in North Dakota. Note the dense shelter belt protecting the farmhouse

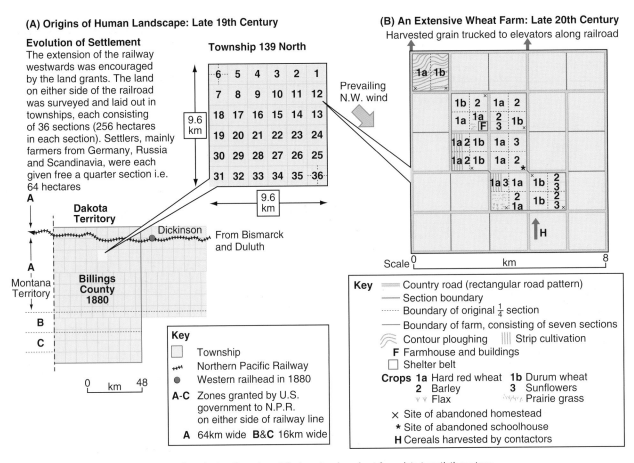

(A) Origins of Human Landscape: Late 19th Century

Evolution of Settlement

The extension of the railway westwards was encouraged by the land grants. The land on either side of the railroad was surveyed and laid out in townships, each consisting of 36 sections (256 hectares in each section). Settlers, mainly farmers from Germany, Russia and Scandinavia, were each given free a quarter section i.e. 64 hectares

Township 139 North

6	5	4	3	2	1
7	8	9	10	11	12
18	17	16	15	14	13
19	20	21	22	23	24
30	29	28	27	26	25
31	32	33	34	35	36

9.6 km

9.6 km

Prevailing N.W. wind

Dakota Territory

Dickinson

From Bismarck and Duluth

Billings County 1880

Montana Territory

A

A

B

C

Key
- ☐ Township
- ⊶ Northern Pacific Railway
- ● Western railhead in 1880
- **A-C** Zones granted by U.S. government to N.P.R. on either side of railway line
- **A** 64km wide **B&C** 16km wide

0 km 48

(B) An Extensive Wheat Farm: Late 20th Century

Harvested grain trucked to elevators along railroad

1a	1b				
	1b 2	1a 2		
	1a 1a F	2 3 1b	
	1a 2 1b	1a 3	
	1a 2 1b	1a 2	
		1a 3 1a	1b 2 3
		2 1a	1b 2 3

H

Scale 0 km 8

Key
- ══ Country road (rectangular road pattern)
- — Section boundary
- ···· Boundary of original $\frac{1}{4}$ section
- — Boundary of farm, consisting of seven sections
- ≋ Contour ploughing ‖‖ Strip cultivation
- **F** Farmhouse and buildings
- ☐ Shelter belt
- **Crops 1a** Hard red wheat **1b** Durum wheat
  2 Barley **3** Sunflowers
  Flax Prairie grass
- × Site of abandoned homestead
- ∗ Site of abandoned schoolhouse
- **H** Cereals harvested by contactors

Fig 8.11 (A) Origins of human landscape: late nineteenth century. **(B)** An extensive wheat farm: late twentieth century

Great Plains, farms continued to have to cope with recurring drought years. Where accelerated soil erosion was a serious problem, it was tackled by

◆ contour ploughing

◆ strip cultivation (see Figure 8.12)

◆ shelter-belt planting.

Farm businesses, whether family- or corporate-owned, became increasingly dependent on agricultural contractors, especially at harvest time (see Figure 8.14) and on fossil fuels.

Proponents of modern agriculture often brag that one farmer in the USA feeds 116 people. It would be more accurate to say that one farmer, backed up by huge amounts of fossil fuel, feeds 116 people, for farming is the nation's largest consumer of petroleum products. It is dependent on oil for an amazing variety and quantity of farm inputs, from pesticides to fertilisers to diesel fuel used by tractors, combines and irrigation pumps ... All across the country, millions of Farmer Browns have been replaced by millions of barrels of crude ... The problem is

that while Farmer Brown is a renewable resource, oil is not, and a system of farming dependent on the latter is not sustainable ...

Source: O.G. Davidson: *Broken Heartland.* Un. of Iowa Press 1996

Fig 8.12 Strip cultivation: broad, high rows of sunflowers, aligned north–south, give shelter to the intervening strips of spring wheat, as well as improving the soil and diversifying farm output

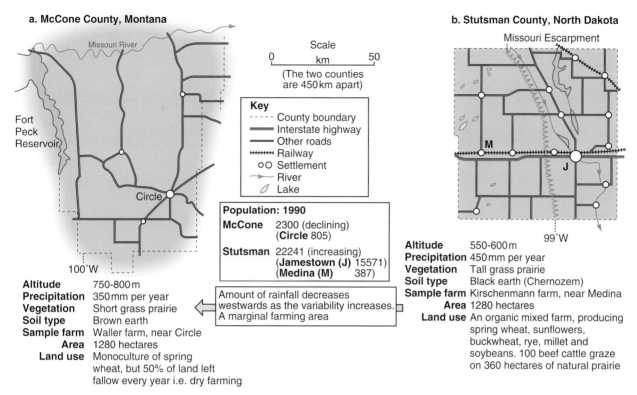

a. McCone County, Montana

Missouri River

Fort Peck Reservoir

Circle

100°W

b. Stutsman County, North Dakota

Missouri Escarpment

M

J

99°W

Scale

0   km   50

(The two counties are 450 km apart)

Key
- - - - - County boundary
━━━━ Interstate highway
──── Other roads
┼┼┼┼┼ Railway
oo Settlement
→ River
◿ Lake

Population: 1990

McCone   2300 (declining)
     (**Circle** 805)

Stutsman   22241 (increasing)
     (**Jamestown (J)** 15571)
     (**Medina (M)**   387)

Altitude	750–800 m
Precipitation	350 mm per year
Vegetation	Short grass prairie
Soil type	Brown earth
Sample farm	Waller farm, near Circle
Area	1280 hectares
Land use	Monoculture of spring wheat, but 50% of land left fallow every year i.e. dry farming

Amount of rainfall decreases westwards as the variability increases. A marginal farming area

Altitude	550–600 m
Precipitation	450 mm per year
Vegetation	Tall grass prairie
Soil type	Black earth (Chernozem)
Sample farm	Kirschenmann farm, near Medina
Area	1280 hectares
Land use	An organic mixed farm, producing spring wheat, sunflowers, buckwheat, rye, millet and soybeans. 100 beef cattle graze on 360 hectares of natural prairie

Fig 8.13 Two counties on the Great Plains, an area of extensive commercial farming

Extensive Farming in North Dakota and Montana Today (see Figures 8.8, 8.13)

As the smaller family farms are sold off, larger farming businesses continue to expand and prosper, but the farming landscape as a whole is changing:

◆ new crops have been introduced, e.g. sunflowers

◆ some land has been taken out of wheat production. For example, a Hutterite colony producers potatoes, eggs and rears pigs on a 1600 ha communal farm, while elsewhere a 640 ha holding was reseeded with native grasses on which 500 bison now graze

◆ part-time farming and the dependence on farm cooperatives have increased. A new cooperative bison processing plant and the Pasta Cooperative plant at Carrington, processing durum wheat, have increased farm incomes significantly

◆ some farms have become organic, while still operating on a large scale, for example, the Kirschenmann Farm (see Figure 8.13)

◆ smaller cattle ranches have doubled their carrying capacity by careful management of their prairie pasture: by fencing it into paddocks and rotating

their use for grazing and mowing. The larger ranches in the semi-arid west are now more dependent on irrigated fodder crops and feed-lots. Some now grow spring wheat, using dry farming methods (e.g. 130 ha on the Diamond A Ranch, Wyoming)

◆ the rural population continues to decline, and this is marked by the presence in the landscape of abandoned homesteads and schools, and of rapidly-declining small towns such as Circle, McCone

Fig 8.14 This vast area of spring wheat is harvested using teams of combine harvesters. The grain is then stored in elevators along the railways.

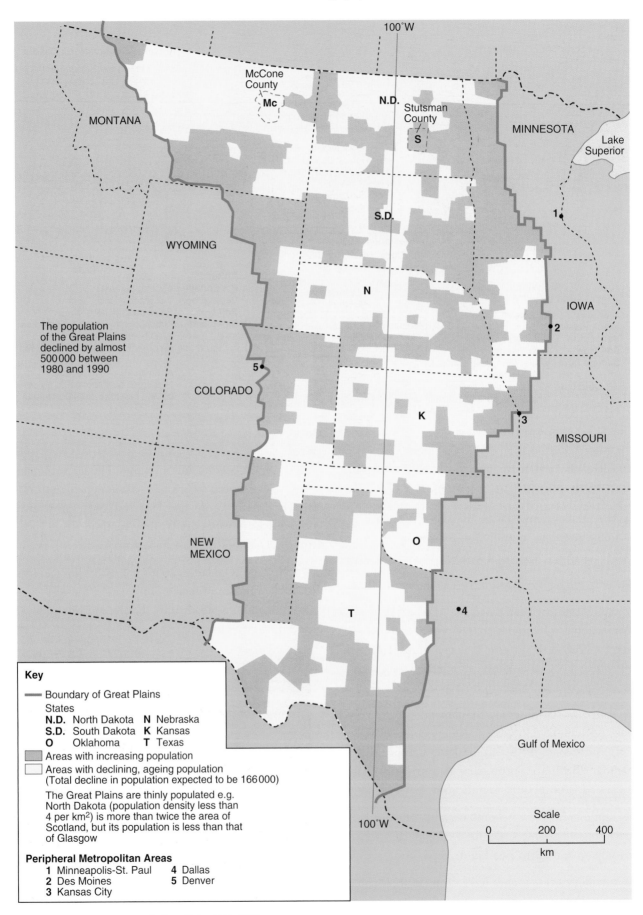

Fig 8.16 Population change in the Great Plains 1990–2010

Fig 8.15 Population is diminishing and services are declining in many small towns in the spring wheat belt. Note the typical gridiron street plan

County and Montana, with diminishing services (see Figure 8.15).

North Dakota is a large, thinly populated area with a population density similar to that of Skye & Lochalsh, that is, less than 4 people per sq km. This sparse, declining rural population is typical of the extensive farming areas on the Great Plains (see Figure 8.16). The migrants are attracted to the peripheral metropolitan areas like Des Moines, or to smaller growth points within the Plains. In North Dakota, in counties such as Stutsman, small towns which have growing populations are located

- near major growth points (e.g. Fargo)
- on the Interstate Highways and railways
- near new irrigation projects
- near areas with strip mining of coal, and gas wells
- within or near Indian reservations

Despite the marked decline in rural population, North Dakota's extensive farming is still a major source of the USA's wheat supply. The summary diagram, Figure 8.17, shows how suitable the area is for extensive farming, but also some of the consequences.

Extensive Subsistence Agriculture: Shifting Cultivation in the Amazon Basin

The Landscape of Shifting Cultivation: Pestaza, Ecuador

As our plane gained altitude, detaching itself from the bluish foothills (of the Andes) and striking towards the blinding morning sun, the orderly arrangement of plantations gave way to a scattering of cleared areas. Here and there the tin roof of a colonist made a splash of brightness. Soon the clearings became increasingly rare and the last traces of the pioneer frontier were eventually swallowed up in a sea of little green hills gently undulating away to an indistinct horizon. Beneath our wings, the forest looked like a huge carpet of broccoli, interspersed with the paler plumes of palm trees.

After just a few minutes airborne, we had left behind a landscape that spoke of human activity, though only skimpily indicated, to plunge into an anonymous, infinitely repetitive world with no reference points at all. There were no holes, no rent in this mantle of greenery, which was occasionally embroidered with a silvery thread by the reflection of the sun on a tiny twisting rivulet. No sign of life on the (riverine) beaches, no solitary wisp of smoke; nothing betrayed any human presence beneath that monotonous canopy . . .

Source: P. Descola: *The Spears of Twilight*

Shifting Cultivation

Once a form of livelihood in the temperate woodlands of Europe at a time when the population density was very much lower than today – from Neolithic to Medieval times – shifting cultivation is now confined to the humid tropics of South America, Africa and South East Asia (see Figure 8.1). However, it has been estimated that it still supports as many as 300 million people. Where the settlement is permanent, perhaps because of higher population density, as in West Africa, and where only the clearings are 'shifted', the system is more properly called **bush fallowing**. In most cases, the settlement is usually small and temporary, its lifespan being

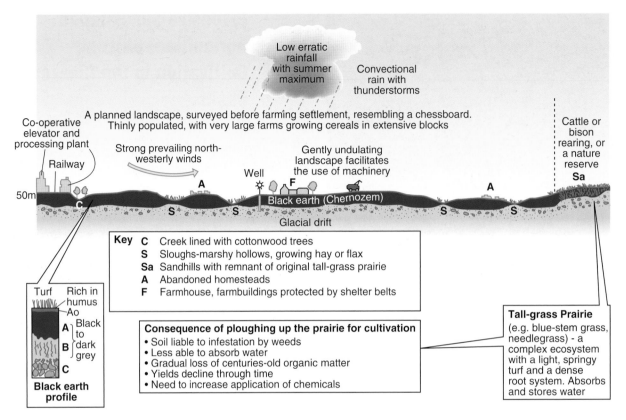

Fig 8.17 Extensive farming on the Great Plains: spring wheat production in North Dakota

determined by the rate of decline in soil fertility and the productivity of the clearings around it (see Figure 8.24 & Figure 8.25). Although there are changes taking place in the more accessible areas of the Amazon Basin, traditional shifting cultivation is still found in the more remote, and more thinly populated parts, for example, part of Pastaza Province in eastern Ecuador described above and shown in Figure 8.18. Look carefully at the large scale map in this figure. The outline of Stutsman County, North Dakota, (see Figure 8.13) has been superimposed on this map. Whereas the American county has a population of over 22 000, there are only a few hundred people in the comparable area (approximately 6000 km²) of the rain forest. This very low density is due to

◆ its isolation, poor communications and lack of economic development

◆ the inability of shifting cultivation to support a larger population, rather than any negative aspects of the ecosystem. Smallpox has been eradicated, although malaria is endemic. Measles, whooping cough and chicken pox occur, but influenza is the

major hazard. However, disease is not the major factor in limiting the population.

The larger scale map in Figure 8.18 shows an area inhabited by the Achuar people, a tribe of the Jivaro Indians, whose territory straddles the disputed frontier region between Ecuador and Peru. Almost entirely unsurveyed and undeveloped, very unlike North Dakota, this dense rain forest is without any roads (see Figure 8.19). The map shows a parallel drainage pattern which provided navigable north-west to south-east waterways, safest to use from September to January. The sketch shows a characteristic river bank site, while the map shows the **dispersed** pattern of settlement (which may range in size from 1–12 houses) along the river courses in what is a very empty landscape (see Figure 8.20):

Marginal people . . . usually settle on the periphery of a region occupied by a network of relatives . . . Several hours' or even a day's journey separates them from the main kernel of the habitat, composed of about six or seven households scattered within a radius of about a dozen kilometres, along a section of the same river or on the

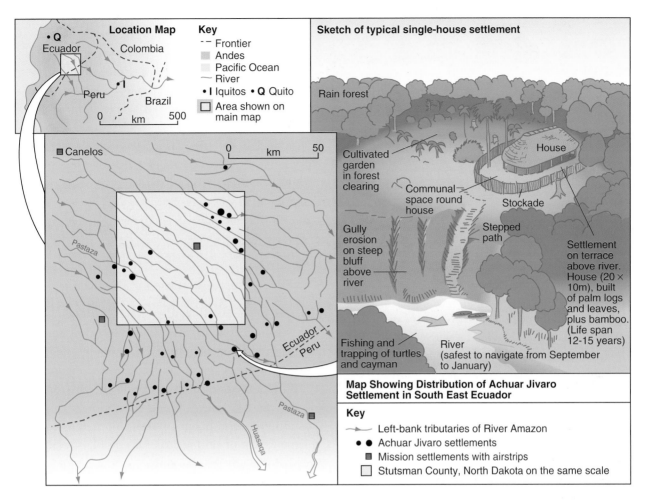

Location Map

Key
- -- Frontier
- ▪ Andes
- ▫ Pacific Ocean
- — River
- •I Iquitos •Q Quito
- ▫ Area shown on main map

Q Ecuador • Colombia

Peru •I Brazil

0 — km — 500

▪ Canelos

0 — km — 50

Pastaza

Ecuador / Peru

Pastaza

Huasaqa

Sketch of typical single-house settlement

Rain forest

Cultivated garden in forest clearing

Communal space round house

Stockade

House

Stepped path

Gully erosion on steep bluff above river

Settlement on terrace above river. House (20 × 10m), built of palm logs and leaves, plus bamboo. (Life span 12-15 years)

Fishing and trapping of turtles and cayman

River (safest to navigate from September to January)

Map Showing Distribution of Achuar Jivaro Settlement in South East Ecuador

Key
- → Left-bank tributaries of River Amazon
- •● Achuar Jivaro settlements
- ▪ Mission settlements with airstrips
- ▫ Stutsman County, North Dakota on the same scale

Fig 8.18 Landscape of shifting cultivation, Upper Amazon Basin

banks of its smaller tributaries. The inhabitants of such a neighbourhood are relatives who intermarry ... and define themselves collectively by reference to the principal water course that provides for them and determines their travels ... The social space of the Achuar can therefore be seen as a network of houses/territories, quite dense at the heart of what might be called a 'neighbourhood', but looser on its outskirts, which eventually dissolve into vast, uninhabited expanses, that separate them from other social entities, identical but potentially hostile.

Source: P. Descola: *The Spears of Twilight*

The most effective way of communication is by light aircraft, owned by government departments and missionaries, but on rare occasions used by Achuar men to travel to a larger settlement to barter head-dresses and blowpipes, for example, for a new axe, machete or pots.

The Farming Process

There is plenty of land for cultivation: it is an extensive system in which labour and output are

Fig 8.19 The dense canopy of the selvas in the Upper Amazon Basin

Fig 8.20 Isolated selva settlement in the Amazon Basin. Note the building materials used and the pitch of the roof

varieties of which constitute the major part of daily nourishment, is a lazy gardener's dream. Each cutting produces from 2 to 5 kilos of roots, unearthed by a shove from a machete. Once these roots are collected, a couple of slashes suffice to reshape the stem into a little rod which, stuck carelessly into the earth, is soon covered with leaves and produces a new set of roots within a space of a few months. Like the yam, this accommodating plant can be left in the earth well beyond the point when it reaches maturation without danger of spoiling. Stockpiling of food supplies is therefore unnecessary, without the dead seasons faced by the cultivators of cereals; here the garden remains a reserve for flourishing, starch-producing crops that will provide for one's needs throughout the year . . .

Source: P. Descola: *The Spears of Twilight*

relatively small in relation to the total area of land available. Cultivation is aided by the year-long growing season, characteristic of the equatorial climate (see Figure 8.21). The description which follows emphasises this, as well as some of the skills involved.

. . . it is true that tropical agriculture does not demand great efforts. Here, there is no need to turn the earth, hoe the clods, water or mulch young seedlings, enrich the soil, prune off unnecessary shoots and battle against parasites. Most plants are propagated vegetatively, either by cuttings, as in the case of manioc, or by planting a shoot – from a banana tree for example – or, as in the case of a yam, by burying part of a tuber. Manioc, the numerous

The use of the word 'garden' is apt, because cultivation is done by hand over a relatively small area cleared from the rain forest, although one settlement may have several gardens (see Figure 8.22). The methods used to exploit the environment do not cause damage: shifting cultivation is a **sustainable** form of agriculture, although it requires a large extent of land to make it possible. Because tropical soils (mainly **latosols**) are basically of low fertility, and even when first used have a limited amount of humus, declining yields lead to a garden being abandoned, and the sequence described below takes place once more, perhaps in a new part of the forest up to two days' march from the original site:

It is of course the men who clear the ground: once they have felled the bigger trees with an axe, and slashed back the residual bushes with their machetes, they call upon the

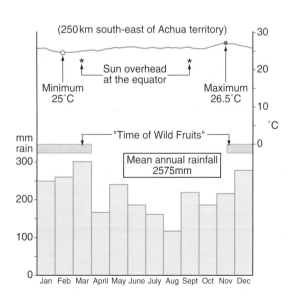

Fig 8.21 Climate graph, Iquitos (3½° S), Upper Amazon Basin

Fig 8.22 A new cultivated 'garden' in a forest clearing in the Amazonian selvas. Note the young manioc (cassava) plant in the foreground

women to burn what is left. Then, once a thick carpet of ash covers the future garden, their last contribution is to plant the rows of banana trees that are to mark out the plots of the various co-wives. After this symbolic gesture, they retire from the scene, leaving it to their female companions.

With the aid of digging sticks of chonta wood, the women first plant manioc cuttings all over their plots, then distribute, in apparent disorder, yams, sweet potatoes, taroes, beans, squashes, groundnuts and pineapples. All that then remains to be done is to plant the trees whose seasonal fruits help to vary the somewhat monotonous everyday diet: chonta palms, avocado trees, sweetsop trees, caimitos, ingas, cacao and guava trees. These tend to be planted along the edge of the area, totally cleared of grass, that surrounds the house, a collective space that is not subject to the exclusive jurisdiction which each wife exercises over her own parcel of land. Here are also the plants which are communally used by one and all: pimento, tobacco, cotton, bushes of clibadium and lanchocarpus – the juices of which asphyxiate fish caught with poison – gourd plants, racou and genipa for face paining, and last, but not least, herbal remedies and narcotic plants . . .

When mature, the garden resembles an orchard set in a vegetable garden standing high with produce. The tall papapaya stems rise above a most impressive tangle of plants: the taroes spread like monstrous sheaves of arum lily leaves, the peeling banana trees intertwine and stoop beneath the weight of enormous bunches of fruit, the squashes swell like balloons at the foot of charred stumps, carpets of groundnuts border thickets of sugar cane, arrowroot flourishes along the great fallen tree trunks left over from the felling operation, and everywhere the bushes of manioc unfurl tentacles of finger-like leaves . . . (see Figure 8.23)

Source: P. Descola: *The Spears of Twilight*

The produce of these 'gardens' (see also Figure 8.24), along with fishing and gathering of wild honey and fruits within a two-hour travel radius, and hunting for peccaries and monkeys in the untouched rain forest beyond that, make each household self-sufficient. The gathering makes for a more varied diet, while the fishing and hunting increase the protein content. The productivity of the gardens can be very high, estimated as ranging from 2000 to over

Fig 8.23 Newly harvested manioc roots. The grated roots then have to be squeezed to extract prussic acid from the juice before cooking

20 000 calories per person per day (where the population concentrations are very small). The amount of time spent on, and the relative importance of, each activity varies from tribe to tribe. In some cases gardening is very time-consuming, for instance, the Achuar weed their gardens:

. . . Entza is busy weeding, using her machete. Every day she attacks the weeds that are competing with her cultivated plants. It is a patient task which takes up most of the time spent in the garden . . . The apparent confusion of the plants that strikes the unversed observer at first glance is in reality the product of the careful equilibrium struck between the groups of plants that vary greatly in their appearance and needs . . . To be sure, the weeding does slightly prolong the life of the garden beyond the three or four years beyond which the exhaustion of the not-very-fertile soil dictates that it be abandoned . . .

Source: P. Descola: *The Spears of Twilight*

There is evidence to suggest that the gardens of shifting cultivators

◆ are best cultivated, using appropriate technology, for two years, then left fallow for 10 years

◆ in which fewer than eight crops have been grown will revert during the fallow period to rain forest (secondary); otherwise the abandoned clearing will degrade to unproductive grassland

◆ can be maintained in continuous cultivation, using a simple crop rotation involving maize, soya bean and rice, for example, but requiring frequent applications of fertiliser. This might be tried after attempts to feed an increasing population by

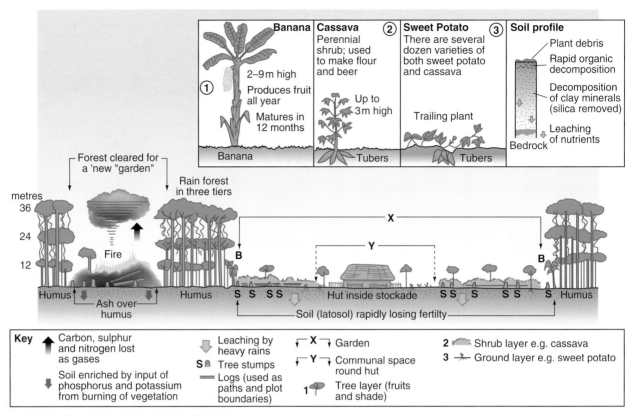

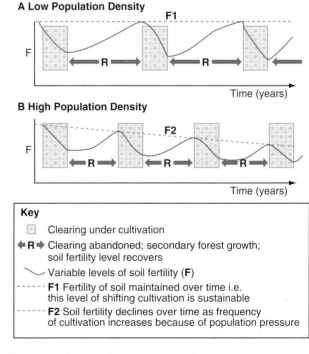

Fig 8.24 Shifting cultivation (Achuar Indians, Ecuador)

reducing the fallow period have failed or where no new land is readily available (see Figure 8.25).

Changes in Farming in the Amazon Basin

Such a change in the very nature of shifting cultivation can only take place where there is a need for continuous cultivation brought about by a shortage of virgin forest, perhaps as a result of population pressure, and dependent on improved contact with the outside world. In such cases, self-sufficiency may be maintained at the expense of sustainability. Great population pressure and land-hunger in areas beyond the rain forest, such as North East Brazil or on the Andean Altiplano in Peru, have resulted in immigration by 'colonists' on a large scale to take up holdings along the new roads within the Amazon Basin, especially within the Rondonia federal unit in Brazil (see Figure 8.26). Such colonists are from very different environments, and lack the knowledge and farming skills of the native peoples in sustainable land management. Extensive areas of rain forest have also been cleared:

Fig 8.25 Shifting cultivation population density

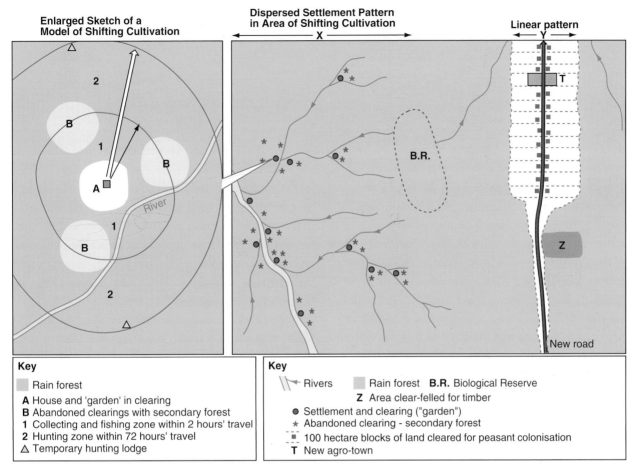

Enlarged Sketch of a Model of Shifting Cultivation

Dispersed Settlement Pattern in Area of Shifting Cultivation
← X →

Linear pattern
← Y →

Key

	Rain forest
A	House and 'garden' in clearing
B	Abandoned clearings with secondary forest
1	Collecting and fishing zone within 2 hours' travel
2	Hunting zone within 72 hours' travel
△	Temporary hunting lodge

Key

	Rivers		Rain forest	**B.R.** Biological Reserve
			Z	Area clear-felled for timber
●	Settlement and clearing ("garden")			
⋆	Abandoned clearing - secondary forest			
▪	100 hectare blocks of land cleared for peasant colonisation			
T	New agro-town			

Fig 8.26 Rural settlement, Rondonia, Brazil

for large-scale cattle ranching to provide beef for the vast urban population of South East Brazil

where mineral and timber resources are being exploited

where large-scale hydroelectric schemes are being developed,

and so native peoples have been displaced and shifting cultivation has vanished. The chain saw and bulldozer have replaced the axe and the machete, and clear-felling of the rain forest has made way for heavily fertilised fields growing a single crop, instead of 'gardens' with their tree stumps, fallen logs and profusion of crops.

However, on a world scale, shifting cultivation is still very important as a livelihood. Globally, it is estimated that shifting cultivators clear between 20 and 60 million ha of forest and scrub each year, and then burn between 1 and 2 billion tonnes of dry matter, thus contributing to global air pollution. It

therefore might be argued that while on a microscale, shifting cultivation is sustainable, overall it may not be environmentally friendly, for instance, in areas of very high population density where it takes the form of bush fallowing.

8.5 Intensive Peasant Farming: Kedah State, Malaysia

We have now looked at two very different forms of extensive farming in North and South America. In contrast, the monsoon lands of South East Asia are predominantly areas of intensive peasant agriculture. Climatically, there are some similarities with the Amazon Basin, but the biggest difference between Kedah on the one hand, and the Amazon Basin and the Great Plains on the other, is the population density.

Consider the sets of population figures in the following table:

	State/province	Area (km^2)	Population
(Great Plains)	N Dakota, USA	183 000	639 000
(Amazon Basin)	Pastaza, Ecuador	30 000	41 000
(Amazon Basin)	Rondonia, Brazil	243 000	1 131 000
	Kedah, Malaysia	9 000	1 413 000

Malaysia is a federal state in South East Asia, just north of the equator, consisting of 13 states, one of which is Kedah, and two federal territories, in two widely separated areas – the Malayan Peninsula south of Thailand, and the northern part of the island of Borneo (see Figure 8.27A). Plantations in the Peninsula are the world's main source of both rubber and palm oil; the rain forest is a valuable source of tropical hardwoods, and the large resources of both tin and oil have helped to make Malaysia one of the new industrialised economies of South East Asia. The population is predominantly Malay, who are Muslims, but there is a significant Chinese minority.

Kedah is by far the most densely populated of the agricultural areas studied in this chapter. Its population is essentially youthful; there is a very low death rate and a long expectation of life. There is a high level of both education and medical care although the farming community is among the poorer sector of the Malay population, and there is some evidence of rural–urban migration. However, the countryside is still crowded and the intensive nature of the farming is both a consequence of, and a reason for, the dense rural population. Here, land is at a premium; holdings are small, perhaps, amounting to only two hectares, in 10 or 12 scattered plots; the human input is considerable, but both the capital invested and the output are increasing (see Figure 8.36). Figure 8.28 clearly illustrates the characteristic distribution of the rural population in Kedah with settlements in the form of

◆ villages ('**kampongs**'), long and linear in shape, found along the raised banks of rivers, or along embanked roads, sometimes for a distance of several kilometres

◆ other kampongs on small 'islands' in the middle of the rice fields ('**sawahs**').

The landscape has a geometrical appearance, with small square or rectangular fields dotted with isolated

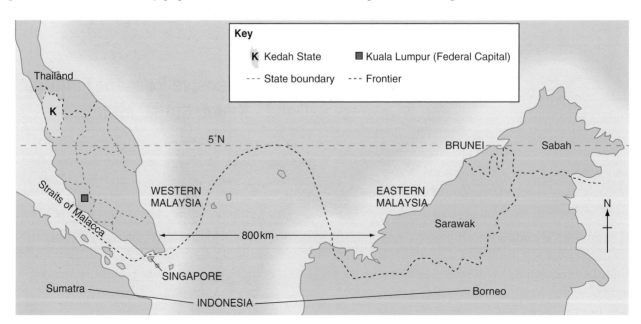

Fig 8.27a Location map: South East Asia

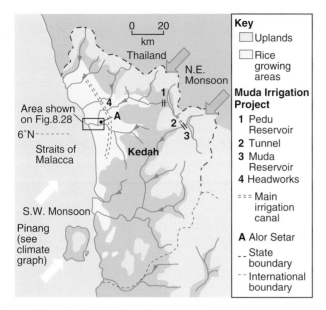

Fig 8.27b Location map: North West Malaysia

deposits. Paths, tracks and seasonal roads were more common than all-weather roads, but there is now an efficient communications network with a north–south motorway, new hard-topped roads, reinforced farm roads, as well as the railway.

There are some similarities with the Amazon Basin:

◆ the climate is equatorial, although here there is a monsoonal influence

◆ the natural vegetation was rain forest, but many centuries of cultivation and a high density of population have resulted in its disappearance. Along the coast and estuaries fragments of mangrove forest remain. This consists of trees with stilt roots growing on silt flooded by brackish water.

The graph (Figure 8.30) shows that the growing season lasts all year, with uniformly high temperatures and consequently a very small range. The dry months of January and February traditionally allowed the rice to ripen, but lack of rain had to be overcome when agriculture became more intensive and farmers grew two crops of rice each year instead of one. This has been made possible by the construction of reservoirs in the mountains to the east as part of the Muda Irrigation Project (see Figure

wooden shelter sheds, canalised river courses, and long, straight irrigation/drainage canals. The meandering course of the River Kedah is an exception to the regularity of the landscape. Figure 8.27B shows that the land consists of a wide, very flat plain, mostly below 5 m, extending for 20 km between the sea and the mountainous spine of the peninsula, and made largely of alluvial and marine

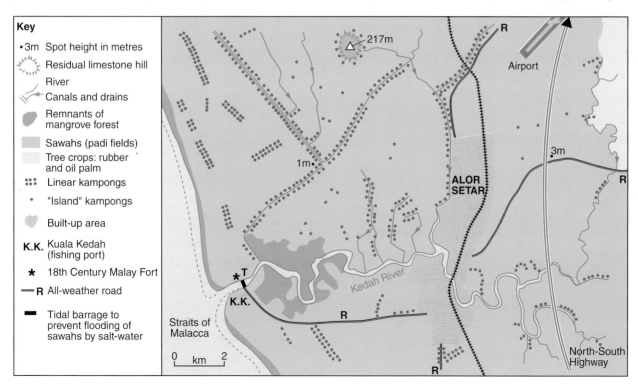

Fig 8.28 Landscape of intensive rice cultivation, Kedah, Malaysia

Fig 8.29 Padi ripening in a drained sawah on the Kedah plain. Note the line of trees marking the site of a kampong in the background

8.27B) to provide irrigation water to flood the rice fields in the dry season (see Figure 8.31). The output of rice (**padi**) has greatly increased, and farming has changed from intensive peasant **subsistence** agriculture to intensive peasant **commercial** agriculture. The changes have involved:

◆ the use of more productive and faster maturing varieties of rice which mature twice as quickly (see Figure 8.38), and thus allow two crops per year

◆ planting rice seed directly into the flooded padi fields, rather than transplanting seedlings (see Figure 8.32) from nursery beds

◆ using tractors instead of draught animals (see Figure 8.33)

◆ a great increase in the use of chemical fertilisers, instead of using guano gathered from limestone caves in residual hills

◆ using hired machinery instead of hired labour – rice is now harvested and threshed by half-tracked combines rented out by Chinese contractors instead of being done by hand

◆ spending less time on other farming activities, since there is no longer the time or the land available: there cannot be grazing for cattle and buffalo in dry padi fields after harvest, if there is double-cropping of rice

◆ becoming part-time farmers or farm-workers, with jobs in the six new industrial estates scattered about the countryside.

Many of these aspects are illustrated in the following description:

*The date is early October and the time 6.45 pm. Beneath the clear sky, the temperature of the early evening, after the heat and the humidity of the afternoon, is pleasantly comfortable. In the flat landscape of near shadeless and mostly flooded fields, the rays of the rapidly declining sun brilliantly light the different tints of green indicating **padi** at different stages of growth. Within the fields and also on some of the little kampong islands, with house and fruit trees casting long shadows, there is a great deal of activity. Farmers, nearly up to their knees in mud and water, walk behind and guide their little two-wheeled tractors with rotovators attached to the front. Behind them, youths follow with their cone-shaped fish traps. Beside some of the recently flooded fields women with wide-brimmed hats squat, fishing with long bamboo lines, whilst beneath a fruit tree and close to her newly-built house with its cement block walls and louvred windows, a farmer's wife guts fresh fish. Nearby, another farmer walks along a low, narrow **bund** of mud. At irregular intervals, he stops, bends down, and besides a small hole in the bund's side, lays a spoonful of baited rat poison. Half an hour later, the sun drops below the horizon to set the western sky ablaze with orange and yellow hues. As the light rapidly fades, a farmer leads his two cows back from an embankment, between track and main irrigation canal, where they have been tethered to graze, to the security of the kampong. In the fields, activity rapidly ceases. Some workers slowly return to their houses. Others mount their little Japanese motor cycles and, with wife mounted on the pillion seat, quickly make the short journey home.*

Source: J.A. Hocking

Farming in Kedah has thus become less diverse and more specialised, focusing more than ever on rice production, with a strictly controlled allocation of irrigation water (see Figure 8.35). Farmers now have more capital to invest in their farming enterprise, their houses and transport. Before mechanisation became common, a kampong usually formed a labour cooperative on a voluntary basis to ensure that everyone's rice seedlings were transplanted in time. Although farmers may now buy or rent farm

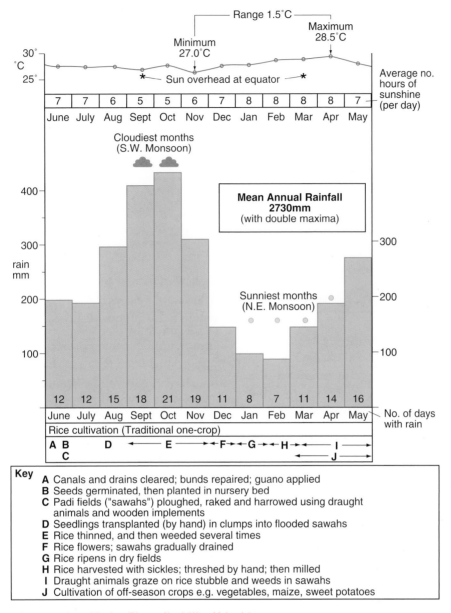

Fig 8.30 The climate of traditional rice cultivation (Pinang, North West Malaysia)

machinery, immigrants from Thailand may be hired to do some seasonal manual work. This might involve partial transplanting to even out the distribution of the young plants in the sawahs.

The Malay farmers now have a better quality of life, but there have been some adverse effects on the environment, including the build-up of agrochemicals in the sawahs, contaminating both plankton and fish. Nevertheless, fishing in the sawahs, canals and drains is still important in some areas of the state, especially among the poorer farmers (see Figure 8.34). In general, the transformation in Kedah of what was intensive

peasant farming for subsistence into even more intensive peasant commercial farming on a small scale is a good example of the beneficial effects of changes in agricultural practice in the last 40 years. The modern landscape of rice cultivation in Kedah is shown in the summary diagram Figure 8.36.

8.6 The Green Revolution

The dramatic expansion in rice production in Kedah mirrors the changes which have taken place throughout the rest of Monsoon Asia, as well as other regions, mainly tropical, within the Developing

Fig 8.31 Flooded padi fields in Kedah. Note the bunds around the fields and the kampong 'island' with an orchard round the houses

Fig 8.33 Using a mini-tractor and rotovator to plough the flooded fields. The young boy is carrying fish traps

World during this time. This became known as the **Green Revolution**. In simple terms, this involved the development and use of improved short-stem, high-yielding cereals (rice, maize, wheat), increased and improved mechanisation, irrigation and drainage, and the increased use of agrochemicals. The result of these changes was that food production more than kept up with population increase. In 1960 it was estimated that about half of the world population did

not have sufficient food (at that time reckoned to be 2000 calories per day) but by 1990, 80% of the world's population had at least 2500 calories per day.

The scientific breeding of new strains of rice at the International Rice Research Institute (IRRI) in the Philippines has continued to the present day (see Figures 8.37 and 8.38), and the new strains have been introduced to the sawahs of Kedah as they have become available. Kedah was ideal for the innovations since it

◆ was a most favourable agroclimatic region in terms of relief, soil, growing season, hours of

Fig 8.32 Traditional collecting of bundles of rice seedlings from nursery beds for transplanting in the padi fields

Fig 8.34 Gutting fish caught in the padi fields. Fish are now a declining part of the diet because of agro-chemical pollution

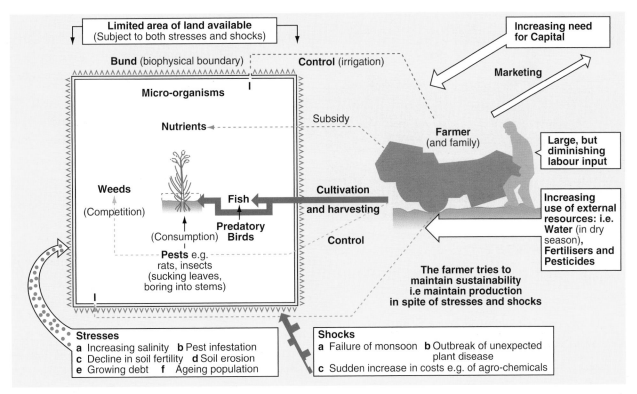

Fig 8.35 Rice production in Kedah as a system

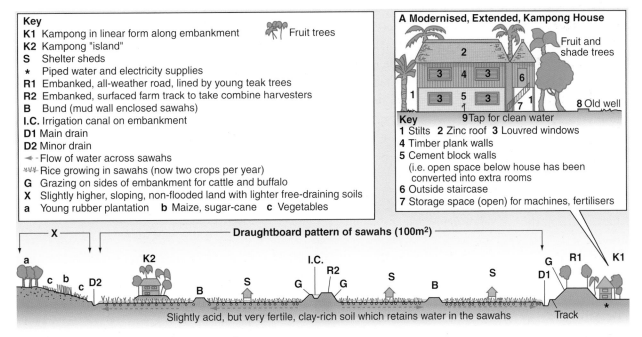

Fig 8.36 The landscape of rice cultivation: summary diagram

Fig 8.37 Researcher assessing the disease-resistance of new strains of rice at the IRRI in the Philippines

sunshine, rainfall, and the provision of irrigation water

◆ had an educated and progressive farming population willing to change traditional practices

◆ had the infrastructure already in place to handle the distribution of seed, fertiliser, pesticides and herbicides as well as the equitable distribution of irrigation water.

The first Green Revolution evolved from research into the breeding of high-yielding staple food crops, the benefits of which to the poor and hungry were then established. It has been suggested now that in the next thirty years the approach should be reversed, so that the needs of the poorest, least well-fed people in the world are assessed before identifying research priorities. The outcomes would be to achieve both secure supplies of food and **sustainability**. Critics have pointed out that the Green Revolution has had its limitations:

◆ in some parts of the Developing World (e.g. sub-Saharan Africa), there is still not an adequate supply of food

◆ natural resource degradation and environmental problems have remained or increased, and there are signs of diminishing returns (ever-increasing inputs are not matched by an equivalent increase in output)

◆ the poorest people have not all benefited from the changes and an increasing number of landless labourers have migrated to cities. In many areas, it has been the wealthier farmers who have gained most from the Green Revolution.

The critics suggest that there will now have to be a greater rate of increase in food production than in recent years, without further damaging the environment, and that the increased food supply will have to be accessible to all needy people in all parts of the Developing World. In other words, the future Green Revolution must be equitable, sustainable and environmentally friendly. Sustainable agriculture is not just the ability to produce sufficient food, but, in the view of Gordon Conway, an eminent agro-ecologist, should be seen as the ability to maintain agricultural productivity while minimising the stresses and shocks which can affect the farming process (see Figure 8.35). Sustainability can be achieved by a variety of agricultural technologies (see Figure 8.39). One is green manuring, such as the growing in padi fields of **azolla** fern, the leaves of which contain blue-green algae, and when harvested and ploughed into the soil, can greatly increase the nitrogen content and thus the yields.

Some of these innovations will affect the rice growers of Kedah State. As mechanisation and capital investment increase, along with productivity, there will be an increase in the size of the holdings, and in the number of people commuting from the kampongs to jobs in the secondary and tertiary sector, especially in the state capital of Alor Setar.

A. Early Developments at the International Rice Research Institute (I.R.R.I.)

Founded in 1962 in the Philippines: Early sucess with IR8 rice.
Participating farmers were given a 'package' consisting of

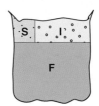

S Seed
I Insecticide
F Fertiliser

  • IR8 rice seed   • insecticide   • pesticide

IR8 rice   ∗ matured for 130 days   ∗ could grow at any time of the year
              ∗ yielded 9 tonnes per hectare

Problems   □ increase in plant disease and pest problems
              □ poor cooking quality □ consequent low prices

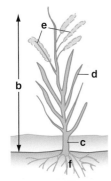

B. Aims of Recent Rice Breeding Programmes at the I.R.R.I.

To produce a directly-seeded, irrigated rice (hybrid, semi-dwarf) with

a a short growing season of 100-130 days   **b** a limited height i.e. 90cm
c sturdy stems   **d** thick, erect, dark-green leaves (rich in chlorophyll)
e only three or four panicles (seed clusters) per plant, but yielding about 1kg per plant
f a vigorous root system   **g** multiple resistance to pests and diseases

Results   very high yields, up to a maximum of 75 tonnes per hectare

C. Results of the Introduction of New Strains of Rice

Rice Production in Asia (million tonnes)
1970: 287
1995: 501

Reasons for the increase of 75%
  • increase in irrigated area
  • new strains of rice
  • increase in use of chemical fertilisers

Effects of the Green Revolution

Productivity
∗ higher yields   ∗ higher food production   ∗ higher farm income in most-favoured areas
∗ more employment and higher wages, except where mechanisation coincided with an increased labour supply

Equability
∗ increased profits for land owners and factory owners
∗ decline in real income for farmers
∗ increased number of landless labourers
∗ unequal distribution of benefits e.g. persistence of malnutrition in Sub-Saharan Africa

Sustainability
∗ greater resistance to pests and diseases   ∗ greater reliance on agro-chemicals
∗ increase toxicity of soil   ∗ danger to workers' health from chemical sprays
∗ loss of soil structure   ∗ increase in water-logging and salination of the soil

Fig 8.38 The Green Revolution and rice production

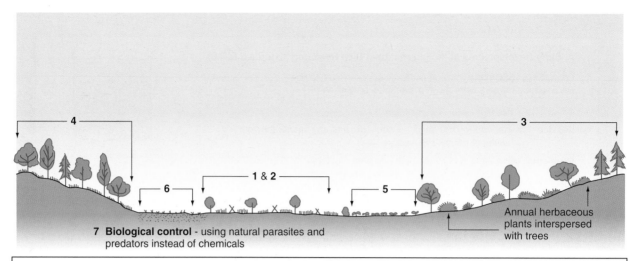

7 Biological control - using natural parasites and predators instead of chemicals

Annual herbaceous plants interspersed with trees

Possible inputs of agricultural technologies to make agriculture sustainable

1 **Inter-cropping**

2 **Rotation**

3 **Agro-forestry**

4 **Sylvo-pasture** with livestock grazing in clearings

5 **Green manuring** i.e. growing plants, for example azolla fern, cowpeas, lupins or velvet beans which are ploughed
into the soil to fix the nitrogen

6 **Conservation tillage** i.e. minimal or no ploughing. The seed is placed into the soil without turning over the surface.
This reduces the loss of nutrients; helps control run-off, and limits the likelihood of soil erosion

Fig 8.39 Sustainable agriculture

 ASSIGNMENTS

1 The Farm Systems Diagram (Figure 8.5) is incomplete: there are no details of the inputs. **Copy** the diagram, and **insert** the inputs (one for each bullet point) selecting from the following list:
Farm size; Growing season; Precipitation; Government and EU quotas and subsidies; Farm tenure; Transport; Soil; Capital; Altitude; Markets; Farm buildings; Aspect; Technologies: 1. Mechanical, 2. Chemical, 3. Biological; Exposure; Bank credit and loans; Soil.

2 Study Figure 8.6 carefully and with the aid of an atlas
   a) **Name** the states shown by their initial letters.
   b) **Measure** the distance between points a–b and c–d on the map.
   How do these compare with the distances between
   i) Paris and Budapest; **ii)** Palermo and Stockholm?
   What are the approximate latitudes of points c and d?
   Comment on your answers.

   c) **Describe** and **explain** the ways in which farming on the Great Plains changes from a. south to north and b. east to west.

3 Study Figures 8.8 and 8.11A
   a) **Describe** and **explain** the pattern of communications in North Dakota, noting the difference between the alignments of the roads and the railways.
   b) **Describe** the distribution of the main towns in the state.
   c) **State** three possible reasons for the damming of the Missouri River.

4 Study Figure 8.11b carefully, noting the evidence it gives of a survey of land use on a typical farm in North Dakota, then
   a) **Describe** the layout of the farm, the pattern of roads and farmtracks, and the shape and size of the fields. **Suggest reasons** for the patterns you describe.
   b) **Calculate** the size of the farm (see also Figure 8.11A), and the area under each crop.

c) State two ways in which the farmer has tackled the problem of soil erosion.

5 Study Figure 8.13 carefully and **suggest reasons** why

a) The Kirschenmann Farm is so different in its land use from the Waller Farm, although both are the same size. Figure 8.5 may give you some ideas.

b) The communications network in Stutsman County is so much better developed than in McCone County.

c) The population of McCone County (6828 sq km) is so very different from that of Stutsman County (5884 sq km).

6 Study Figure 8.16, and **locate** the two counties on the map. Which county appears to be typical in its population change of the Great Plains as a whole?

7 Study Figure 8.10 which shows a farming scene on the Great Plains. Draw a simple sketch of the scene, and annotate it to identify the main features of the landscape and the farming.

8 a) Study the field sketch of the single-house Achuar Jivaro settlement (Figure 8.18), and using the information provided (and the text if necessary), **describe and explain** the choice of site for the settlement.

b) Study the map of the distribution of such settlements in south-east Ecuador, and then **describe and explain** this distribution, and the low population density, in the area shown on the map.

9 Read the descriptions of shifting cultivation, study Figure 8.24 and then

a) List the different activities carried on by men and women.

b) Explain the significance of the burning process.

c) Classify the crops mentioned according to **(i)** height and **(ii)** uses.

d) Explain why the gardens are eventually abandoned.

10 Study Figure 8.26 which shows a typical area in Rondonia, Brazil, and then **describe** the **differences** in settlement pattern and population shown on the map.

11 Study Figure 8.25 and **describe** the impact of differences in population density on

a) the frequency of cultivation of clearings.

b) length of recovery period between cultivation.

c) the level of soil fertility.

12 Look at the table showing the area and population of North Dakota, Pastaza and Kedah. **Calculate** the population density for each state, and **link** each with the appropriate type of farming: **shifting cultivation; intensive peasant agriculture; extensive commercial cultivation**.

13 Study Figure 8.28 and **explain** why rural settlement in Kedah

a) is found either in linear form or as 'islands',

b) is not found along the coast.

14 Study Figure 8.30 and Figure 8.35

a) List the advantages of the climate of Kedah for rice growing.

b) List the ways in which the cultivation of rice could be said to have been 'intensive'.

c) What problem of the climate had to be overcome when the switch was made from one to two crops per year? How was this done?

d) What do you consider has been the biggest change in methods of rice cultivation since the onset of the Green Revolution?

e) Why do rice farmers in Kedah now use many more chemicals than before? What have been the results?

f) In what ways has rice cultivation become more commercial, while still being intensive?

15 Study Figure 8.33 carefully. Draw an outline sketch, and annotate it to show:

Flooded sawah; shade and fruit trees; bund; kampong; rotovator; fish traps.

16 a) Draw a graph to illustrate the relative yields of the crops using the data in the table below.

b) Identify the crops which are grown by
(i) the shifting cultivators in south-east Ecuador.
(ii) the peasant farmers in Kedah.

c) Suggest reasons why rice is the highest-yielding cereal.

Yields and protein content of some important tropical crops: 1991

	Avg yields (tonnes/ha)	Protein content (%)
Rice	2.69	7.5
Maize	1.59	9.5
Guinea Corn	0.87	10.5
Millet	0.65	10.5
Cassava	9.63	1.6
Sweet Potato	5.86	1.6
Yams	7.00	2.0
Soybeans	1.34	38.0
Groundnuts	0.89	25.5
Bananas	13.00	1.1

d) What are the essential differences in yield and dietary value between

   (i) cassava (manioc) and sweet potato on the one hand and

   (ii) soyabeans and groundnuts on the other?

e) The figures above are average for all tropical areas. Study Figure 8.38. What effect has the scientific breeding of new rice strains had on yields in the most favoured areas of cultivation?

17 Study the text, Figure 8.35 and Figure 8.38. To what extent has the Green Revolution been a success? What problems have apparently not been solved? What new problems have arisen?

Key terms and concepts

- agro-forestry (p. 255)
- aquifer (p. 234)
- arable farming (p. 230)
- biological control (p. 255)
- biological technology (p. 256)
- black earth (p. 238)
- bund (p. 249)
- bush fallowing (p. 240)
- chemical technology (p. 255)
- chernozem (p. 238)
- commercial farming (p. 230)
- conservation tillage (p. 255)
- contour ploughing (p. 237)
- crop rotation (p. 244)
- dispersed settlement (p. 241)
- dry farming (p. 236)
- Dust Bowl (p. 236)
- extensive farming (p. 231)
- fallow (p. 236)
- farm cooperatives (p. 238)
- fossil fuel (p. 237)
- 'gardens' (p. 243)
- Great Plains (p. 232)
- Green Revolution (p. 251)
- green manuring (p. 253)
- growing season (p. 236)
- guano (p. 249)
- homestead (p. 236)
- hunting and gathering (p. 233)
- intensive farming (p. 231)
- inter-cropping (p. 255)
- IRRI (p. 251)
- 'kampongs' (p. 247)
- latosol (p. 243)
- marginal farming (p. 236)
- mechanical technology (p. 255)
- mixed farming (p. 230)
- monoculture (p. 238)
- organic farming (p. 232)
- 'padi' (p. 249)
- pastoral farming (p. 230)
- peasant farming (p. 246)
- plantations (p. 247)
- prairie (long grass & short grass) (p. 233)
- 'sawahs' (p. 247)
- sections (p. 235)
- sedentary cultivation (p. 231)
- shelter belt (p. 237)
- shifting cultivation (p. 231)
- strip cultivation (p. 237)
- subsistence farming (p. 230)

- ◆ sustainable agriculture (p. 243)
- ◆ townships (p. 235)
- ◆ transhumance (p. 231)

Suggested reading

Broadley, E & Cunningham, R (1991) *Core Themes in Geography – Human* Oliver & Boyd

Conway, G (1997) *The Doubly Green Revolution: Food for All in the 20th Century* Penguin

Descola, P (1996) *The Spears of Twilight: Life and Death in the Amazon Jungle* Harper-Collins

Guinness, P (1998) *Brazil: Advanced Case Studies* Hodder & Stoughton

Guinness, P & Nagle, G (1999) *Advanced Geography: Concepts and Cases* Hodder & Stoughton

Hocking, JA (1997) *Changes to Small-Scale, Intensive Rice Farming in N.W. Malaysia* SAGT Journal, No. 26

Maclean, K & Thomson, N (1988) *Landscapes and Peoples of Western Europe* OUP

Nagle, G & Spencer, K (1996) *A Geography of the European Union – A Regional and Economic Perspective* OUP

Price, B & Guinness, P (1997) *North America: An Advanced Geography* Hodder & Stoughton

Waugh, D (1994) *Geography: An Integrated Approach* Nelson

Waugh, D (1997) *The U.K. and Europe* Nelson

Waugh, D (1998) *The New Wider World* Nelson

Witherick, M (1995) *Environment and People* Stanley Thorne

Internet sources

Agricultural Resources: www.educationindex.com/ag/
Canadian Wheat Board: www.CWB.ca/
Food & Agriculture Organisation (FAO): www.fao.org/
Green Revolution: International Food Policy Research Institute: www.cgiar.org/ifpri/2020/backgrnd/25years.htm_top

Shifting Cultivation
1: www.forest.gov.my/cultivat.html
2: http://www.sru.edu/depts/artsci/ges/lamerica/d-5-13.htm

INDUSTRIAL GEOGRAPHY

As Figure 6.5 has shown, most developed countries in the world have reached the post-industrial stage of the economy, in which secondary (manufacturing) industry has lost its pre-eminence. Not only are more people employed by the tertiary sector, but a growing quaternary sector is emerging (see p. 295).

In Chapter 9, this change will be examined on a variety of scales through case studies of: industrial regions – the coalfields of the Sambre-Meuse and North-East England; particular places – Saltaire and Cambridge; particular industries; and individual companies – Nokia and Quintiles. These follow a summary of the changing locational factors, and changing industrial landscapes through time, since the creation of a particular industrial landscape has long-lasting implications for both population and the environment

9.1 Industrial Location

The factors which influence the location of a particular factory or a particular industry through

time, that of an industrial area, are varied. It is necessary to look back in time to understand how the reasons for industrial location have changed; how the locations might have shifted, and how the resulting landscapes have evolved. Some industrial landscapes have virtually disappeared and may be only of interest to industrial archeologists, for

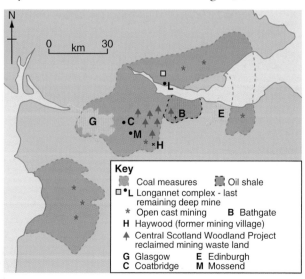

Fig 9.1 Former mining areas in Central Scotland

Fig 9.2 The eighteenth century cotton mills at Stanley on the River Tay have recently been converted to provide high quality housing, workshops and a museum

9.2 The Changing Location of Industry with Special Reference to Scotland

The earliest industry was small-scale, used manual power, for example in spinning and weaving cloth, and was carried on by individuals in houses or workshops attached to them. This is known as the **domestic** system of manufacturing. **Water-power** was used to grind grain in small mills on riverside locations. At a time when the economy was based on **self-sufficiency** (producing everything required) and most people were either working on the land, quarrying, or fishing (working in the **primary sector**), the manufacturing industry (the **secondary sector**) developed anywhere where there were clusters of population in villages (e.g. Kilbarchan in Renfrewshire) and small towns. **Labour** was supplied by a small number of skilled craftsmen in each location. Industry was therefore widely **dispersed**, making use of local **raw materials** at a time when **transportation** was slow and difficult because of the

example, the former coal mining village of Haywood in North Lanarkshire (see Figure 9.1), while other landscapes have survived, and have become elements in our **industrial heritage**, for example, Bonawe, New Lanark and Saltaire. These are industrial sites which have been restored, and now attract many visitors.

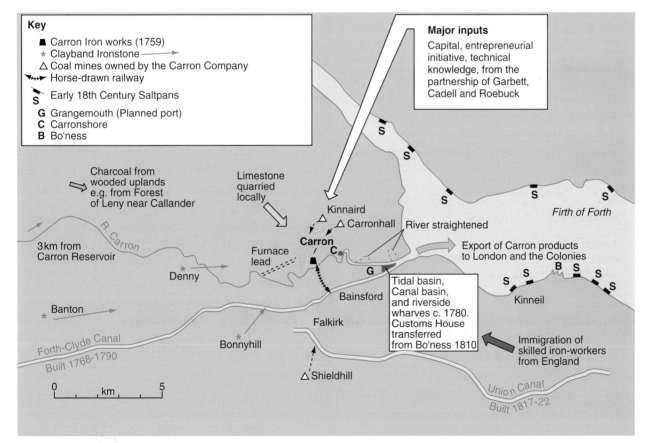

Key
- ■ Carron Iron works (1759)
- ★ Clayband Ironstone ———→
- △ Coal mines owned by the Carron Company
- ⊷┉┉ Horse-drawn railway
- S̶ Early 18th Century Saltpans
- **G** Grangemouth (Planned port)
- **C** Carronshore
- **B** Bo'ness

Major inputs
Capital, entrepreneurial initiative, technical knowledge, from the partnership of Garbett, Cadell and Roebuck

Charcoal from wooded uplands e.g. from Forest of Leny near Callander

Limestone quarried locally

Kinnaird
△ △ Carronhall

River straightened

Firth of Forth

3km from Carron Reservoir

R. Carron

Carron
C

Furnace lead

Export of Carron products to London and the Colonies

Denny

G

Bainsford

Tidal basin, Canal basin, and riverside wharves c. 1780. Customs House transferred from Bo'ness 1810

Kinneil

Banton

Falkirk

Forth-Clyde Canal Built 1768-1790

Bonnyhill

Immigration of skilled iron-workers from England

△ Shieldhill

Union Canal Built 1817-22

0   km   5

Fig 9.3 Carron iron works

Fig 9.4 The former Kvaerner yard at Govan in Glasgow is now owned and operated by BAe Systems, together with the former Yarrows yard downriver at Scotstoun

lack of roads. The largest numbers of craftsmen were to be found in Edinburgh and Glasgow.

The second stage was at the beginning of the **Industrial Revolution,** when industry became less dispersed with the development of the **factory** system, and the construction of large textile mills using **water-power**, and often an associated village (e.g. at Stanley on the River Tay, and New Lanark on the River Clyde, see Figure 9.2). **Bonawe**, as will be seen later, was a different example. The money required (**capital**) for the new enterprises came from:

◆ businessmen (**entrepreneurs**) already involved in industry, and willing to take risks, that is, to invest in a new enterprise in the hope of making a **profit**

◆ landowners who were trying to maximise the **income** from their estates

◆ **loans** from the new banks

◆ merchants who had made fortunes in trading with the colonies.

Riverside locations were vital, but industry was still dispersed in river basins, and in rural, rather than urban, locations. Self-sufficiency was no longer a factor and the **market** expanded as transport improved with the opening of **turnpike** (toll) **roads** and the building of canals. The demand for labour increased and people, made jobless and landless by the Agricultural Revolution, migrated to the new

mills. Production in the textile industry was still **labour-intensive** (in relation to the capital invested large numbers of workers were employed, despite the introduction of new **technology**). There was thus a marked switch of labour from the primary to the secondary (manufacturing) sector.

The third stage came with the large-scale exploitation of the coalfields in the nineteenth century and the development of **steam power** (dependent on coal) with factories and ironworks being built, and new industrial towns and large villages rapidly and haphazardly growing up in close proximity. Industry became **concentrated** in these new **urban** areas where the fuel and raw materials – coal, blackband iron ore (ore found within the coal seams), limestone, clay (for brick making), building stone, sands and gravels (from glacial outwash deposits) – could be extracted. Industrial expansion was aided by the application of new **technology**, for instance, the steam engine, the pump and the blast furnace. The ironworks at Carron near Falkirk (see Figure 9.3) is a good example of an early industry becoming established because of

◆ local supplies of coal and raw materials

◆ the application of new technologies

◆ the importance of entrepreneurs, who encouraged innovation and supplied the necessary capital.

Rural–urban migration and immigration from Ireland increased as the demand for labour multiplied. Miners' rows and tenement blocks were hurriedly built in places such as Coatbridge and Airdrie, as the canal network was expanded and railways were introduced, thus facilitating transport and widening the market as self-sufficiency ended. Glasgow became Scotland's premier industrial centre because of its improved port facilities and imports of cotton, tobacco and sugar; easy access to a wide **hinterland** through its network of canals and railways; readily-available coal supplies, and accumulated wealth from trade with the colonies which became available for investment. Local **innovations**, (e.g. the first steamship, Bell's 'Comet', and later the first iron ship built by Napier), led to the rapid development of the shipbuilding industry along the Clyde, so much so that by the end of the century there were 50 yards employing 100 000 people (see Figure 9.4).

Fig 9.5 The 450 ha site of the Ravenscraig integrated steel works (1964–1992), cleared and decontaminated with funding from the RESIDER programme. The infrastructure is now being put in place for a new industrial village and business park

There was thus a great increase in the workforce employed in the secondary sector, but there was also a large increase in the number of coal miners (primary sector). **Heavy industry** (coal mining, manufacturing iron and steel and chemicals, heavy engineering and shipbuilding) became **concentrated** (or **fixed**) on the coalfields or in ports on their edge.

The impact on the environment was cumulatively disastrous, as waste material from coal and shale mines and iron furnaces was piled up close to slum housing, and emissions from factories, furnaces and gas works polluted the rivers and the atmosphere, as the following contemporary description of Coatbridge (see Fig. 9.1) makes clear:

Dense clouds of smoke roll over it incessantly, and impart to all the buildings a particularly dingy aspect. A coat of black dust overlies everything and in a few hours the visitor finds his complexion considerably deteriorated by the flakes of soot which fill the air ... To experience Coatbridge it must be visited at night when it presents a most extraordinary spectacle ... From the steeple of the parish church no fewer than 50 blast furnaces may be seen.

Source: D. Bremner: *The Scotsman* 1869

Similar problems were to be found in other areas of heavy industry elsewhere in Europe, for example, the woollen manufacturing areas around Bradford in West Yorkshire (see pp. 268–271) and the coal, iron and steel area in the Sambre-Meuse Valley in Belgium (see pp. 272–280). In all these areas, the concentration of heavy industry was creating the **'old' industrial landscape** in inner city areas along railway lines. The landscape was to last, while deteriorating further, well into the middle of the twentieth century (see Figure 9.18).

The fourth stage came in the second half of the twentieth century. At the end of the economic depression of the 1930s, the first **'new' industrial landscapes** were created with the creation of **industrial estates** on the periphery of large urban areas (e.g. Hillingdon in Glasgow and Sighthill in Edinburgh). These made use of electricity. There were no smokestacks, most buildings were one-storey, and were surrounded by landscaped open spaces. Road transport was becoming more important than rail transport, and electricity was becoming more important than steam power. Industry was becoming more dispersed again, less fixed, and indeed **'footloose'**, since power could be brought to any suitable location served by the road network. As the old industrial areas in the inner city areas declined further, other new industrial locations were created as the planned **New Towns** (e.g. East Kilbride, Cumbernauld) were built after 1945 to provide

◆ better housing for the **'overspill'** of population from the large cities, especially Glasgow, where the inner city housing was being cleared

◆ employment in the new **'light'** industries – clothing, foodstuffs, electrical goods, pharmaceuticals,

in what was very much a new industrial landscape.

Government policy was therefore becoming more important than decisions taken by entrepreneurs in locating industry. Later in the twentieth century government agencies such as the **Scottish Development Agency** (later to become **Scottish Enterprise**) and **Locate in Scotland**, working with the local authorities, became the main driving force in attracting **inward investment** from overseas, especially in the electronics industry. The

Fig 9.6 IBM near Greenock, which now produces PCs, opened in 1954, making accounting machines. Note the linear, greenfield site between the A742 and the railway line at the foot of the steep slope

concentration of such firms, with parent companies in such countries as the USA and Japan, was such that Central Scotland was designated '**Silicon Glen**'. Their choice of location, often in the New Towns, was influenced by

◆ government subsidies, grants, loans

◆ the successful example of IBM at Greenock (see Figure 9.6)

◆ the expansion of the motorway network and of the airports

◆ the proximity of the universities which could provide graduate trainees

◆ the availability of a youthful workforce who could be trained to do the delicate but routine assembly work.

The concentration of the electronics industry in Silicon Glen might be seen as either an example of

◆ the '**screwdriver economy**', basic assembly without research, or

◆ the '**knowledge**' economy, with a firm base in research carried on in Central Scotland (see Figure 9.7).

As will be seen later, the electronics industry has undergone several major setbacks, and is much changed in character, but it is a major employer and the main exporter in Scotland.

While air and motorway transport were key locational factors for the electronics industry, sea transport or a tidewater location were still of importance (although traditional port industries had declined in Glasgow and Dundee) in two sectors of the manufacturing industry: (i) the petro-chemical industry at Grangemouth (see Figure 9.40 and pp. 288–290), and more recently, (ii) the engineering yards set up at various locations to provide structures for the North Sea oil and gas fields.

The latter part of the twentieth century also saw a very rapid decline in primary sector employment as

◆ work in farming shrank because of mechanisation, amalgamations, and economic problems

◆ coal mining virtually ceased to exist.

There was a major decrease in employment in the secondary sector as the heavy industries disappeared, along with some of their replacements, as **de-industrialisation** accelerated. As in most mature economies, the vast majority of the Scottish workforce is now engaged in the **tertiary sector** (working in the service industries in education,

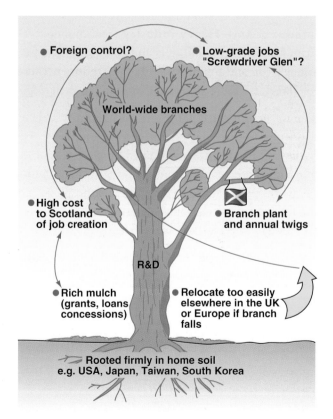

Fig 9.7 The transnational tree – some criticisms

health care, retail, finance, insurance, leisure and tourism). As the old industrial landscapes are finally razed, reclaimed and renovated, a new kind of industrial landscape has evolved, sometimes in the regeneration of urban cores (e.g. the Merchant City in Glasgow), or in peripheral **business parks**, similar in appearance to earlier industrial estates. In such parks, manufacturing is virtually non-existent, and the low buildings are essentially modernistic office blocks of concrete and glass, with aluminium cladding used for warehousing.

Scotland has now reached a fourth stage in industrial development, employment and landscape. A **quaternary sector** has emerged, more firmly based on the knowledge economy and modern information technology – in biotechnology and making use of telecommunications. The **science parks** which are found in association with the universities contain laboratories, often funded by **transnationals** (major firms based in the USA, Japan and South Korea amongst others, with many branches in other parts of the world). In these parks, research in biotechnology, of global importance for heathcare, is carried out. The Wellcome Trust Building in Dundee (see pp. 294–295) and the Roslin Biotechnology Park are examples, but the largest concentration of such work is at Cambridge (see Figure 9.46). In the business sector, the most striking development in the last decade of the twentieth century was the proliferation of **call centres** ('phone factories') associated, but not exclusively, with banking and insurance. They are sometimes in remote locations, but employ more than 100 000 people in Glasgow alone (see Figure 9.50). The landscapes of the quaternary sector are varied, ranging from converted buildings in the CBD or Inner City, to business parks, university campuses, and custom-built 'parks' in rural areas.

The Motorola factory makes mobile phones in a new industrial landscape in Bathgate – see page 282 and Figure 9.30

9.3 CASE STUDY 1: BONAWE FURNACE

Situation

Bonawe is located on the south shore of Loch Etive in Argyll (see Figure 9.8). It had a long life as an iron furnace, starting production in 1753 in the early stages of the Industrial Revolution, although it was not typical of that period of industrialisation. It finally closed in 1876, by which time it was only one of two charcoal-burning furnaces left in Britain. It is worth considering why the furnace was located here, and how it came about, although it was not the only iron furnace located on the shores of a sea loch in the Western Highlands of Scotland. Furnace on the west side of Loch Fyne was the best known of the others.

Charcoal

Bonawe was established at a time when charcoal was the fuel used in blast furnaces, and when even in mid-eighteenth century England there was a growing shortage of deciduous woodland to supply the necessary charcoal. About eight tonnes of wood were needed to make one tonne of charcoal. Iron founding was of major importance in the Furness district of north-west England, based on local supplies of iron ore and limestone, and what proved to be inadequate supplies of charcoal. It was not economic to transport charcoal long distances because, although it was light, there was a very high wastage rate (25%) during transport, in what was a very expensive commodity. The Ford family, iron masters at Ulverston in Furness, therefore sought an alternative and assured sources of charcoal, preferably at as low a cost as possible. They found them in the deciduous woodlands of the Loch Etive area, especially in one of the tributary glens, Glen Kinglass, about 9 km up the lochside from Bonawe. One hectare of coppiced oak and hazel could

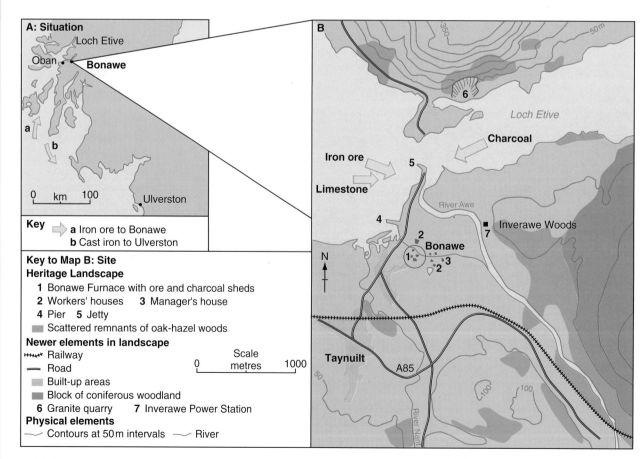

Fig 9.8 Bonawe furnace: situation and site

A: Situation

Loch Etive
Oban
Bonawe

a
b

0 km 100

Ulverston

Key
a Iron ore to Bonawe
b Cast iron to Ulverston

Key to Map B: Site
Heritage Landscape
1 Bonawe Furnace with ore and charcoal sheds
2 Workers' houses   3 Manager's house
4 Pier   5 Jetty
▓ Scattered remnants of oak-hazel woods
Newer elements in landscape
╋╋╋ Railway
— Road
▓ Built-up areas
▓ Block of coniferous woodland
6 Granite quarry   7 Inverawe Power Station
Physical elements
⌣ Contours at 50 m intervals   ⌣ River

Scale
0   metres   1000

B

350   50 m
6
Loch Etive
Charcoal
Iron ore   5
Limestone
River Awe
4
7 Inverawe Woods
2
1 Bonawe
2 3
N
Taynuilt
A85
50   100   100
River Nant

Fig 9.9 Bonawe Heritage Site. Notice the renovated charcoal and iron sheds in the foreground with the furnace and the tenement building beyond

provide enough charcoal to fuel a day's production of iron (about two tonnes) at Bonawe, and it was actually cheaper to produce iron there than at Ulverston, despite having to import heavy iron ore (haematite ore from Cumbria, as well as carboniferous ore from Central Scotland) by sea. This was therefore a most unusual example of the bulky raw materials being taken to the source of the fuel required. The charcoal used at Bonawe eventually had to come from an area within a 65 km radius of the furnace.

Sea transport, cheaper and quicker than land transport in the eighteenth century, was a vital factor, and Loch Etive provided a navigable, sheltered waterway. Ships arrived at the jetty at the mouth of the River Awe, carrying cargoes of iron ore from Furness, limestone from the island of Lismore near the mouth of Loch Etive, and charcoal from Glen Kinglass. The ships from Furness took back bars of pig iron to be forged at Ulverston. Figure 9.10 shows the operation at Bonawe as a systems diagram.

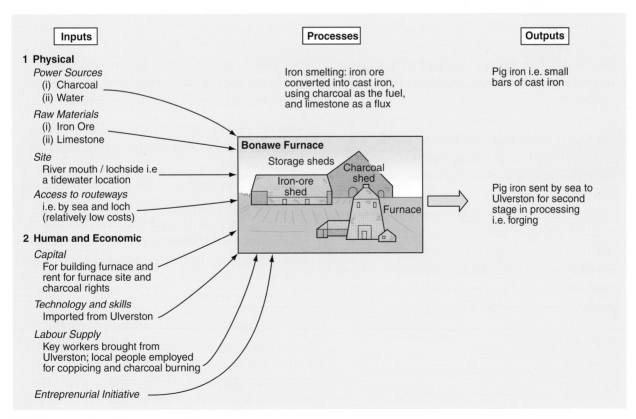

Fig 9.10 Iron smelting at Bonawe as a system

Site

The site was as important as the situation, and the iron masters were able to rent a sloping area of land close to the River Awe and a short distance from the shore of the loch. The river was vital because the bellows in the furnace were water-powered, and a lade was constructed from the river through the furnace (see Figure 9.8). The slope from south to north gave the necessary head of water in the furnace lade. The buildings in the furnace complex (charcoal sheds, the iron-ore shed and the furnace itself) were substantial, of stone and slate, and built into the hillside. All the buildings, storage sheds and furnace could thus be loaded from the back (up-slope), and emptied from the front (down-slope). In addition, houses were built for workers, including a tenement block, and there was also a school and a church. Relatively few people were needed for the furnace itself, with about six or seven men working under the direction of the furnace master. Others worked for the wood agent, but it is thought that as many as 600 extra people were engaged seasonally in the making and transporting of charcoal.

Bonawe is therefore an example of:

◆ **entrepreneurial initiative** in seeing the development potential of a remote rural area
◆ an isolated industrial location based on the fuel source, to which the raw materials were brought
◆ **inward investment** from outside the industrial location
◆ the import of skill and technology into a rural area
◆ the rise of associated industries on a small scale (a limited example of the **multiplier** effect). Examples are the export of oak bark, a by-product of the charcoal industry, to tanneries in Lancashire and on the Clyde, and the spinning of wool by the wives and daughters of the workers, again for export to industrial areas
◆ a **heritage** site, since the furnace buildings have been restored by the government and now attract many visitors (see Figure 9.9).

Impact on People

For over a century, Bonawe provided alternative employment in a remote rural area for the local people, although as in most modern examples of foreign, inward investment, key personnel were brought in to do the highly skilled work and carry out the managerial roles. Some of the key workers were provided with houses: terraced cottages on the higher ground to the south of the furnace, and tenements to the north. There is still evidence today in the traces of runrig (a method of dividing land using mounds of earth) that they were encouraged to cultivate small areas of land near the tenements, as well as leasing grazing rights for cattle. There was also payment in kind and the families became dependent on company shipments of oatmeal.

The impact on the local economy of the final closure of the furnace, even after years of declining production and temporary closures, must have been considerable. Superior technology – the **hot-blast furnace** using coking coal – eventually made Bonawe redundant.

Impact on the Environment

Inevitably in iron-smelting there is waste, but fortunately the amounts at Bonawe were relatively small. There are no unsightly **spoil heaps** because the waste was quarried for use as road metal after the closure, and the residue has long since been colonised by plants. The industrial buildings were so well constructed, that after restoration they both blend in with, and enhance, the landscape. Most importantly, the use of charcoal as the fuel did not destroy the oak woodland. Coppicing involved

◆ the cutting of trees in a small area of a woodland down to stumps, and using the initial wood for charcoal
◆ moving on to another patch of the woodland, and repeating the process
◆ allowing the original coppiced trees to regrow from the stumps, usually in the form of many thin poles
◆ returning to the first coppiced area, perhaps after as along a period as 30 years, and harvesting the poles for charcoal.

Coppicing was therefore an example of careful, **sustainable**, forest management (look at Figure 5.3 again). Essentially, the furnace at Bonawe had no detrimental effect on the environment, although admittedly the development there was on a very small scale.

9.4 CASE STUDY 2:
SALTAIRE, West Yorkshire

The description 'small scale' could not be applied to Saltaire, which in 1853 contained the largest mill in Europe, and which today is seeking **World Heritage Site** status as a planned, industrial village, like New Lanark.

Titus Salt

The establishment of Saltaire was thanks to the vision of Titus Salt, who wrote:

I looked around for a site suitable for a large manufacturing establishment, and I fixed on this as offering every capability for a first-rate manufacturing and commercial establishment. It is also, from the beauty of its situation, and the salubrity of its air, a most desirable place for the

erection of dwellings. Far be it from me to do anything to pollute the air or the water of the district . . . I hope to draw around me a population that will enjoy the beauties of the neighbourhood – a population of well-paid, contented, happy operatives.

Bradford 1850

Salt was a typical Victorian entrepreneur who became a millionaire through his innovative use of alpaca, angora and mohair in the manufacture of worsted cloth in the woollen manufacturing town of Bradford, West Yorkshire (see Figure 9.11). By the mid-nineteenth century, the population of the Bradford area had mushroomed, with an increase of more than 600% between 1811 and 1851. As the population soared, so too did Salt's capital, and by 1851, he employed some 3000 workers in six scattered spinning and weaving mills in the town, as well as many outworkers in surrounding small towns and villages. As the description above suggests, Salt

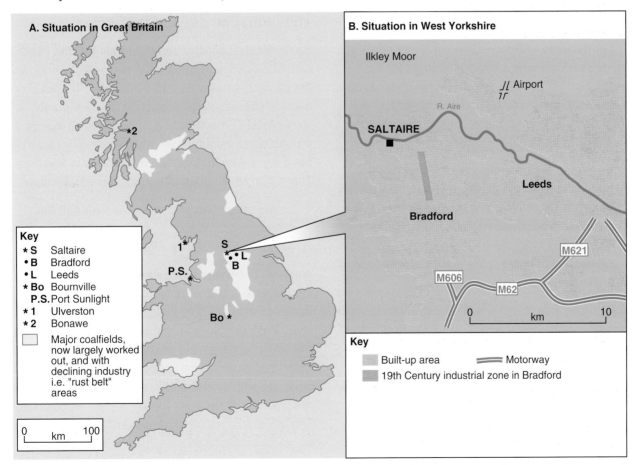

Fig 9.11 Saltaire. **(A)** Situation in Great Britain. **(B)** Situation in West Yorkshire

was an early environmentalist, in his awareness of the damage which was being done by very rapid industrialisation and urbanisation. Bradford suffered from severe atmospheric pollution, while the canal and the river had become open sewers. The water supply was inadequate and contaminated; housing for the incoming workers was both substandard and inadequate, with appalling overcrowding (e.g. eight people working and living in a small cellar). As a result, endemic diseases were widespread, and epidemics occurred frequently – there had been an outbreak of cholera in 1849. Mortality rates in general were high, especially among infants, and expectation of life was short. The consequences of the Industrial Revolution were grim for the workers and their families in the fast-growing **conurbations** throughout Europe, not least in Glasgow and Dundee.

Saltaire

Salt was all too aware of these human consequences of the industry which had made him wealthy, and he was perhaps influenced by the earlier example of Robert Owen at New Lanark in his decision to make a decisive change for the better. Salt was already known for his philanthropy in Bradford, stemming from his deep religious faith, but he was also an astute business man. He had problems to solve:

◆ business was booming, new technology was becoming available, but he needed a large area of land on which to concentrate and expand his business. There was no such area of land in Bradford
◆ he needed to improve the quality of life of his workforce.

His solution was to move to a near-greenfield site in the valley of the River Aire, some 6 km north of Bradford, in a main communications artery (see Figure 9.11B). The chosen site was 'greenfield' in some important respects, but it already had a major part of an infrastructure in place:

◆ the Leeds–Liverpool Canal, in use since 1774, had been built south of the river course. Bulky cargoes of coal for the steam engines could be

brought in cheaply, as could the bales of wool and other fibres
◆ a railway line, originally linking Leeds to Bradford, had been extended north, and then west through the Aire valley, just south of the canal. This provided a faster means of access to suppliers and customers, as well as links to ports for imports and exports.

There was a small mill beside a weir, which had at various times been used as a woollen mill or for milling flour. It was demolished as part of the development of the site.

The River Aire itself was another locational factor in that it could provide the necessary water, after the reconstruction of the weir, for the steam engines and for the scouring, dyeing and finishing of the cloth. Salt was able to buy 20 ha of land to build a planned industrial village, zoned by function, which was to be called Saltaire, in what was then largely open countryside.

The Industrial Zone: 4 hectares

Salt's mill was constructed in two stages: the major part was opened in 1853 (after two years' work), and the second, smaller part 15 years later. The four hectare site was large enough to house the new wool-combing machinery, and integrate all the processes of manufacture which previously had been carried on in six different factories in Bradford. The major advantages of the site were that it was

◆ on the valley floor, between the river and the railway
◆ astride the canal (see Figure 9.12).

Although there was a large area of land available, the mill was five storeys high, built of sandstone (rather than brick), and designed in a very ornate style. For example, the chimney stack in the 'new' mill was concealed in a structure in the style of an Italian bell tower. More importantly, the mill was equipped with smoke filters, and the area around the mill was preserved in as 'green' a state as possible. The mill employed 3000 people, even though it was so highly mechanised.

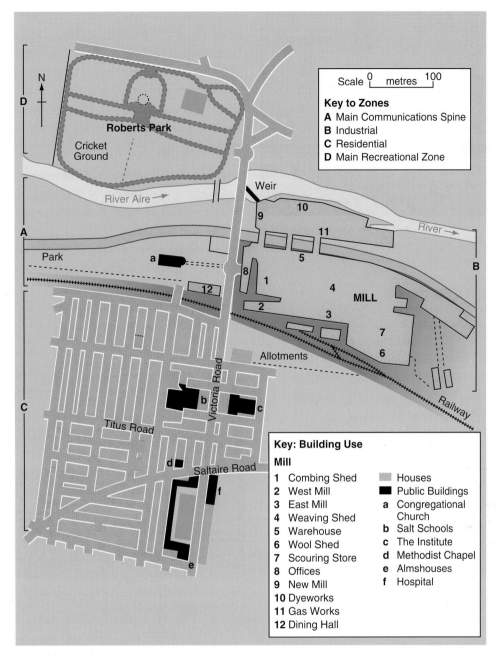

Fig 9.12 Saltaire: plan and functional zones

The Residential Zone: 10 hectares

This was laid out in stages in rectangular form (see Figure 9.12), varied only by the curving Saltaire Road, on gently sloping ground immediately south-west of the mill. Most of the houses, available at very low rents, varied in size from four to six rooms; all had a yard in which the lavatory was located; some had gardens but all had a supply of gas from the mill and a piped water supply, as well as access to an efficient drainage system. There were public baths and a wash house. Some of the 800 houses, like the mill, were ornately built – three-storey houses for senior employees. Shops were allowed to develop along the main streets but public houses were banned. A school, a hospital, an institute (community centre) and almshouses, where pensioners were given free accommodation, were added as the population grew to 4400 and the village spread up the slope. Roads were relatively wide, trees were planted, and everywhere there was light. Saltaire was not a place of 'dark satanic mills' so characteristic of the Industrial Revolution, and provided a model for later planned industrial villages, such as Bournville and Port Sunlight.

Fig 9.13 Saltaire: the 'new' mill with the bell-tower chimney, renovated as a centre of the electronics industry, the arts and the heritage industry

Recreational Open Space

West of the mill, there was a large area of parkland in which the Congregational Church and works dining room were located; south of the mill, beyond the railway, allotments were provided, while the extensive Roberts Park was laid out on the other side of the river, north-west of the mill.

Saltaire Today (see Figures 9.13 and 9.14)

A Site Transformed – Power Looms to Microchips

The mill had a long and productive career but there was a final, slow decline, and by the 1980s it had closed down, as had many others throughout the West Yorkshire '**rust belt**', as result of de-

industrialisation – the decline and eventual disappearance of traditional heavy industry. Saltaire itself, which had ceased to be company-owned in 1933, became something of a 'ghost' village, but its fortunes were revived by another local businessman, Jonathan Silver in 1990s. His efforts resulted in the mill being transformed. Today it contains:

◆ an art gallery housing a major permanent collection of works by David Hockney, the Bradford-born artist
◆ electronics workshops of various kinds employing over 2000 people
◆ specialised shops, boutiques and one of the largest restaurants in the north of England, all catering for the 'heritage industry', in this case the million or more people who now visit Saltaire each year.

Fig 9.14 Saltaire: renovated terraced housing, now privately owned, and a surviving corner shop

9.5 CASE STUDY 3:

THE SAMBRE-MEUSE COALFIELD:

A 'Bruise on the Banana'.

This is an example of an old industrial landscape, originally based on local coal measures, a European example of a **rust belt**. It is the eastern part of the extensive Franco-Belgian coalfield, which stretches 220 km from northern France into the area around Liege (see Figure 9.15). The geographer, F J Monkhouse, described it in the following terms as long ago as 1965 when the number of collieries had declined from 113 to 47 in only eight years . . .

Derelict collieries, overgrown spoil banks, a chaos of pit shafts, blast furnaces, and steel works, chemical factories, long rows of small, drab,

garden-less dwellings built in irregular rows – all these are typical of the crowded and haphazard industrial development of the nineteenth century (see Figure 9.16).

By the time a visitor from another rust-belt area, the north east of England, recorded his impressions in the 1990s, the coal mines had all closed, either because they had been worked out or because of the increasing difficulty and costs of mining the intensively folded and faulted coal seams.

In Philippeville we parked the car, and caught the train into Charleroi. Charleroi was the centre of the Pays Noir, the Belgian Black Country. The Sambre Valley in which it lay had been a rich coalfield, and the hub of the Wallonian steel industry. Artists who visited the area at the turn of the century, such as Maximilien Luce, painted scenes of smoke and raging flame, confined inside a loose picket of brick chimneys, beneath sky the colours of a well-established bruise. Nowadays the coal has gone, the steel business, once the heart of all industrial production, is in need of a pacemaker and the sky

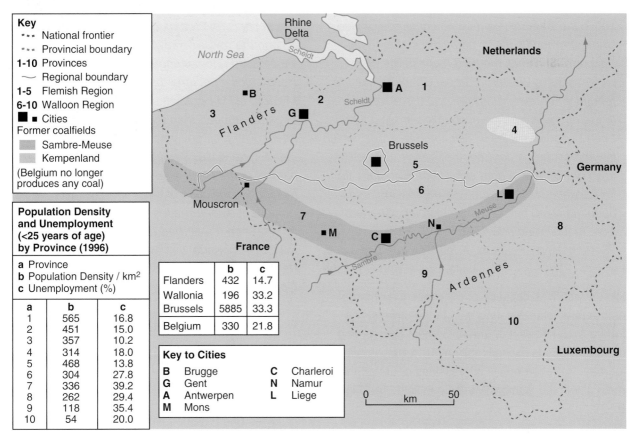

Key

- - -	National frontier
- - -	Provincial boundary
1-10	Provinces
~	Regional boundary
1-5	Flemish Region
6-10	Walloon Region
■ ▪	Cities

Former coalfields

- Sambre-Meuse
- Kempenland

(Belgium no longer produces any coal)

Population Density and Unemployment (<25 years of age) by Province (1996)

a Province
b Population Density / km²
c Unemployment (%)

a	b	c
1	565	16.8
2	451	15.0
3	357	10.2
4	314	18.0
5	468	13.8
6	304	27.8
7	336	39.2
8	262	29.4
9	118	35.4
10	54	20.0

	b	c
Flanders	432	14.7
Wallonia	196	33.2
Brussels	5885	33.3
Belgium	330	21.8

Key to Cities

B	Brugge	C	Charleroi
G	Gent	N	Namur
A	Antwerpen	L	Liege
M	Mons		

Fig 9.15 Belgium: base map

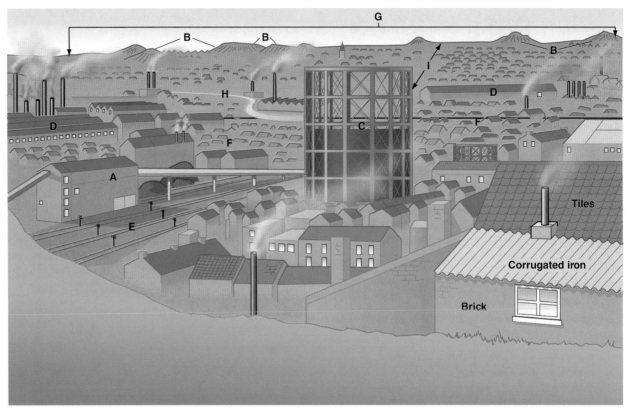

Fig 9.16 Field sketch: old industrial landscape – Liège c. 1960

above the Sambre is a less violent shade of blue. Unemployment is rife, even by the standards of Belgium, which has the third-highest rate of joblessness in Western Europe. Charleroi has the reputation of being run-down, but there was more to the atmosphere than economic hardship. I had been to poorer places in Belgium. Mouscron, which lay across the border from the French rust-belt town of Roubaix, a grim satellite of Eurostar Lille, for instance ... Mouscron was depressing, sad, haunting even, but it wasn't disturbing. Charleroi was, though, and how. (See Figure 9.18)

Source: Harry Pearson: *A Tall Man in a Low Land.*
Little, Brown and Co. 1998

His unease was because of the presence of large numbers of jobless young men in the streets. (As Figure 9.15 shows, the unemployment rate among the under-25s is exceptionally high). Later on his journey, he recorded his impressions of Liège, the main city in the Sambre-Meuse coalfield (see Figure 9.17):

The train from Tongeren to Liège rumbled through the farming land of the Haspengouw, past rolling wheat fields and teams of toiling Sikhs, crouching among the rows of strawberry plants. Near Glons, we crossed another linguistic boundary, and the Haspengouw became the Hesbaye ... After Milmort, tall water towers and bulky relay stations began to appear, then the first of the conical spoil heaps with their spiny covering of tiny pine trees (see Figure 9.19). Soon the train was running parallel to the River Meuse, and we were into the industrial dereliction of Liège itself, with its redundant blast furnaces and unattended coal quays. Even the air had a gritty texture. The hop town of Poperinge was twinned with Hythe in Kent. Liège looked like it had a suicide pact with Consett.

Source: Harry Pearson: *A Tall Man in a Low Land.*
Little, Brown & Co. 1998

Origins of Heavy Industry in The Sambre-Meuse Valley

The valley was noted for its iron industry in the Middle Ages, several hundred years before the Industrial Revolution. There were some similarities with Bonawe in that the iron smelting was dependent on charcoal and water power, but the blackband iron ore and limestone were extracted locally (see Figure 9.22).

Belgium was the first country on mainland Europe to undergo an Industrial Revolution. The process started in 1823, seven years before Belgium was created as an independent state, when the first coke oven was installed at an iron works in Seraing, about 8 km upstream from the centre of Liège (see Figure 9.17), by John Cockerill. By this time, coal had replaced charcoal as the fuel source, and Cockerill, a Yorkshireman, provided the technical knowledge, a skilled immigrant workforce and the capital required for this industrial innovation. This was an early example of foreign investment. SA Cockerill-Sambre, in which the majority stake is now held by Usinor, the French steel company, is still the principal steel producer in Belgium, although production has declined and the original locational factors have long since ceased to matter.

Locational Factors

In the nineteenth century, the iron and steel industry in the Sambre-Meuse Valley developed because of

◆ **availability of the energy source and raw materials** in close proximity to each other – coal

seams (although contorted) close to the surface, blackband iron ore and limestone

◆ **the industrial tradition** of the valley, and the availability of a workforce skilled in coal mining and metal manufacture, both of which had been important since the thirteenth century

◆ **a strategic position** which facilitated the 'pulling-in' of migrant labour from Flanders, where cloth production had been important since the Middle Ages, and from the Ardennes and north-east France

◆ **excellent transport facilities**, with, at first, cheap water transport on navigable rivers and canals, and then what was to become the densest rail network in Europe

◆ **inward investment by entrepreneurs** from England and France in transport, mining, iron-working and the application of the latest technology

◆ **close proximity to markets** in northern France and western Germany

Vertical Integration of Industry

By the 1830s, when the railway network was growing steadily, the Sambre-Meuse was the largest producer of coal in Europe and the Cockerill company was the largest industrial company, not

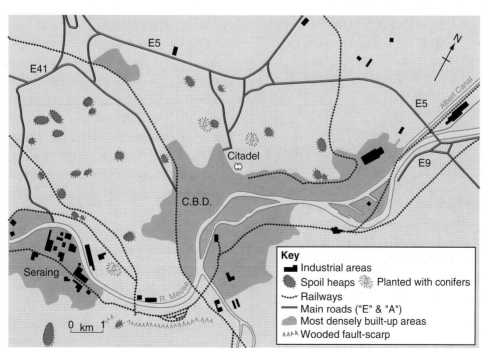

Fig 9.17 Industrial landscape: Liège

only in Belgium, but in the whole of Europe. It employed as many as 2500 workers in its core enterprise, the iron works at Seraing, but shrewd investment had resulted in Cockerill gaining control of all the stages of production from mining to final processing – the company was an example of **vertical integration** of industry (see Figure 9.20).

The Multiplier Effect

The expansion of the iron and steel industry, the improvements in the transport network and the rapid increase both in the total population and the skilled workforce, resulted in the attraction of other industries (see Figure 9.21). Some were associated very closely with existing industries such as chemical manufacture, using the by-products of the coking process; the manufacture of locomotives and rolling stock; engineering; gas works (and later power stations). Other industries were more loosely connected, for example, glass-making and zinc smelting, while the expanded local market brought about an increase in the local manufacture of food

Fig 9.19 One of the many coal spoil heaps on the Hesbaye Plateau, now planted with conifers

products, clothing, and household goods. Such a 'snowball' expansion in industry is an example of the **multiplier effect**.

De-industrialisation and Industrial Inertia in the Sambre-Meuse Valley

Almost inevitably, as one of the pioneering areas in the Industrial Revolution, the Sambre-Meuse coalfield was one of the first to suffer from the effects of de-industrialisation, long before the terms '**rust-belt**' and **smokestack** were coined. This involved a substantial reduction in the **heavy** industries – coal mining, iron and steel, chemicals and heavy engineering. The iron and steel industry is still found in the valley, admittedly in a contracted form, although the original locational factors have ceased to apply. This is an example of what is known as **industrial inertia**: an industry remains in its original location, relying on imported raw materials and fuel; its skilled workforce, and perhaps government subsidies to keep it going. By the middle of the twentieth century, the coking coal needed was being brought into the area from the Campine coalfield to the north, and imported from the Ruhr, while all the iron ore was being imported from Luxembourg, France and Sweden, and from even further afield, Brazil, Liberia and Mauritania. Since then, the industry has declined further through **rationalisation** – the closure of uneconomic plants, investing in expensive new technology, and

Fig 9.18 An old industrial landscape in Charleroi in 1986: coking ovens loom above terraced housing

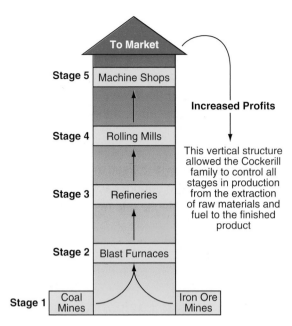

Fig 9.20 Vertical integration of the iron industry

shedding a large proportion of its workforce, while concentrating on the production of high quality steel.

The **de-industrialisation** of the Sambre-Meuse began in the second decade of the twentieth century and its subsequent acceleration can be attributed to several factors:

◆ **lack of investment in the steel industry**, especially during the depression of the 1930s. The result was an antiquated industrial structure which could not cope with the changing demands of the market

◆ **a rapid decline in coal production**, so much so that by 1984 all the coalmines had closed. Apart from the lack of coking coal, the high costs of mining because of the geological difficulties meant that locally produced coal could not compete with cheaper imports of coal, even from as far away as the USA. There was also competition from alternative sources of electricity generation, especially nuclear energy

◆ **an obsolete infrastructure**, especially in terms of transport – the railway network had to be modernised, canals had to be widened and deepened, and motorways (E-Routes) constructed

◆ **its inland location**, which increased its costs of production by its total reliance on imported coal and iron. The higher cost of labour and social

security in the Sambre-Meuse made it even less competitive. It faced increasing competition from coastal steel works such as the Sidmar plant near Ghent

◆ **the loss of both home and export markets** because of increasing imports of steel in the second half of the twentieth century, first from Japan, and later from South Korea and Brazil, both examples of **NICs** (Newly Industrialising Countries)

◆ **a negative external image** which discouraged investment. This arose because of the accumulated 'smokestack' elements in the landscape (see Figure 9.16), militant industrial relations, and the inequalities within Belgium itself. The Sambre-Meuse coalfield is in Wallonia (French-speaking), and a recent survey of regional preferences showed that Flanders (Flemish-speaking, and with a more varied industrial base) was four times more attractive for new industrial investment than the Sambre-Meuse coalfield (see Figure 9.15).

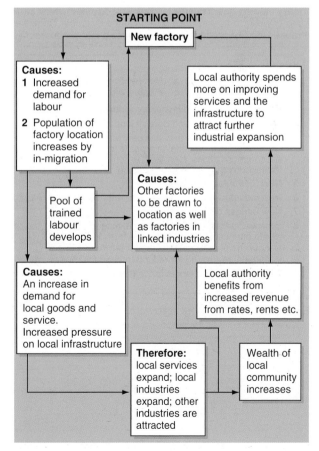

Fig 9.21 A simple version of the multiplier model

Consequences of De-industrialisation

a) For the Environment

The legacy of 150 years of heavy industry, and the subsequent de-industrialisation, is very evident in the landscape, with derelict industrial buildings, usually along the riverside, and a high frequency of spoil heaps and subsidence hollows on the Hesbaye plateau above the valley floor (see Figure 9.22). Pollution of the atmosphere (see the extract above) and the rivers are still major problems. The

concentration of heavy industry on the valley floor more than 100 m below the plateau also resulted in frequent temperature inversions and the formation of a distinct heat island, both over the centre of Liège and above Seraing. Efforts were made to repair the damage in the last decades of the twentieth century.

b) For the People

Unemployment rates increased and remain high, especially among the young. Lack of opportunity within the Sambre-Meuse has resulted in young

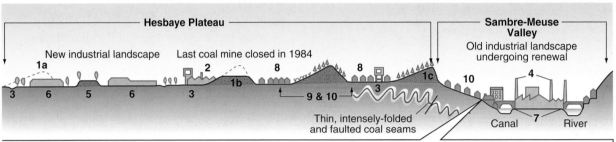

"Heavy industry has created a landscape of chimneys, blast furnaces, cooling towers, pithead frames, and slag heaps...a disorderly mass of workers' houses grew up around the factories. There was no overall plan...some are grouped in mining villages, others are arranged in alleys, courtyards or passages...a row of houses may end in a wooded slag-heap."

Key

1 "Terrils" (spoil-heaps) **a** Removed to clear site ("brownfield") for buildings **b** Top levelled
c Planted with conifers, and in some cases, vines

2 Derelict pit-head buildings either renovated as industrial museum, or demolished

3 Subsidence hollows and derelict land reclaimed

4 Surviving steel works now use new technologies (computers and robotics). Atmospheric and river pollution now firmly controlled. New industrial prosperity at Liege based on work related to the space programme, software engineering, telecommunications. Charleroi: aeronautical industry, petro-chemicals, computer graphics. Namur: major science parks (biotechnology) and food processing

5 New "E" route (motorway)

6 New industrial estates, using road, rather than rail or canal, transport

7 New bridges over river and canal

8 Improved infrastructure in former mining villages e.g. the renovation of the model village of Grande Hornu near Mons, where bituminous coal was mined from 1778 to 1953. The 400 houses have been renovated and the main buildings are now museums, galleries and a business park

9 Vocational re-training for redundant workers/miners, many of whom were immigrants from Southern Europe and North Africa

10 Early retirement schemes for coal miners and steel workers

Fig 9.22 Sambre-Meuse Valley: improvements and re-industrialisation

people with the necessary qualifications commuting to work in the cluster of science parks around Brussels, or to jobs in the new '**sunrise**' industries in Flanders and neighbouring areas in France, Luxembourg and Germany. Unemployment may also have resulted in permanent migration. A major consequence of the economic decline has been substandard housing. The small 19th Century miners' cottages – '*corons*' – built round a courtyard, and the equally inadequate rows of houses on the valley floor mentioned in the first extract above have not always been rebuilt or replaced.

Solving the Problems of the Sambre-Meuse Valley: Re-industrialisation

The attempts to resolve the problems of this area were not helped by the economic and political rivalry between Wallonia and Flanders, both of which have their own regional government, but the national government and the European Union have both played a major part in the economic regeneration of the Sambre-Meuse. For example, the **European Regional Development Fund** granted the Liege area **Objective 2** status, by which funds, through the **Resider** programme, were made available to improve the economic development of regions in economic decline and with high unemployment. The **European Coal and Steel Community**, through its **Rechar** programme, made substantial financial provision for the regeneration of the degraded industrial environment (see Figures 9.22 and 9.23), as well

Fig 9.24 This vertical aerial photograph shows a new industrial estate on a greenfield site on the Hesbaye Plateau

as providing money to fund vocational retraining and early retirement schemes.

The reindustrialisation of the area has thus involved:

- a massive modernisation of the transport network
- the beginning of the reclamation of the industrial wasteland
- the building of new industrial estates on **greenfield** sites (rather than on **brownfield** sites) on the periphery of urban areas, and along the new E-Routes (see Figure 9.24)
- the attraction of new **sunrise** industries, such as electronics and pharmaceuticals, perhaps owned by **transnational** companies.

This development of new industries has been favoured by the availability of skilled labour, although retraining has been necessary; the excellent transport network, which gives access to a potential market of some 50 million people within some three hours time-distance; and the provision of a package of economic incentives. This last allows money for labour-retraining schemes, reduced tax bills and loans and grants for land purchase. Local development programmes such as the *'Investir a Charleroi'* have therefore been successful in attracting firms involved in electronics, synthetic fibres, confectionery and printing. Examples of investment by transnationals include pharmaceutical companies (Beecham), an earth-moving equipment firm (Caterpillar), as well as

Fig 9.23 The renovated pithead buildings at this former colliery are now part of an industrial heritage site.

many service companies. The industrial base has therefore been widened and work opportunities have, in theory, been increased, although unemployment remains high. The other major concern is that the new industries are even more susceptible to external influences and changing markets than the old heavy industries; electronics is a modern example of the multiplier effect in industry, but is also subject to cyclical growth and decline. The economic regeneration of the Sambre-Meuse has been partially successful, but Wallonia as a whole has lost out to Flanders in the growth of jobs in the service sector (**tertiarisation**), notably, tourism and the concentration of EU jobs in

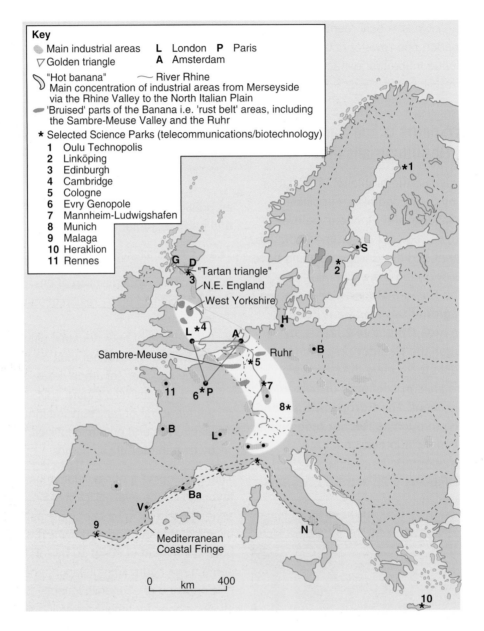

Key
🔘 Main industrial areas   **L** London   **P** Paris
▽ Golden triangle   **A** Amsterdam

🔗 "Hot banana"   ⌒ River Rhine
Main concentration of industrial areas from Merseyside via the Rhine Valley to the North Italian Plain

🔗 'Bruised' parts of the Banana i.e. 'rust belt' areas, including the Sambre-Meuse Valley and the Ruhr

✳ Selected Science Parks (telecommunications/biotechnology)
  1 Oulu Technopolis
  2 Linköping
  3 Edinburgh
  4 Cambridge
  5 Cologne
  6 Evry Genopole
  7 Mannheim-Ludwigshafen
  8 Munich
  9 Malaga
 10 Heraklion
 11 Rennes

Fig 9.25 Industrial location in Europe

Brussels. One major advantage which should always favour the Sambre-Meuse Valley is its favourable location within the northern segment of the '**Euro-banana**', with its north-west–south-east axis of industrial development (see Figure 9.25).

Like other areas of smokestack industries, it might be said that the Sambre-Meuse was very much a bruise on this 'banana' although the measures already taken might ensure that the bruise will gradually fade away.

Other Rust-Belt Areas: Beyond the Banana

The Sambre-Meuse is a typical rust-belt area, but it is not unique, hence the many areas, within and outwith the 'banana', notably the Ruhr coalfield, which have benefited from the ECSC's Rechar programme. Within Britain, two of the rust-belt areas which have suffered most from the effects of de-industrialisation are Central Scotland and North East England.

9.6 Central Scotland

The human consequences of industrial decline in one small community were graphically described in 1995:

Three cheers for once-thriving town

Joe Boyle can remember when Mossend was big and busting enough to support four picture houses ... Nursing a pint at the bar of the Derby Inn on the Lanarkshire village's main street, the 67-year-old former steelworker has plenty of time for reminiscing. He has not worked since being made redundant six years ago when the Clydesdale Steel Works shut down ... However, there was a time when there was too much to do rather than too little. Work was hard but not scarce. Beardmore's steel works and the ICI chemical plant absorbed most of the local labour. Before that there was coal mining. You just had to look at the amenities to know that Mossend was thriving. 'You could get a suit from the tailors. We had ice-cream shops, even a couple of fish and chip shops. People used to come from Bellshill to shop here. Now there's nothing'. Since the bulldozers moved in about 12 years ago, when the economy began to slide, Mossend has become little more than a mile-long ribbon of road joining Holytown and Bellshill, not far from the Ravenscraig site – perhaps the highest-profile casualty in the recent decline of Scotland's manufacturing base ... Locals boast that it once sported 20 pubs, but only two have survived, and a new one, built beside the shopping centre, has already

been boarded up after going out of business. ICI closed long ago, and the collapse of the steel industry in Lanarkshire has hit the village hard ... Small industrial units dominate one side of its road, their squat ugliness partly hidden by trees. A fishing tackle shop and a discount bedding store have survived as they are not reliant on passing trade, but there are few customers. The custom-built shopping centre across the road tells the same story. By mid-afternoon, the shutters are down on ... at least half the retail units ... Local people insist that the community spirit is thriving, and they are confident that yesterday's announcement of a new television tubing plant bringing more than 3000 jobs will mean rich rewards. The tiny dot on the map is now known as the site of Scotland's biggest-ever inward investment.

Source: Lynn Cochrane: *The Scotsman*

The rust-belt village of Mossend (see Figure 9.1) described above is typical of many places in Central Scotland which have faced serious problems in the last 50 years, with the virtual disappearance of the traditional heavy industries – coal mining, steel manufacture, shipbuilding, heavy engineering and chemicals. Figure 9.1 shows the former extent of coalmining, and although there are several large areas

Fig 9.26 The coal-fired power station at Longannet on the opposite side of the Forth from Grangemouth uses coal from the last deep mine in Scotland

Fig 9.27 The new Chunghwa Tube works at Mossend benefits from both the improved road network and the nearby Eurocentral rail freight terminal

where opencast mining is carried on, especially in Lanarkshire, there is only one deep mine left in production, serving the power station at Longannet in West Fife (see Figure 9.26). The closure of the steel mill at Ravenscraig mentioned above was a major blow to heavy industry in Scotland; now there are only one or two small steel works left, while

shipbuilding on the Clyde is similarly down to the last remnants, struggling to survive.

Finding the Solution

As these traditional industries have declined, there have been many attempts to attract new industries to Scotland. The government created **development areas** and **enterprise zones**, in which entrepreneurs and transnationals received substantial funding for setting up businesses. EU funding was also sought from the European Development Fund and the ECSC Rechar scheme, which came to the aid of the Sambre-Meuse Valley. Before these measures were taken, and before the influx of electronics companies such as the new Taiwanese Chunghwa cathode ray tube plant at Mossend (see Figure 9.27), mentioned above, hopes of economic revival were based on the expectation that the motor vehicle industry would become re-established in Central Scotland, but in turn four major enterprises have failed (see Figure 9.28)

◆ Hillman-Rootes-Citroen (cars) at Linwood near Paisley

The Most Significant Investments for the Knowledge Economy: 1946-1999
Source: *Scotland on Sunday 14/12/97*

1	N.C.R.	Dundee	1946
2	IBM	Greenock	1951
3	Hewlett-Packard	S. Queensferry	1966
4	Motorola	East Kilbride	1969
5	Digital E.C.*	Ayr	1976
6	NEC	Livingston	1981
7	Shin-Etsu Handotai Europa	Livingston	1984
8	Compaq	Bishopton	1987
9	Sun Micro	Linlithgow	1988
10	Chunghwa	Mossend	1995
11	Hyundai	Dunfermline	1996
	(unfinished; mothballed; bought by Motorola in 2000)		
12	Cadence	Livingston	1997

*Compaq took over Digital in 1998 which resulted in the closure of plants at Stirling and Irvine

There are major clusters of electronics firms in some of these locations e.g. East Kilbride, as well as other places e.g. Glasgow, Prestwick, Cumbernauld and Glenrothes.
Electronics firms in minor outlying centres e.g. Haddington, Selkirk, and Galashiels, were closed down in the late 1990s

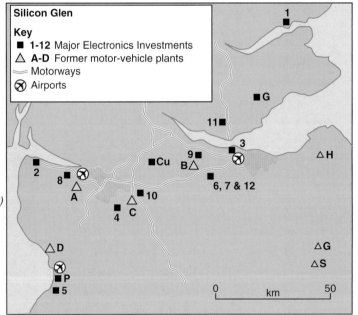

The map above also shows the locations of former motor vehicle manufacturing / assembly plants i.e.

A Linwood
B Bathgate
C Uddingston
D Irvine

Fig 9.28 Silicon Glen – 50 years of inward investment

Ford, and is therefore another example of a transnational firm. Production of Volvo buses is now concentrated in larger plants in Sweden and Poland, while trucks for the European market are made in Sweden and at Antwerp in Belgium. The closure of the Volvo plant, and of the other vehicle plants earlier, had very serious effects on the local economy, and they are examples of why the Scottish economy has been described as

an unsettling rollercoaster of achievements and setbacks, competing for investment and sales in a world transformed by technology.

Figure 9.28 shows, in sequence, the establishment of what might be considered the most significant examples of inward investment in Scotland in the second half of the twentieth century, some of which will be considered later.

They do not include the four vehicle plants listed above, since they did not survive, despite receiving large subsidies from the government and the high expectations that the motor vehicle industry would be the keystone of the re-industrialisation of the Central Scotland rust-belt. There were problems both with the management and trade unions, while

Fig 9.29 Bathgate: the BMC–Leyland Light-Medium Vehicles Division, set up to provide work for redundant coal miners, relied on strip steel from Ravenscraig during its short existence

◆ BMC (trucks and vans) at Bathgate in West Lothian (see Figure 9.29)

◆ Caterpillar (earth-moving equipment) at Uddingston

◆ Volvo (buses and trucks) at Irvine in Ayrshire.

Truck production at the Volvo plant ceased in mid-2000, two years after the bus division closed down. Volvo, originally a Swedish firm, is now owned by

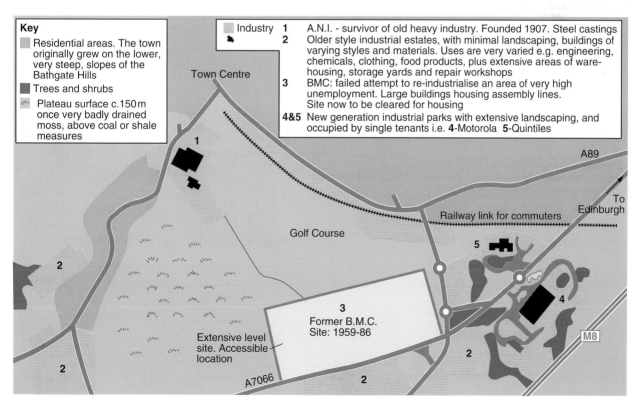

Key

Residential areas. The town originally grew on the lower, very steep, slopes of the Bathgate Hills

Trees and shrubs

Plateau surface c.150m once very badly drained moss, above coal or shale measures

Industry **1**
2

1 A.N.I. - survivor of old heavy industry. Founded 1907. Steel castings
2 Older style industrial estates, with minimal landscaping, buildings of varying styles and materials. Uses are very varied e.g. engineering, chemicals, clothing, food products, plus extensive areas of warehousing, storage yards and repair workshops
3 BMC: failed attempt to re-industrialise an area of very high unemployment. Large buildings housing assembly lines. Site now to be cleared for housing
4&5 New generation industrial parks with extensive landscaping, and occupied by single tenants i.e. **4**-Motorola **5**-Quintiles

Town Centre

A89

To Edinburgh

Railway link for commuters

Golf Course

5

3 Former B.M.C. Site: 1959-86

4

Extensive level site. Accessible location

A7066

2

2

2

2

M8

Fig 9.30 Bathgate – a multiple industrial landscape

another reason put forward for their failure was that they were too distant, both from the suppliers of the components, and the markets. Another is that in three cases, they were more vulnerable to closure because they were foreign-owned.

Impact on the Community: Bathgate on the Rollercoaster (see Figure 9.30)

When the BMC plant closed in 1986, there was a great increase in unemployment. It had employed 6000 at its peak; after it closed more than a quarter of the workforce in the Bathgate area was looking for work. Long before then, the traditional extractive industries in the area, oil shale and coal mining, had ceased, and a new growth industry, Plessey Electronics, had failed. There were several plans put forward for the use of the disused 80 ha BMC site, including a giant theme park, but eventually it was used for car storage. After 13 years of neglect, it was finally announced that the brownfield site would be cleared to build 1000 houses. In the meantime, another wave of industrial development had brought two major new employers to Bathgate – Motorola (electronics), employing 4000 people, and Quintiles (pharmaceuticals), employing 1000 (see Figure 9.31). As a result of all the changes, there are very varied industrial landscapes to be found within a very small area.

9.7 North East England: The Nissan Car Factory

Foreign ownership and distance from both suppliers and customers have not mattered in the case of the Nissan factory at Washington New Town near Sunderland (see Figure 9.32), located in a classic rust-belt/smokestack industrial area where coal mining, steel-making, shipbuilding and associated industries have been in rapid decline. The last deep coal-mine in North East England at Ellington (see Figure 9.33), which supplied the nearby Alcan smelter, was expected to close in 2000 because of increasing geological difficulties mining under the North Sea. Now highly mechanised open-cast mining is all that remains. Defence cuts have had an adverse effect on both shipbuilding and tank manufacture on Tyneside.

The unemployment figures in the Sunderland-Washington area were so high in the 1980s that it was granted **enterprise zone** status by the government, as well as qualifying for aid from the EU under the Rechar and Resider schemes. Location theory suggests that the optimum location in Britain for motor vehicle manufacture is in the West Midlands where the multiplier effect has resulted in a cluster of car factories surrounded by their suppliers of components, so that transport costs are low. However, government policy was to direct inward investment to areas of chronic de-

Fig 9.31 Bathgate: the Quintiles plant opened in 1997 on a greenfield site immediately east of the cleared site of the vehicle factory

industrialisation and consequent high unemployment. The grants, loans and concessions were a very large carrot to foreign firms, especially as a location in North-East England gave non-European companies free access to the large EU market for their products.

Figure 9.32 summarises the other factors which influenced Nissan's choice of a location. Low labour costs were another critical factor as was the availability of a large area of flat land, a former airport, to build the extensive assembly lines required (see Figure 9.34).

The choice of location has been a great success, so much so that the Washington plant survived the massive rationalisation which took place within the parent company in 1999. Nissan was rescued from a financial crisis by a merger with Renault, which

became the dominant partner, and announced the closure of four factories, including two in Japan.

The reasons for the Washington plant being allowed to continue were attributed to a strong Japanese management and a well-trained, well-treated workforce of 5000 which had achieved a very high level of productivity as the table below reveals:

The world's most productive car plants: 1997		
Car plant location	Maker	Cars per worker per year
Mizushima, Japan	Mitsubishi	147
Suzuka, Japan	Honda	123
Washington, Durham	Nissan	98

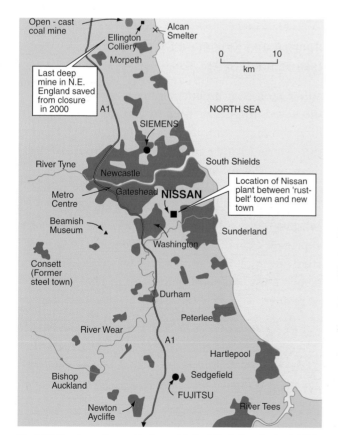

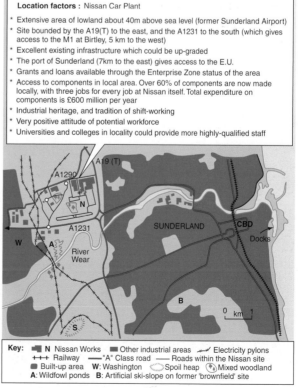

Fig 9.32 Sunrise and sunset in the rust-belt of North East England. **(A)** Location of Nissan Plant. **(B)** Site of Nissan.

Fig 9.33 Ellington colliery, the last deep mine in North-East England and one of only 17 left in Britain, was saved from closure in 2000 by government intervention

Fig 9.34 This aerial view of the Nissan site shows the extensive area of flat land required for the assembly lines

Another major factor in Nissan staying at Washington was the announcement of a massive government grant to help fund a major expansion programme in which a new model, the Almera, would be built. **Government intervention** continues to be a major factor in maintaining the Nissan plant at Washington. However, its success must be seen against a background of

◆ large-scale overproduction in the European car industry

◆ less frequent demand for new cars as the quality has improved

Fig 9.35 The final assembly line producing the Nissan Almera. In 1998 the Nissan plant at Sunderland was by far the most productive car factory in Europe

◆ the decline in the number of workers needed to work on assembly lines (see Figure 9.35).

With the decision to build the new Micra there rather than in France, the success of Nissan at Washington seems assured, unlike two other attempts to revive and broaden the industrial base in North-East England.

Sunrise and Sunset on Silicon Fell
(see Figure 9.32A)

While motor car production in this area has succeeded, there have been two major blows to the local economy, with the loss of about 1800 jobs in the electronics industry. In 1999, the closure of the Siemens plant in North Tyneside, after only 15 months of production, was followed by that of the Fujitsu plant in Newton Aycliffe, 34 km south of Washington (see Figure 9.36). These were both very expensive examples of inward investment with very large government contributions, which gave the local people hope of a permanent boost to the local economy:

A lot of people were encouraged to move here (Newton Aycliffe) and buy houses with £99 deposits. Two years later, they're out on their ear. Microchips were supposed to be the best thing since sliced bread; now they're worth nothing.

Source: Mark Rowe: *The Independent on Sunday* 6/9/98

Fig 9.36 The Fujitsu microchip plant at Newton Aycliffe, heavily subsidised by the government, closed after only two years' production

Unfortunately, it would seem that the North-East of England is not immune to the roller-coaster economy which, as we have seen, has also affected Central Scotland. External factors, in this case over-capacity in the world production of microchips, often have a negative affect on re-industrialisation. North-East England now has the larger proportion of its workforce employed in the tertiary and quaternary sectors; in retailing, for example, the Metro Centre at Gateshead; leisure, the heritage industry, for example, Beamish Open Air Museum, and in call centres – Barclaycard employs 2000 in Sunderland (see Figure 9.50).

Fig 9.37 The mothballed Hyundai plant on the M90 east of Dunfermline was taken over by Motorola in 2000 to produce silicon chips for a new generation of mobile phones to be made at Bathgate

9.8 SILICON GLEN IN THE GLOAMING?

Hyundai Plant Stays on Hold: Korean Firm warns no decision on Fife semi-conductor factory until 2001

The planned Hyundai semi-conductor fabrication plant at Dunfermline which could have created up to 5000 jobs will stay mothballed until 2001 at the earliest. The Hyundai announcement . . . continues the wait for what would have been Scotland's biggest inward investment for years . . . the group did not plan to resume construction on the Dunfermline plant and another one in Wales until at least late next year . . . Work was halted in November 1997 as problems hit the South Korean economy.

Source: Ken Symon: *The Scotsman* 22/10/99

The extract above confirms two points:

◆ the continuing attraction of Central Scotland as a location for inward investment by electronics firms

◆ the vulnerability of such investments, despite substantial government aid, to external economic factors.

Hyundai was one of South Korea's largest conglomerates, with as many as 39 subsidiaries across a wide range of manufacturing (**horizontal integration**) including car manufacture – it was considering setting up a car factory at Robroyston in Glasgow earlier in 1997. Since the financial crisis, it has been restructured and merged with a rival, and is now the world's largest manufacturer of dynamic random access memory chips. While Hyundai considers its future, the plant at Dunfermline, its first in Europe, remains unfinished, an unusual example of a completely unused industrial landscape on a greenfield site with all the infrastructure in place (see Figure 9.37). As has been said already, there have been several examples of electronics companies failing to survive, both in Central Scotland and the Borders, because of problems in the global market or the home economy, or simply because of the rapid pace of change and innovation in the industry.

The original reasons for the clustering of electronics firms in Silicon Glen still apply (see p. 263), despite

Fig 9.38 The Alba Centre on a greenfield site at Livingston which will become the core of the new industrial landscape

the failures. The electronics industry employs as many as 60 000 people; it still has a very obvious presence in the new industrial landscapes of Central Scotland (see Figure 9.28), and its products have overtaken whisky as Scotland's major export. Its strengths lie in the diversity of its products and, more importantly, in the increase in research. The emphasis has moved away from the assembly of products (the screwdriver economy) to a better balance between manufacture and research (the knowledge economy). Scotland is:

◆ a major producer of printed circuit boards, e.g. Prestwick Circuits, which exports 70% of its production to Europe, South Africa and North America

◆ the largest producer in Europe of personal computers and workstations.

There have also been major successes in fibre-optics, digital phones, semi-conductors (despite the Hyundai delay) and software, while the arrival of Cadence Design Systems at Livingston in 1999 was seen as being significant in several respects i.e.

◆ it should eventually create up to 2000 jobs

◆ it should put Scotland in the forefront of research in semi-conductor design, having all the systems in an electronic product on a single chip

◆ it could attract other advanced technology companies into Scotland

◆ in collaboration with Scottish Enterprise and the universities, there should be an increase in the number of graduates to help solve the shortage of semi-conductor designers

◆ its presence should help to ensure the survival of existing electronics firms

◆ Cadence is the key tenant and main focus of Scottish Enterprise's Alba Centre Technology Park at Livingston. The hope is not only that other research companies will be attracted to the Park, but that the Internet will link it to similar organisations worldwide which do not have a physical presence in Silicon Glen: the Park will become a global e-village, with as many as 4000 scientists and researchers (see Figure 9.38).

The new industrial landscape in which the electronics

Fig 9.39 The Sun-Micro plant on an isolated greenfield site just south of the M9 on the outskirts of Linlithgow

firms are found in central Scotland has remained essentially the same for the last thirty years, although the companies might have changed. Sun Micro at Linlithgow (see Figure 9.39) which designs, builds and installs Internet systems on a contract basis, is another example of the move away from the screwdriver to the knowledge economy in Silicon Glen. The elements are constant, even if the focus keeps changing, although it is claimed that the new Cadence complex at Livingston is the last word in providing a stress-free working environment (see Figure 9.38).

9.9 The Chemical Industry in Central Scotland

A unique industrial landscape in both Central Scotland and North-East England (on Teeside) is provided by the chemical industry. Grangemouth, on the south side of the upper Firth of Forth (see Figure 9.40A), has been involved with the industries of Central Scotland since it was planned as a port in the early days of the Industrial Revolution.

The present-day landscape is the result of eighty years of evolution within the chemical and petro-chemical industries, so it combines elements more usually associated with old industrial landscapes – multiple stacks, cooling towers, storage tanks – with elements indicating the very complex, high-technological processes which are taking place on the site, for example, the many kilometres of angular pipelines, apparently forming the skeletal shapes of buildings.

It does not resemble a conventional new industrial landscape because of the bulk and height of the industrial buildings, the minimal landscaping, and the untended open spaces. It does not resemble an old industrial landscape in that the processes involved require large areas of land, and the safety factor has meant that the residential area is segregated, if not distant, from the industrial areas. Bounded by the M6 to the south-west, the residential area forms a wedge between the two main industrial zones, with the docks on the north side. Within the residential area, there are parks, sports grounds and tended open spaces, but they are all overwhelmed by the presence of industry, especially that of the BP complex on the north-east side. Such is the vertical scale of the works

on such an open site, although some of the tanks are half-concealed by bunds, that this is an industrial landscape which dominates a very much larger area than the immediate site (see Figure 9.41).

Locational Factors

The first manufacturing process on the site was related to the then thriving oil-shale industry in West Lothian, with Scottish Oils establishing a refinery. Scottish Dyes, making fast chemical dyes from coal-tar, was another pioneer on this site. The attractions of the site (see Figure 9.40B) for heavy industry were as follows:

◆ the tidewater location

◆ water for processing, and tidal water for the disposal of effluent

◆ the extensive area of flat carseland available (at low cost) on the raised shoreline, and later, by reclaiming land from the tidal mud flats at the edge of the Firth, between the mouths of the River Carron to the north-west, and the River Avon to the south-east. (The site is overlooked by the remains of the Roman Antonine Wall, built along the break of slope above the carse.)

◆ the presence of the port facilities and the railway network. By that time, the canal connections in the area were of no consequence.

Today, access to the road network is of much greater importance, while pipelines have largely replaced ships as a means of transporting oil to, and oil products from, the site. Grangemouth was first linked in 1951 by pipeline to Finnart on Loch Long, and much later to Cruden Bay, Mossmorran and Hound Point (see Figure 9.40A).

Industries Today

Although there are several smaller chemical firms, Zeneca and BP are by far the main employees of labour. The former occupies a relatively small area (32 ha) and specialises in agro-chemicals (e.g. fungicides), textile dyes, and fine chemicals with pharmaceuticals forming a very small proportion of its output. BP occupies a vast area (485 ha) on which three different processes are carried out:

◆ crude oil processing

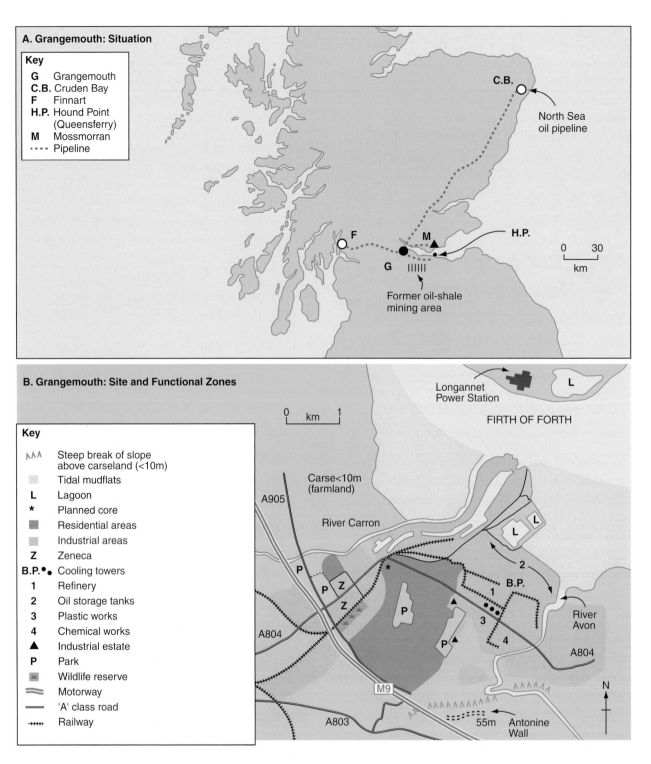

A. Grangemouth: Situation

Key
G Grangemouth
C.B. Cruden Bay
F Finnart
H.P. Hound Point
(Queensferry)
M Mossmorran
---- Pipeline

North Sea
oil pipeline

H.P.

Former oil-shale
mining area

0   30
km

B. Grangemouth: Site and Functional Zones

Longannet
Power Station

FIRTH OF FORTH

0   km   1

Key
ᴧᴧ Steep break of slope
above carseland (<10m)
Tidal mudflats
L Lagoon
* Planned core
Residential areas
Industrial areas
Z Zeneca
B.P.•₀ Cooling towers
1 Refinery
2 Oil storage tanks
3 Plastic works
4 Chemical works
▲ Industrial estate
P Park
🌼 Wildlife reserve
═ Motorway
— 'A' class road
+++ Railway

Carse<10m
(farmland)

A905

River Carron

River
Avon

A804

A804

M9

A803

55m   Antonine
Wall

N

Fig 9.40 Grangemouth

◆ oil refining

◆ petro-chemical production.

Grangemouth is the only place in Europe where all
three processes are found on the same site. BP have
recently invested heavily in a modernisation
programme which will involve expanding

production of polypropylene, ethylene and ethanol,
as well as building a combined heat and power
(CHP) station. The sums involved in such
developments are enormous: Zeneca invested £76
million in producing their new fungicide made from
naturally occurring fungicides found in mushrooms,
to create a total of 140 jobs. Both Zeneca and BP are

Fig 9.41 A unique industrial landscape in Scotland: the BP complex with its cooling towers and stacks on the flat carselands at Grangemouth

examples of **capital-intensive** industries, although they employ over 4000 workers between them.

Impact on the Population

The chemical industry is the major source of employment for Grangemouth in particular, as well as a large part of its hinterland, with many more people employed indirectly. The recent expansion at both Zeneca and BP employed more than 1500 construction workers. Another positive effect is the attraction of working here for graduates of Scottish universities in both chemistry and chemical engineering. On the other hand, almost all the jobs in the chemical industry are filled by males. This made the recent announcement of the closure of two clothing factories on the town's small industrial estates all the more serious for female employment.

Impact on the Environment

The safety factor is a prime consideration in chemical production, but pollution of the atmosphere and of the Firth is more likely to occur than an explosion. Both Zeneca and BP have reduced atmospheric emissions by the use of filters; solid waste from Zeneca is incinerated off-site, while liquid wastes (containing some heavy metals, but not mercury) is piped from Zeneca into the main channels of the Firth. All of these matters are carefully monitored but, as in all such large centres of chemical-production, these methods of dealing with pollution raise further questions. On a more positive note, an urban wildlife park has been created on the Jupiter site from 4 ha of derelict railway sidings close to the Zeneca works (see Figure 9.40B).

9.10 'Footloose' Industries on Science Parks

Figure 9.25 shows, in addition to the 'Euro-Banana' and the main industrial areas in Europe outwith that concentration, the location of selected science parks. The main industrial areas have evolved:

◆ at a power source using coal or hydro-electric power

◆ round a major port, i.e. where transport was a major factor

◆ in and around a major city i.e. where the market was crucial.

Science parks have evolved for very different reasons. In some cases, the science parks, as might be expected, arose from the commercialisation and application of scientific research at the nearby or host university, for example, Cambridge (see Figure 9.46). In some cases, it was an attempt to expand the industrial base of old areas within the Euro-Banana, such as Mannheim-Ludwigshafen. In other cases, it was the climatic attraction of the area (e.g. Malaga) to semi-retired entrepreneurs with capital and time to spare. In one particular case, Oulu in Northern Finland, the reasons are more complex and perhaps illustrate the importance of the **chance factor** in industrial location (see Figure 9.43).

The choice of Oulu as the location of the **technopole** (a cluster of IT and biotech. firms)

Fig 9.42 The new headquarters of Nokia in Salo, a town on the Gulf of Finland

Nokia, a world leader in the telecommunication industry, evolved from

- a merger between two companies which initially had no links with IT
- an area where there was an abundance of HEP
- an area where the main industry had been textiles, although the main local raw material was timber
- an area where the Industrial Revolution began in Finland using local water-power, and imported, tax-free, raw materials and machinery

The major factors in Nokia's recent rapid growth have been

- the demand from the market, both national and international
- the evolution of a knowledge economy in Finland
- its massive recruitment of scientists and research

Key

T	Tampere
N	Nokia
P	Pori
⇨	Early imports of cotton & machinery
St. P.	St. Petersburg
H	Helsinki
T	Turku
*	New headquarters of Nokia
X	Technopole

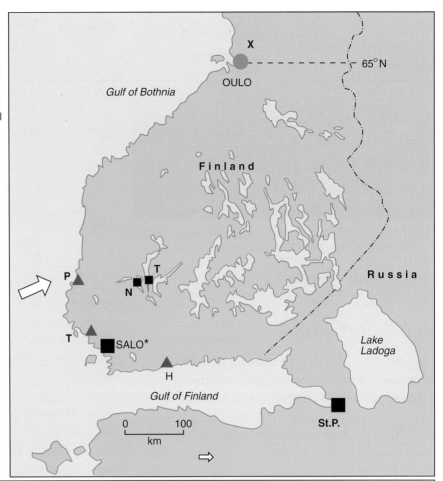

Evolution

1828	James Finlayson, a Scot, established an iron foundry and cotton mill at Tampere
Factors	**a** entrepreneurial skills **b** water power, using the 20m drop between two lakes **c** a concession from the Czar to import, duty free, cotton and machinery
Result	Other textile mills were established in the Tampere region
Changes in power source	Steam power replaced water-power in the late 19th C.; hydro-electric power became available in the early 20th Century
Spread and diversification of industry	Nokia, a small town 10km SW of Tampere, attracted industries, including paper making
1960s	The Nokia group formed by a merger between two companies, one making paper, the other rubber goods
1967	The group then diversified by making conventional telephones
1990	Nokia made the decision to sell off its paper and rubber interests, and concentrate on making mobile telephones
1998	Nokia became the world's main producer of mobile telephones

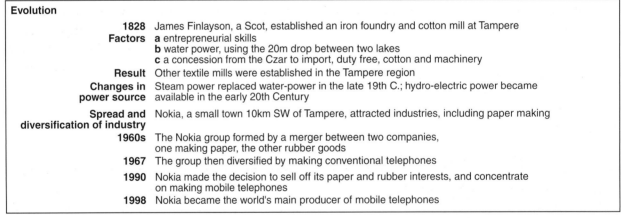

Fig 9.43 Nokia – the origins and evolution of a Finnish transnational

might be traced back to the beginnings of the Industrial Revolution in Finland. James Finlayson, a Scottish Quaker, had been employed in the workshops at St Petersburg, and he used his entrepreneurial skills to obtain the concession from the Czar to establish the first water-powered mills at Tampere (Finland was then part of Russia). Tampere, then a small village, eventually became the main industrial centre in Finland, specialising in textiles. Nokia was then a small settlement on the periphery of Tampere. Figure 9.43 shows the links between these early developments and the astonishing rise to world status of the IT firm, Nokia, in the last decade of the twentieth century.

The chance factors relating to Nokia in the last third of the twentieth century were

1 the formation of a group, Nokia, involving two

disparate industries: the manufacture of toilet paper on the one hand, and wellington boots and tyres on the other (an example of **horizontal**, rather than vertical, integration).

2   the decision of the group to then diversify by making telephones and computers.

3   the crucial decision in 1990 taken by the Nokia Group to sell off its paper and rubber-making divisions, leave the town of Nokia, and later, to sell its ICL computer manufacturing division to Fujitsu. Nokia now concentrates on the manufacture of mobile telephones in several locations between Helsinki and Turku, with its headquarters at Salo (see Figure 9.42).

4   the decision to invest heavily in the technopole at Oulu, the first science park to be established in the north of Europe. There are now over 150 IT firms at Oulu, working with the local university, and supported by government policy which favours economic developments in the far north, and linked by IT to other, distant, science parks including one at Rennes in Brittany. Oulu is perhaps the best example in Europe of employment in the **Quaternary Sector** – large scale tele-working, despite its northern latitude (65°N).

Nokia's success in overtaking Motorola as the world's main producer of mobile telephones in 1998 was firmly based on research. Since it employs more than 13 000 scientists worldwide, Nokia is now

Fig 9.45 Cambridge Science Park has grown in extent and importance since it was established in 1970. Note the varied architecture, the extensive landscaping and the proximity to the A10–A14 junction

Finland's most successful company, supplying 20% of its exports, and is very much the product of a knowledge-based economy. Every graduate in electrical engineering in Finland is guaranteed a job with Nokia, and its ever-more sophisticated products find a ready market within Finland itself as well as abroad. Such products as mobile phones which are also fax machines and palm-top video conferencers, are examples of very high value, low-bulk products, relying on small sophisticated components. They can be made anywhere in small modern factories, they can be transported cheaply, thus, the manufacturing process is '**footloose**', and can be located anywhere a suitable workforce is available (see Figure 9.44).

Science Parks around Cambridge
(see Figures 9.45 and 9.46)

Nokia has a world-wide presence, and one of the locations where it is found is the area round Cambridge, which has the largest concentration of IT and biotechnology firms and scientists in Europe on 'Silicon Fen', both on science parks and individual locations. The first location factor was the quality of research being done at the university, and once the base was established, the multiplier effect came into play, with the arrival of Microsoft perhaps being the most vital. Figure 9.46 suggests some locational factors involved in the concentration of science parks in the Cambridge area, while the newspaper extracts describe the impact on the local people and the environment (see pp. 302–303).

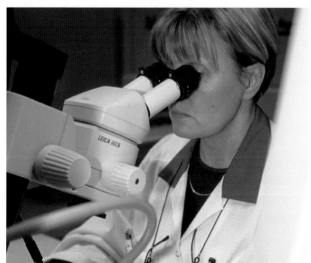

Fig 9.44 Scientific research is the basis for Nokia's success in the worldwide market for their mobile phones

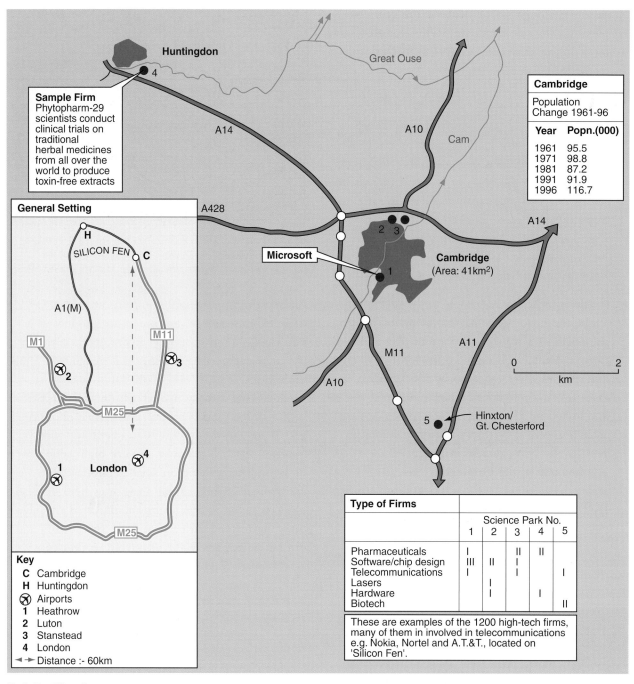

Sample Firm
Phytopharm-29 scientists conduct clinical trials on traditional herbal medicines from all over the world to produce toxin-free extracts

Cambridge

Population Change 1961-96	
Year	Popn.(000)
1961	95.5
1971	98.8
1981	87.2
1991	91.9
1996	116.7

General Setting

SILICON FEN

Microsoft

Cambridge (Area: 41km²)

Hinxton/ Gt. Chesterford

Key
C Cambridge
H Huntingdon
⊗ Airports
1 Heathrow
2 Luton
3 Stanstead
4 London
◄—► Distance :- 60km

Type of Firms	Science Park No.				
	1	2	3	4	5
Pharmaceuticals	I		II	II	
Software/chip design	III	II	I		
Telecommunications	I		I		I
Lasers		I			
Hardware		I		I	
Biotech					II

These are examples of the 1200 high-tech firms, many of them in involved in telecommunications e.g. Nokia, Nortel and A.T.&T., located on 'Silicon Fen'.

Fig 9.46 Silicon Fen

Science Parks in Scotland

The presence of so many universities in Central Scotland with a high reputation, both for research and for producing postgraduates in sciences related to biotechnology, has been a critical factor in the establishment of science parks in Scotland. Biotechnology can be simply defined as the manufacture of useful products from any living system. Brewing and baking would fit this definition,

but in the last 20 years, biotechnology has been at the forefront of major advances in agriculture, preventing environmental pollution, and most notably, in health care. Most companies are involved in research, but also in testing, trialling and validating new drugs and sophisticated new medical kits. The best-known example of biotechnology, the cloning of sheep at PPL Therapeutics at Roslin near Edinburgh, has as its ultimate objective the production of 'milked medicine' to treat diseases such as cystic fibrosis. Figure 9.47 shows the distribution of the main centres of biotechnology in Scotland. There are some science parks which are very similar in appearance to business parks, but with laboratories occupying much of the space, sometimes in 'incubator' units, available at low cost to help small new firms. There are also individual, isolated, units, some on university campuses, some on industrial estates, and some even in town centres, so the landscape associated with biotechnology is very varied. One of the major clusters of biotechnology research is Dundee (see Figure 9.48):

The skyline is now dominated by the gleaming white facade of the Wellcome Sciences Institute, another juicy grape in the nascent cluster of the biomedical and biotechnological institutes and companies springing up in

Dundee. The city of the 'Three J's' (jute, jam, and journalism) is on the road to becoming one of technology and tourism . . .

Source: Rob Stokes: *Scotland on Sunday* 16/11/97

There is also a technology park in the city which is the location for several firms, as well as a medipark in Ninewells Hospital. The Wellcome Trust building houses 250 scientists and 1250 support staff doing research with the aid of substantial funds from various sources, including the Wellcome Trust, which also has a presence in the Cambridge area (see Figure 9.46). It is not the only transnational biotechnology firm involved in Scotland; the American–based Quintiles is a major participant, both in research and testing of the drugs they make (see Figure 9.49). This illustrates the fact that

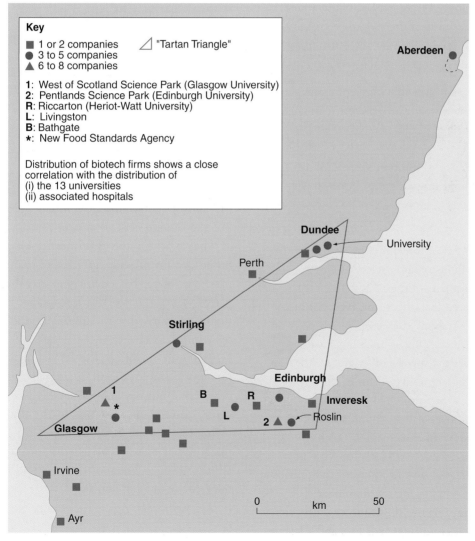

Fig 9.47 Clusters of biotech firms in Scotland

Fig 9.48 One of the main tenants on The Technology Park, Dundee, is Axis-Shield, specialising in the development and manufacture of medical diagnostic products

biotechnology is part of both the secondary sector, although on a limited scale, and the quaternary sector. Although most of the workforce is very highly skilled, there is also a need for unskilled labour to pack, store and despatch products.

The distribution of biotechnology companies in Scotland has a very strong correlation with that of

◆ the electronics industry

◆ the 13 universities, some of which are associated with large teaching hospitals (e.g. Ninewells in Dundee).

It has been predicted that in the first decades of the twenty-first century, biotechnology will have the same effect on industry and society as electronics and IT had in the last part of the twentieth century. It has been estimated that at present, biotechnology in Scotland employs about 10 000 people directly and as many again indirectly, but it has a long way to go before it matches the success of 'Medical Alley', a cluster of 600 biotechnology firms in the Minneapolis area, or San Diego in California. San Diego, with just over half the population of Scotland, has eight times as many biotechnology firms. The future success of biotechnology in Scotland will depend on the ability of these firms and government agencies to attract substantial investments from financial institutions, entrepreneurs and transnational companies. For example, despite their successful cloning experiments, PPT at Roslin

were in severe financial difficulty at the end of 1999 because of insufficient funding. Following their Alba Centre initiative for the semiconductor industry, Scottish Enterprise announced late in 1999 a major funding programme for the biotechnology industry, aimed at doubling both the number of firms and the number of employees.

9.11 The Quaternary Revolution

The largest change in Scotland's employment structure, as a part of a UK-wide revolution at the end of the twentieth century, was the rapid spread of **call-centres**, which are sometimes called 'phone' or 'white-collar' factories' (see Figure 9.50). Their distribution is dispersed and random at first sight, but the map shows that they are to be found

◆ generally outwith the south-east of England

◆ in areas where there is land for building or, more usually, there are suitable premises for conversion

◆ in regions where there is high unemployment, especially among young people, that is, rust-belt areas.

The first centres were Direct Line Insurance, followed by the First Direct Bank in Leeds. In Scotland, more than half the jobs in call centres at the end of the twentieth century were in Glasgow, but there were centres also in remote locations such as Thurso where the BT Internet Helpline is based. Some centres provide advice and many sell products, but it has been suggested that many are places where the work is low-grade, low-paid, highly pressurised with shift work and compulsory overtime. However, in some cases the workers are more highly qualified: the IBM Euro-Helpline at Greenock employs foreign language graduates, as does the Quintiles helpline at Livingston. While call centres owe their existence to the use of modern technology, it might be that if 'e-commerce' using the Internet were to be as widely adopted as it has already been in Finland, there would be a reduced need for many call centres.

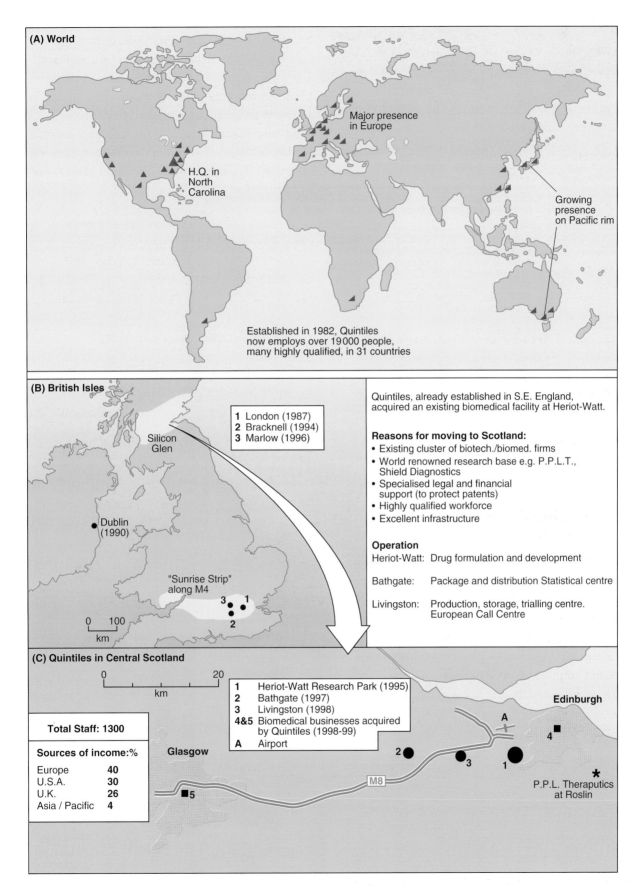

(A) World

Major presence
in Europe

H.Q. in
North
Carolina

Growing
presence
on Pacific rim

Established in 1982, Quintiles
now employs over 19 000 people,
many highly qualified, in 31 countries

(B) British Isles

Silicon
Glen

1 London (1987)
2 Bracknell (1994)
3 Marlow (1996)

Dublin
(1990)

"Sunrise Strip"
along M4

0   100
km

3   1
2

Quintiles, already established in S.E. England,
acquired an existing biomedical facility at Heriot-Watt.

Reasons for moving to Scotland:
• Existing cluster of biotech./biomed. firms
• World renowned research base e.g. P.P.L.T.,
  Shield Diagnostics
• Specialised legal and financial
  support (to protect patents)
• Highly qualified workforce
• Excellent infrastructure

Operation
Heriot-Watt: Drug formulation and development

Bathgate: Package and distribution Statistical centre

Livingston: Production, storage, trialling centre.
               European Call Centre

(C) Quintiles in Central Scotland

0   20
km

1   Heriot-Watt Research Park (1995)
2   Bathgate (1997)
3   Livingston (1998)
4&5 Biomedical businesses acquired
    by Quintiles (1998-99)
A   Airport

Edinburgh

A

4

Total Staff: 1300

Glasgow

Sources of income:%

Europe	40
U.S.A.	30
U.K.	26
Asia / Pacific	4

2

3

1

M8

5

*
P.P.L. Theraputics
at Roslin

Fig 9.49 Quintiles: a biotech transnational from North Carolina to West Lothian

Conclusion

The industries and the industrial regions of Europe, and particularly of Scotland, have undergone a great transformation in the last 50 years. Manufacturing industries have become more specialised and more dependent on advanced technology, and have to be able to respond to rapid changes in the market, and perhaps to a limited lifespan for their products.

Unless there is a strong research and development base for their manufacturing process, they are increasingly vulnerable to innovations made and decisions taken elsewhere in the global economy. They are constantly faced with the need to cut costs and they are perhaps at the mercy of, on the one hand, government decisions on defence policy (which damaged the shipbuilding industry), or on the other, of competition from producers in Eastern

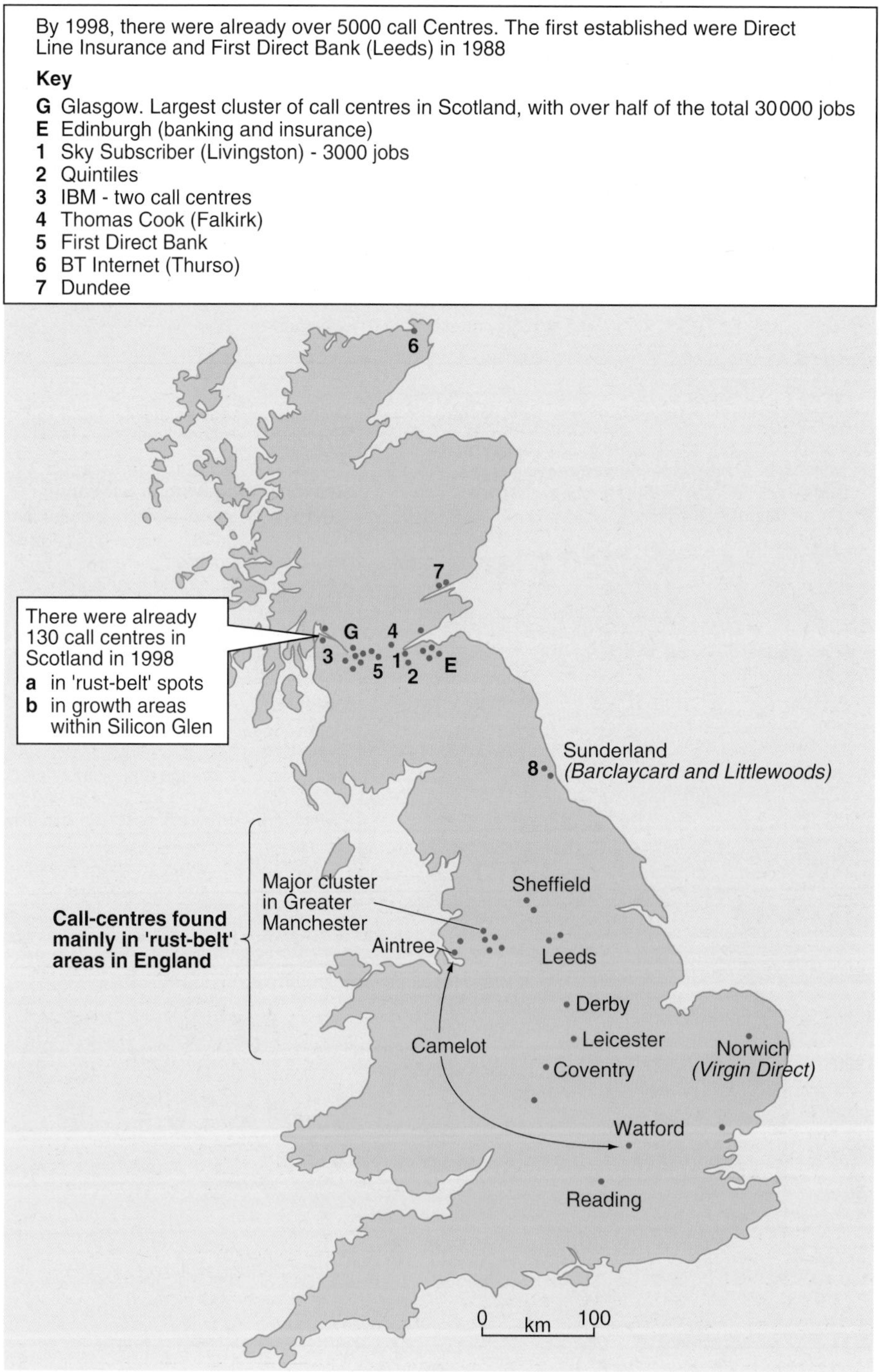

By 1998, there were already over 5000 call Centres. The first established were Direct Line Insurance and First Direct Bank (Leeds) in 1988

Key

G Glasgow. Largest cluster of call centres in Scotland, with over half of the total 30 000 jobs
E Edinburgh (banking and insurance)
1 Sky Subscriber (Livingston) - 3000 jobs
2 Quintiles
3 IBM - two call centres
4 Thomas Cook (Falkirk)
5 First Direct Bank
6 BT Internet (Thurso)
7 Dundee

There were already 130 call centres in Scotland in 1998
a in 'rust-belt' spots
b in growth areas within Silicon Glen

Call-centres found mainly in 'rust-belt' areas in England

Sunderland (Barclaycard and Littlewoods)

Major cluster in Greater Manchester

Aintree

Camelot

Sheffield

Leeds

Derby

Leicester

Coventry

Norwich (Virgin Direct)

Watford

Reading

0 km 100

Fig 9.50 'Phone factories' – selected call centres

297

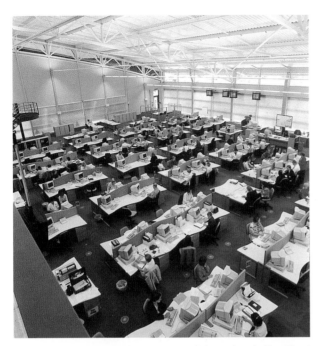

Fig 9.51 The massed ranks of operatives at their monitors in First Direct Bank's custom-built call centre in Leeds

Europe and the ELDCs with much lower manufacturing costs, and changes in fashion, (e.g. the clothing industry). One way of cutting costs might be to reduce the amount of stock of components held on site (which also saves space) and rely on a small number of reliable suppliers, clustered within a very short distance, who can delivery quality-controlled parts/materials as and when they are required – the JIT (Just-in-Time) model. One example of such a supplier is the Scottish firm, Murray International Metals, which buys in, stores and sells steel products to a variety of industries.

The processes and the products are changing, and in general fewer workers are required in manufacturing. There has been a pronounced move away from blue-collar to white-collar work on business and science parks, which have very different landscapes from the rust-belt areas and even the industrial estates. The increasing occurrence of teleworking, that is, working

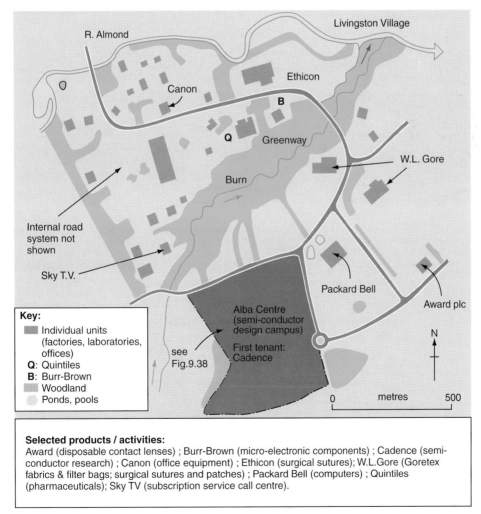

Key:
- Individual units (factories, laboratories, offices)
- **Q**: Quintiles
- **B**: Burr-Brown
- Woodland
- Ponds, pools

Selected products / activities:
Award (disposable contact lenses) ; Burr-Brown (micro-electronic components) ; Cadence (semi-conductor research) ; Canon (office equipment) ; Ethicon (surgical sutures); W.L.Gore (Goretex fabrics & filter bags; surgical sutures and patches) ; Packard Bell (computers) ; Quintiles (pharmaceuticals); Sky TV (subscription service call centre).

Fig 9.52 A new industrial landscape: Kirkton Campus, Livingston

1. c.1750: Stage 1:Domestic Industry

1 Flax field
2 Bleach field
3 Weavers' cottages

2. c.1800: Stage 2: The Factory System
The Industrial Revolution (A)

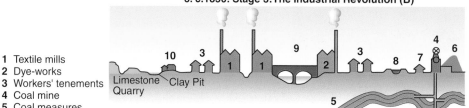

1 Cotton mill
2 Mill-lade
3 Water-wheel

4 Workers' houses
5 Bridge
6 Coal measures unexploited

3. c.1850: Stage 3:The Industrial Revolution (B)

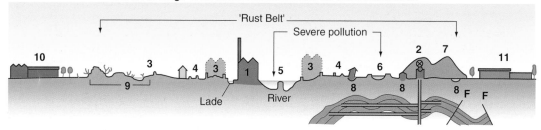

1 Textile mills
2 Dye-works
3 Workers' tenements
4 Coal mine
5 Coal measures

6 Pit spoil-heap
7 Miners' rows
8 Canal
9 Railway bridge
10 Brick works

4. c.1975: Stage 4: De-industrialisation and Re-industrialisation

De-industrialisation:

1 Derelict mill
2 Coal mine closed - early seams worked out; increasing mining problems
3 Mills & works demolished, leaving derelict, 'brownfield' sites
4 Tenements demolished
5 Railway closed; bridge demolished
6 Disused canal

7 Greatly enlarged spoil heaps
8 Serious problems with subsidence
9 Disused quarry & pit
10 Motor vehicle assembly plant
11 Industrial estate (clothing & food products)

Present Day: Stage 5: Re-industrialisation,restoration & expansion

Restoration

1 Heritage centre (restored mill)
2 Industrial museum (former coal-mine)
3 Spoil-heaps lowered & planted
4 Subsidence areas made safe
5 Canal re-opened
6 River cleaned
7 Wildlife park

Re-Industrialisation and Expansion

8 New housing
9 Business park
10 Science park
11 Expanded Industrial estate (a clothing factory facing closure b food products)
12 Electronics firms (research & production)
13 Call centre (banking, insurance, finance)
   - vehicle production ended

Fig 9.53 Evolution of an industrial landscape in Central Scotland

from home, may produce yet another landscape to consider in the future. Although there may be some environmental concerns about the high technology and biotechnology industries, for instance the need for large amounts of clean water to rinse acids in the semi-conductor industry, the impact of industry on the environment is now very carefully monitored.

Most industrial regions have been subject to constant change in the last 200 years. It is now largely forgotten that in the nineteenth century, West

Lothian, now a thriving part of Silicon Glen, was of great importance in the oil-shale industry. The evolution of such a Scottish industrial landscape, from secondary to quaternary, is summarised in the model (Figure 9.53), while an actual example of high technology landscape is shown in Figure 9.52. This small area, containing both high-tech and biotech industries, research and manufacturing, as well as a major example of quaternary employment, might be the keystone of the industrial future of Scotland.

 ASSIGNMENTS

Annotating and analysing field sketches and photographs of industrial landscapes.
1. See Figure 9.16
   a) **Study** the sketch of the industrial landscape round Liege as it was in 1960. **Identify** the features labelled A–I, selecting from the following list: **Steep valley side of the River Meuse; Iron and steel works; High density 19th Century housing; Hesbaye Plateau c 120 m; Railway sidings; River Meuse; Coal hoist and storage area; Gas works; spoil heaps**.
   b) The two main physical elements in this landscape are the valley floor and the plateau. In which of these two areas were **(i)** primary industry and **(ii)** manufacturing industry located?
   c) **Comment on**: **(i)** the height of the buildings, **(ii)** the density of the buildings, **(iii)** the relationship between residential and industrial areas, **(iv)** building materials (if possible), **(v)** the presence/absence of open space, **(vi)** any evidence of pollution.
   d) **Suggest** ways in which this industrial landscape should have changed since 1960. Study Figure 9.17 for help as this shows the Liege area in the 1980s.
2. This chapter contains many photographs of new industrial landscapes e.g. Motorola on p. 264, and Figures. 9.27, 31, 34, 38, 39, 42, 45 and 48. Study all of them carefully and **referring to specific examples,**
   a) **describe** the buildings in these landscapes, mentioning height and extent,
   b) **describe** the landscaping around these landscapes,
   c) **assess** the importance of road transport in such landscapes,
   d) **comment** on the working environment such landscapes provide.
3. **Study** Figure 9.52, which shows a composite

industrial landscape of secondary, tertiary and quaternary activities on an industrial campus at Livingston, West Lothian.
   a) **Describe** the setting in which the industrial and commercial units have been built. You should mention the water and vegetation features.
   b) **Comment** on the layout and the density of the buildings. N.B. you should note that there is a complex internal road system, which is not shown.
   c) Look at the list of products and activities provided. On this basis, **divide** the firms into two categories **(i)** Electronics/IT and **(ii)** Biotechnology. One firm fits into neither category. How should it be classified?
4. **Study** Figure 9.30. **Describe** the **changes** you might observe on a field trip in the Bathgate area in
   a) the character of the industrial landscapes from west to east.
   b) the types of industries carried on.
   Suggest reasons why so much use, past and present, has been made of this area for industrial development.

Analyse the results of surveys of industry and employment:
1. | Percentage of work force employed in the Secondary Sector: Mid-1990s |
   | --- |
   | UK: 17.0   Scotland: 15.5   West Lothian: 37.5 |

   a) **Draw** a graph to show the above employment figures.
   b) **Suggest reasons** for the much higher importance of manufacturing in West Lothian (in which Livingston and Bathgate are located) than in the rest of the country.
2. **Study** the unemployment figures given in Figure 9.15
   a) **State** the apparent correlation between population density and unemployment among the under-25s in Belgium.

b) Which of the two regions has the higher rate of unemployment? **Suggest** the main reason for this.

c) **Suggest** a reason why the Brussels area has such high unemployment among young people, despite its varied employment opportunities.

3

a) Percentage employed in manufacturing and b) Hourly labour costs in EU											
	a	**b**		**a**	**b**		**a**	**b**		**a**	**b**
Aus.	33	12.8	Fr.	27	11.0	It.	32	9.4	Swed.	25	12.8
Bel.	29	13.1	Ger.	37	16.3	Neth.	23	11.7	U.K.	28	8.9
Den.	26	14.0	Gre.	24	5.3	Port.	32	2.9			
Fin.	27	12.4	Ire.	28	7.7	Sp.	30	6.9			

(i) With the aid of an atlas, **identify** and **locate** the member countries of the EU named above.

(ii) On a map, **plot** the five most expensive and the five least expensive countries in terms of labour costs. **Comment on** the location pattern this reveals.

(iii) Is the cost of labour the most critical factor in locating new industry in Europe? **Give two reasons** to back up your opinion.

(iv) Is the importance of manufacturing in these countries in providing employment likely to increase or diminish in the future? **Give reasons** for your answer.

4

Changing industrial production 1991–95 in selected countries					
(Based on an index of 100 for 1990)					
	1991	**1992**	**1993**	**1994**	**1995**
Belgium	98.2	98.1	93.0	94.7	98.7
Finland	91.0	92.1	96.9	107.3	115.3
UK	96.4	96.0	98.4	103.5	106.0
Ireland	103.3	112.7	119.1	133.3	158.5

a) **Draw** a graph to illustrate the above figures.

b) **Describe** the differences in industrial production in the four countries.

c) **Suggest two reasons** for the trends you have noted (information given in Task 3 might be helpful).

Other general assignments:

1 **Study** Figure 9.3. The most famous product of the Carron Iron Works was the carronade, a cannon used by the Royal Navy. Look carefully at the layout of Figure 9.10. **Draw** a similar systems diagram to show the industrial process at Carron.

2 **Study** Figure 9.7. From evidence in the text:
   a) compile a list of transnational firms which have closed 'branch' factories in Scotland.
   b) name at least two transnationals which are more likely to stay in Scotland because of their commitment to research and development.

3 **Study** the text and Figure 9.12. **List** the benefits of building the mill on this site and of a planned village for the workforce at Saltaire.

4 **Study** Figure 9.21. **Find out** to what extent the microelectronics industry in itself illustrates the multiplier effect. In other words, what kind of 'twigs' might grow from the presence of a branch microelectronics firm in an area?

5 **Study** Figure 9.25.
   a) **Describe** the distribution of the main industrial areas shown and give at least two reasons to explain this distribution.
   b) **Describe and explain** the very different distribution of the science parks shown on the map.

6 **Study** Figure 9.28 and **read** the text about Silicon Glen. **Draw** a simple summary diagram to show the main locational factors which continue to attract

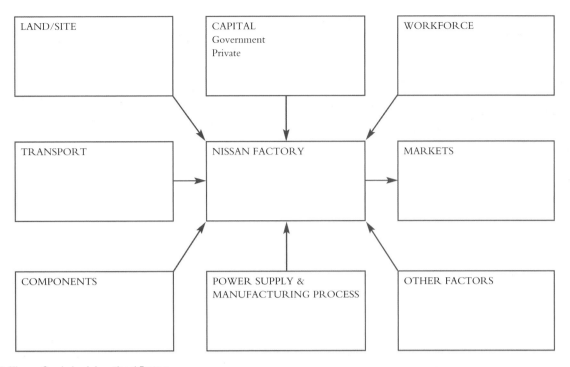

Fig 9.54 Nissan, Sunderland: Locational Factors

electronics firms to this area. Which is now the most important factor?

7. a) **Calculate** the cost per job created if £1 billion is spent in creating 1100 jobs (i.e. Siemens semi-conductor plant on North Tyneside).
    b) **Explain** the difference between 'capital-intensive' and 'labour-intensive' industry. Give an example of each type.

8 **Study** the text and Figure. 9.32B of the Nissan plant. **Make** your own summary diagram of the locational factors involved by copying and completing Figure 9.54. Use one side of A4 paper.

9 **Study** Figure 9.40 and the text. **List** the advantages of the site for the chemical industry and the possible disadvantages to the environment and the local population.

10 **Study** Figure 9.46, the newspaper extracts and Figure 9.49.
    a) **Give at least four reasons** for the rapid growth of the clusters of high-tech. and biotech firms in Silicon Fen.
    b) **State at least four consequences** of this growth for the population of the area in and around Cambridge.
    c) How might the problems mentioned be overcome?
    d) **List** the attractions of Central Scotland, and West Lothian in particular, for a transnational biotech. firm such as Quintiles.

Interpreting newspaper extracts.

1   Techno gold-rush boost for Cambridge:

Cambridge is set to become the boom city of twenty-first century Britain, with its population growing by more than 40%, along with an influx of high-tech industry led by Microsoft, the giant international computer firm founded by Bill Gates. Mr. Gates is planning to set up a multi-million-pound research campus as a joint project with Cambridge University. It will be Microsoft's first research facility outside the USA. . . . The expansion has alarmed sections of the community who have formed a 'Cambridge is Full' campaign against increasing traffic and crowded schools . . . Since Cambridge Science Park was set up in 1975, the city's high-tech sector has grown to 600 firms and 19 000 jobs. Cambridge now has only 3% unemployment.

2   Let's build the village right here.
      Source: Caroline McGhie: *The Sunday Telegraph 30/5/99*

Outside are the first few houses, raw upturned earth and signposts in bare fields. Thus was Cambourne, the new village outside Cambridge launched this month . . . It has taken twelve years to reach this point. Eventually, 8000 people will live here in 3300 houses on 40 ha, currently bristling with a miniature forest of seedlings . . . The houses will be built over the next 10 years. Cambourne will be laid out as three distinct villages, all with greens, plus a 20 ha business park, a church, two primary schools, playing fields, a sports centre, health centre, police and fire stations. When the 700th property is built, the main arterial route linking it to Cambridge will become a dual carriageway . . .

Cambridge, where the pressures of modern life on a small and delicate medieval city wreak havoc, has become the victim of its own success . . . The Cambridge area now has the greatest concentration of high technology firms anywhere in Europe . . . The 'Cambridge Phenomenon' now spreads from Stansted to Peterborough, from Bury St Edmonds to Bedford . . . There Cambridge sits, like a child with too many toys. It has the M11 to whisk drivers to London in an hour, a rail service that takes commuters to King's Cross or Liverpool Street in similar time; the east–west A14 link from the Midlands to the east coast ports, and a reputation like no other for pure research and development. House prices are so high that many choose to live in Huntingdon or Ely, and commute to work. 'We have 35 000 to 40 000 commuters now' says Brian Human, the City Council's Assistant Director of Planning, 'and the rush hour has extended from 7.30 to 9.00 am'. One result of this is that Park and Ride schemes are springing up on the outskirts. Cambridge Futures has come up with a number of possible solutions. Options include necklace development in the nearby villages; swapping areas of the Green Belt for development in return for areas of greenfield further out, improving public transport, creating a virtual superhighway for tele-working, and even building a new town. Some of these would increase commuting, pollution and social imbalance. Others threaten to increase the cost to industry or spoil the area.

3 Net Effect in Silicon Fen
Source: Dan Roberts: *The Daily Telegraph* 16/9/99

. . . this is Silicon Fen, where the bleak windswept flatlands stretch eastwards to the North Sea. There are many places in the world where a higher than average concentration of geeks prompts comparisons with California's Silicon Valley . . . Uniquely, East Anglia's ancient university town is finally showing signs of an industry worthy of the name. Unfortunately, not everyone is welcoming these signs of rising prosperity. Cambridge can now boast permanently gridlocked traffic, soaring house prices and planning decisions which look incomprehensible to outsiders. The latter was demonstrated dramatically when the South Cambridgeshire planners refused an application by the Wellcome Trust to extend its Hinxton biotechnology campus. Hundreds of world-class research jobs were turned away. However bizarre the decision looked from a distance, it was endorsed by John Prescott's department, and may force Wellcome to take its development overseas. Despite this decision, Silicon Fen is on a roll. There are about 1200 high-technology firms there, and they have reached critical mass.

Key terms and concepts

- brownfield site (p. 277)
- business park (p. 264)
- call centre (p. 264)
- capital (p. 261)
- capital-intensive (p. 290)
- coppicing (p. 267)
- development area (p. 278)
- de-industrialisation (p. 263)
- domestic industry (p. 260)
- enterprise zone (p. 281)
- 'euro-banana' (p. 279)
- entrepreneurs (p. 261)
- fixed industries (p. 262)
- footloose industries (p. 262)
- greenfield site (p. 269)
- heavy industry (p. 262)
- heritage industry (p. 267)
- horizontal integration (p. 286)
- industrial estate (p. 262)
- industrial heritage (p. 260)
- industrial inertia (p. 275)
- industrial landscapes (p. 262)
- infrastructure (p. 276)
- inward investment (p. 262)
- 'knowledge' economy (p. 262)
- labour-intensive (p. 261)
- light industry (p. 262)
- multiplier-effect (p. 267)
- primary sector (p. 260)
- quaternary sector (p. 264)
- rationalisation (p. 275)
- Rechar (p. 278)
- Resider (p. 278)
- rust-belt (p. 272)
- science park (p. 264)
- 'screw-driver' economy (p. 263)
- secondary sector (p. 260)
- self-sufficiency (p. 260)
- smokestack industries (p. 275)
- sunrise industries (p. 278)
- technopole (p. 290)
- tertiary sector (p. 263)
- transnational (p. 263)
- turnpike (p. 261)
- vertical integration (p. 275)

Suggested reading

Broadley, E & Cunningham, R (1991) *Core Themes in Geography – Human* Oliver & Boyd

Corney, G (Ed.) (1985) *Geography, Schools & Industry* Geographical Association

Dickens, P (1992) *Global Shift* Paul Chapman Publishing

Guinness, P & Nagle, G (1999) *Advanced Geography: Concepts and Cases* Hodder & Stoughton

Maclean, K & Thomson, N (1988) *Landscapes and Peoples of Western Europe* OUP

Nagle, G & Spencer, K (1996) *A Geography of the European Union – A Regional and Economic Perspective* OUP

Raw, M (1993) *Manufacturing Industry: The Impact of Change* Collins

Waugh, D (1994) *Geography: An Integrated Approach* Nelson

Waugh, D (1997) *The U.K. and Europe* Nelson

Waugh, D (1998) *The New Wider World* Nelson

Witherick, M (1995) *Environment and People* Stanley Thorne

Internet sources

Biotechnology Group: Scottish Enterprises: www.sebiotech.org.uk/biotech

Department of Trade & Industry: www.dti.gov.uk/

Invest in Britain Bureau News: www.dti.gov.uk/IBB/nrews.htm

Finland: www.stat.fi/ff//home.html

Lanarkshire Department Agency: www.lda.co.uk.

Locate in Scotland: www.scotent.co.uk/Locate_in_Scotland/index.html

Meuse in Europe: www.meuse-capem.com/location.htm

Scottish Enterprise: www.scotent.co.uk/home.html

After working through this chapter, you should be aware that:

◆ urban settlements developed in distinctive sites or situations

◆ urban areas have evolved because of the interaction of geographical factors (site and situation) with historical and cultural factors

◆ the functions of urban settlements change as they evolve

◆ urban settlements today provide a range of housing, work and services which are usually found in distinctive zones

◆ urban areas are subject to frequent change in the type and location of housing, industry and services (especially shopping)

◆ traffic is now a major urban problem

You should also be able to:

◆ annotate and analyse field sketches and photographs of different parts of urban landscapes

◆ analyse land use maps and data provided on movements of people and traffic to and within an urban settlement

Edinburgh has been chosen as the main case study, but the principles can be applied to any city.

10.1 Urban Growth: Scotland

*A city is both geography and history; it is the expression of what people have done with a place through time. Geographically, it springs from **situation** and **site**. More precisely, a city is the site at which the situation has significance; it is the choice of site which successfully exploits its situation.*

Source: J W Watson

Site and Situation

Towns have evolved for different reasons through history, but their **situation** and **site** are both critical. Towns may evolve at a favourable situation (location) in relation to existing (or potential) trading routes and natural resources and at a particularly beneficial site, perhaps in terms of defence, or in the early days, water supply. The quotation above is making the point that a particularly good site might never be exploited if its location is insignificant. A **crag and tail** might have offered the best possible defensive site, but it would only have become the core of a major town if it had been strategically placed in relation to major trade routes, both by land and sea, and in a position where it could best control

the country. The original site of a town is limited in area, but as a town expands, so too does the site, perhaps then taking in a variety of physical features, as we shall see later in the case of Edinburgh.

10.2 The Origin and Growth of Towns

The earliest towns also had to be in an area which could produce a food surplus to feed the townspeople. The first towns evolved therefore in the basins of the great rivers of the Middle and Near East – the Nile, the Tigris and the Euphrates, and the Indo-Gangetic Plain, all of which are today regarded as being in the Developing World. Urban settlement eventually spread into Europe through Greece and the Roman Empire. Many cities in Western Europe, as far north as Hadrian's Wall, originated as Roman towns and camps, perhaps on the site of what had been an Iron Age fort. Despite all the changes these cities have undergone, their **townscapes** still contain recognisable elements which date from different periods in their history, and are part of their **inherited urban fabric**, for example, the city wall, the cathedral and the town house.

Typical Town Sites

Not all urban settlements in Western Europe originated as Roman cities (there were none at all in Scotland), and there was a great variety in the choice of sites for the old **core**, or **pre-urban nucleus**. Favoured sites included

- a crag and tail
- above an incised meander
- a dry-point above a marsh
- at a hill-foot/on a spring-line
- on an alluvial fan
- at a loch-side
- on a river terrace
- on a raised beach
- at a river confluence
- on an isthmus
- at the lowest bridge point/limit of navigation
- on a headland protected by cliffs
- at or in a gap between hills
- on a sheltered, navigable inlet or river mouth
- on a river meander in a forest clearing
- above or beside a mineral source.

Examples of almost all of these sites can be found in Scotland. In some exceptional circumstances, **aspect**, that is, position in relation to the sun, might have been a important factor in upland valleys.

The Growth of Towns in Scotland

Although there were some large settlements in earlier times (e.g. the fortified Iron Age settlement on top of Traprain Law in East Lothian), it is accepted that the first towns or burghs date from the twelfth century, by the end of which there were just 38 small **burghs**. In contrast, there were many European cities by the twelfth century with populations measured in thousands, and often with cathedrals. Most of the Scottish burghs were situated in areas of better agricultural land, and almost all of

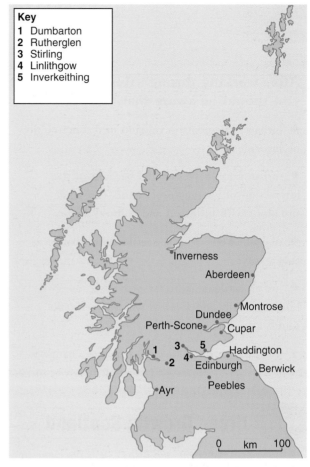

Fig 10.1a Main burghs in fourteenth century Scotland

them were to be found in the east of the country, that is, facing Europe, and also on the routes most often taken by invading English armies. Within this wider situation, the burghs grew up at or on natural

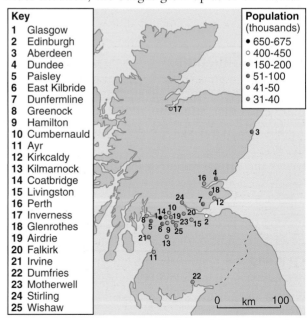

Fig 10.1b Urban hierarchy in 2000 AD

◆ crossing places

◆ natural harbours

◆ defensive sites.

Burghs were established by royal authority; some were actually given a royal charter, while others were established with royal approval by the church and local nobles. They were established to develop trade (and thus increase revenue for the Crown) and for strategic reasons to extend the control of the King in Scotland. These burghs were all small, both in area and population, but in the following centuries, a **hierarchy** developed in terms of their economic importance to the Crown. Figure 10.1A shows the most important towns in the early fourteenth century, and is of interest because of the inclusion of Berwick-on-Tweed (before it became part of England) and the absence of Glasgow, which was then only a minor ecclesiastical burgh. The highest position in any such hierarchy would have been held by the capital, but at that time the royal court (and sometimes the parliament) moved from town to town, from as far north as Aberdeen to Ayr in the south, depending on the state of relations with England, and the physical condition of the royal residence in a particular place.

10.3 Functions of Towns

Some of the burghs in medieval Scotland, for example, Scone and Dunfermline, thus had an administrative function; all had a trading function, with royal burghs such as Haddington and Linlithgow having a monopoly of foreign trade through their ports (at Aberlady and Bo'ness respectively); some had a religious function because of their abbey or cathedral (e.g. Arbroath and Glasgow); some had a defensive function because of their castle (e.g. Stirling and Dumbarton), while four eventually had an educational function because of the establishment of a university (e.g. St. Andrews and Aberdeen). Many burghs were multi-functional: in addition to their trading function they were also centres of workshop industry. Later, in the sixteenth century, the small burgh of Culross was almost unique in that it had an industrial function, both primary and secondary, as it extracted coal, and manufactured salt (from sea water)

and iron utensils. Through time, the functions of Scotland's early burghs changed, and in some cases ended. As the functions changed, the towns concerned came to occupy different positions in the urban hierarchy, the prime position in which was taken by Edinburgh as capital and the largest urban centre in pre-industrial Scotland.

10.4 Urban Morphology

Burghs in medieval Scotland had a distinctive, though limited, townscape and all contained the same elements. Even then, there was a simple zonation of functions and services with administrative and trading functions clustered near the town house; workshops were located in the narrow strips of cultivated land (**tofts**) on either side of the linear residential zone. The religious function, apart from the town kirk, was often located on the periphery of the burgh where an abbey, a monastery, a convent or a friary might be found. As the burgh grew, these buildings were incorporated in the townscape. Some fragments of such medieval townscapes have survived to the present day where they might be both an attraction to tourists and a problem for traffic where they have determined the pattern and the width of streets.

10.5 Urbanisation

In later centuries, many more burghs were created; existing burghs grew rapidly in population, and the population density increased as buildings were enlarged upwards to form **tenements** and as new buildings were built into the tofts. Urban settlement spread to western and northern Scotland. In the eighteenth century, the first '**new towns**' were planned and built (e.g. Inveraray), and old burghs expanded beyond their existing boundaries (e.g. Edinburgh). In the nineteenth and early twentieth centuries, towns mushroomed where coal and oil shale were being exploited (e.g. Coatbridge and Bathgate). The rural-urban migration started by the Agricultural Revolution accelerated as steam-powered factory industry developed rapidly in towns, especially in the west of Scotland, and as a result of this, Glasgow eventually became the core of a vast

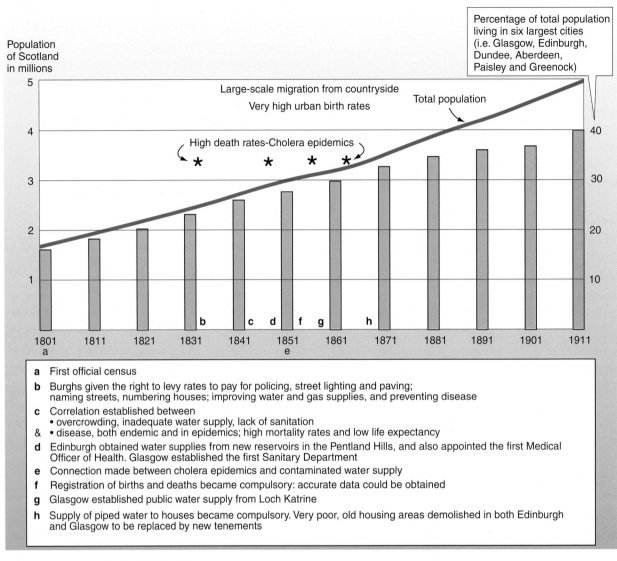

Fig 10.2 Urbanisation in Scotland 1801–1911

conurbation on Clydeside. Glasgow had gone through several stages in its evolution, with its function changing from a religious and educational centre with a local trading role, to a new function as a centre of international trade with the American colonies in the eighteenth century, before becoming the most important industrial centre and the largest city in Scotland. Together with its commercial importance, these elements took Glasgow to the highest place in the urban hierarchy, although Edinburgh remained the capital in name, but without a parliament.

One of the best examples of rapid urban growth was that of Clydebank, which before 1881 was a small village of several hundred people. The establishment of the Thomson shipyard and the Singer Sewing Machine factory (a very early example of American investment in Scotland) created a demand for labour, and workers initially commuted daily from Glasgow by train before tenements were built for them. By 1911, there were 37 000 inhabitants. Such rapid urban growth in the nineteenth century brought problems relating to poor housing, overcrowding and the lack of public utilities, and consequently there were serious health problems in the by then clearly defined residential zones for the factory workers. Figure 10.2 shows population growth in the six largest cities in Scotland and the associated problems (which today might be found in cities in the Developing World), as well as some of the attempted solutions.

The twentieth century saw the appearance of

◆ low-cost, local authority housing

◆ multi-storied urban housing

◆ sprawling peripheral housing and industrial estates in the four largest cities

◆ planned **New Towns**, such as East Kilbride, Cumbernauld and Livingston.

These New Towns were planned to house the overspill from the overcrowded cities of Glasgow and Edinburgh, and to create a favourable environment for the location of the new light industries such as electronics. The second half of this century was characterised by the continuing decline, and in some cases the disappearance, of traditional industries which brought about the creation of large areas of derelict land with Scotland's towns. Urban renewal, the provision and location of services, and traffic congestion became matters of great concern. Most notably, the population of Glasgow shrank dramatically while that of Edinburgh, even before it had its function restored as the home for Scotland's parliament, began to increase again after an initial decline. However, in terms of population, Glasgow is still Scotland's premier city (see Figure 10.1B).

10.6 Edinburgh: Site and Situation

Edinburgh has a distinctive, if not a unique, **site** (see Figure 10.3A), which together with a strategic **situation**, made the city which evolved there the obvious choice as Scotland's capital. The site is the result of a combination of past volcanic action, glacial scouring and moulding, changes in sea level, and to a

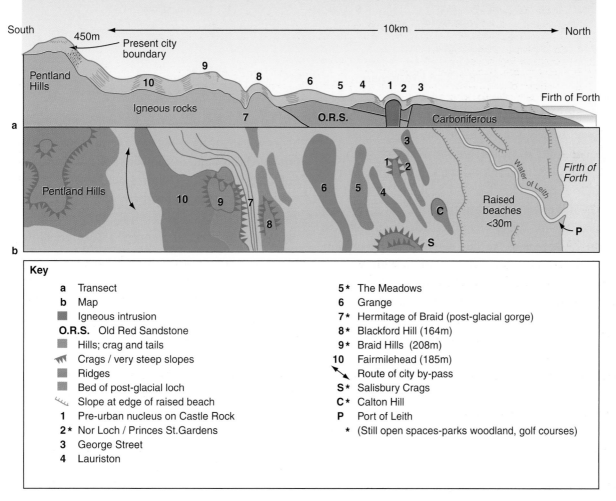

Key

a	Transect	**5***	The Meadows
b	Map	**6**	Grange
▪	Igneous intrusion	**7***	Hermitage of Braid (post-glacial gorge)
O.R.S.	Old Red Sandstone	**8***	Blackford Hill (164m)
▪	Hills; crag and tails	**9***	Braid Hills (208m)
⋙	Crags / very steep slopes	**10**	Fairmilehead (185m)
▪	Ridges	↘	Route of city by-pass
▪	Bed of post-glacial loch	**S***	Salisbury Crags
⤻	Slope at edge of raised beach	**C***	Calton Hill
1	Pre-urban nucleus on Castle Rock	**P**	Port of Leith
2*	Nor Loch / Princes St.Gardens	*****	(Still open spaces-parks woodland, golf courses)
3	George Street		
4	Lauriston		

Fig 10.3a The site of Edinburgh

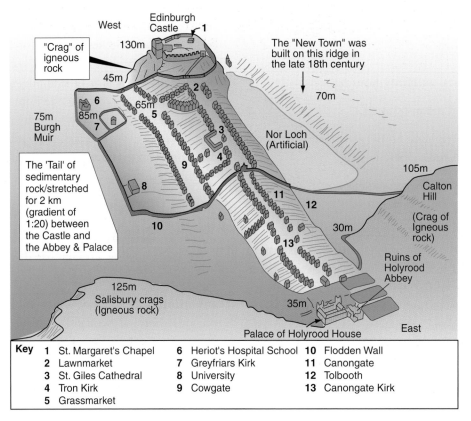

Fig 10.3b Early Edinburgh – late seventeenth century

limited extent, river erosion. The **pre-urban nucleus** was located on the **crag** (where the castle was built) and **tail** (along which the Royal Mile evolved) between the coastline and the hill edge, although some distance from the Water of Leith. The Romans did establish camps in the area, near the mouths of the Rivers Esk and Almond, rather than on the hill or crag tops, but these did not give rise to Roman towns. Water was not a factor in the choice of the castle rock as the site: defence was paramount in the initial stages, although the chosen site was only one of several igneous intrusions rising above the lower areas of sedimentary rock. What made it more attractive were the two deep troughs to the north and south which made defence easier, although the tail meant that the site was very restricted, especially after the town wall was built. Edinburgh actually consisted at one stage of two different burghs, the walled royal burgh of Edinburgh, which grew down the tail from the castle, and the unwalled burgh of the Canongate which grew up the lower part of the tail from Holyrood Abbey and the royal palace of Holyrood House (see Figure 10.3B). In the case of the Canongate, the proximity of the abbey at one

end and the royal burgh at the other was more important than the physical nature of the site.

This choice of Edinburgh's site was influenced also by its wider situation, where travellers and trade were funnelled through the gap between the Pentland Hills and the Firth of Forth. Later, when the town was established, so eventually was its port at Leith which evolved as a separate walled town at its river-mouth site. By then, Edinburgh's situation allowed it

◆ to trade with Europe (the only market before the Americas were discovered and colonies established)

◆ to extend the king's rule over the rest of Scotland

◆ to prevent, if possible, invading English armies penetrating further into Scotland.

After 1767, when defence was no longer a critical matter, and the overcrowding within the wall became too severe, the city expanded with the building of the planned New Town. This had its axis on the George Street ridge, and then spread down the long, steep slope northwards to the Water of Leith; eastwards

Fig 10.4 Central Edinburgh – note the castle on the west-facing crag, the upper part of the New Town on the tail, the planned layout of the New Town to the north and Calton Hill (top right)

round the flanks of Calton Hill, and westwards above and around Dean Village in its deep gorge. This gorge, and the deep hollows on either side of the Royal Mile, were major obstacles which had to be bridged as the city expanded (see Figure 10.4).

Subsequent extensions of the city occupied a variety of sites:

◆ river terraces along the Water of Leith, the right bank of the Almond, and minor streams

◆ the dried-out beds of former glacial lochs to the north, west and south of the Old and New Towns

◆ raised beaches along the shore of the Firth of Forth

◆ flatter ridges (e.g. Lauriston and Grange) to the south

◆ the lower slopes of the igneous hills – Corstorphine, Blackford, the Braids, and even the Pentlands

◆ areas of glacial deposition and moulding in the western extension of the city which took place in 1975.

What, therefore, had been a simple site became eventually a very complex one as the city grew, and expansion took place regardless of physical characteristics or problems. Only very steep rocky slopes and official restrictions have prevented building on the sides of Arthur's Seat which lies within a royal park. It therefore stands as a green island (a fine example of an **urban 'lung'**) rising high above the

surrounding residential and former industrial areas. Figure 10.5 shows the stages in the expansion of Edinburgh as the boundary was successively extended by official legislation.

10.7 Urban Models

All cities have unique sites and histories, and no two cities are absolutely identical in their development. However, it is possible to see common characteristics in the evolution of cities and this has been the basis for the formulation of models of urban growth, mostly American in origin. These were devised on social and economic criteria, based on work done in the early decades of the twentieth century. These different models (see Figure 10.6) proposed that cities grew spatially

◆ in concentric circles from the core, that is, one nucleus, or

◆ in sectors along main routes radiating from the core, or

◆ by the growing together of several built-up areas from different cores, that is, multiple nuclei, or

◆ concentrically, but in contrasting sectors (a combination of the first two models, based on a study of three English cities).

Figure 10.7 shows a newer model by Whitehand which was also based on studies of English cities, especially Birmingham. It also differs especially from the three American-based models in that it concentrates on the ways in which the townscape has evolved through time as a city expands outwards from the Old Core. It takes into account the varied demand for housing through the last hundred years or so and the consequent variations in building type and density. The zones which have evolved during periods of greater demand (greater prosperity, increases in population, deterioration of existing housing stock or even war-damage) are also related to innovations in transport which greatly improved accessibility within the city. Thus, initially, when people walked, cycled or were carried by horse-drawn trams to work, high density housing was constructed in close proximity to the factories. Thereafter, with the successive introductions of

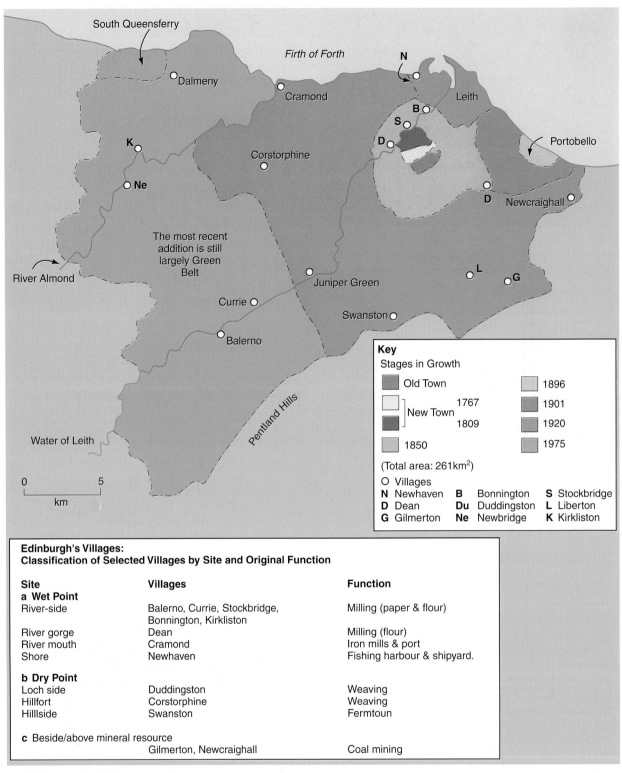

Fig 10.5 The growth of Edinburgh

electric tramways, suburban railways, motor buses, and finally the motor car, it became possible to travel longer distances to work, and the city expanded centrifugally. The model suggests that the Old Core and these three zones which developed during

periods of rapid urban expansion are separated from one another by **Fringe Belts** (fossilised semi-green belts) which evolved during periods of much slower urban expansion. It must be recognised that,

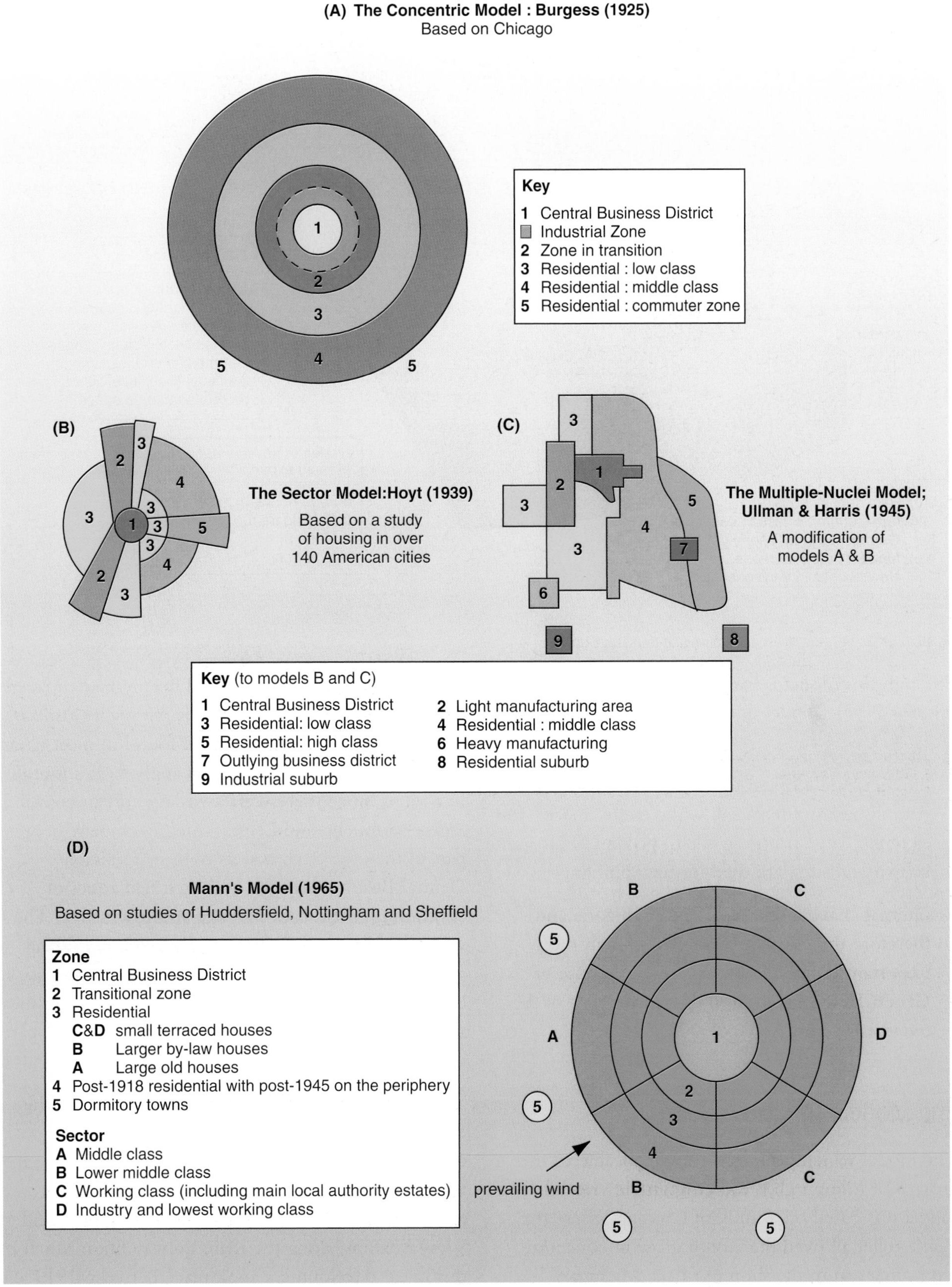

(A) The Concentric Model : Burgess (1925)
Based on Chicago

Key

1 Central Business District
   Industrial Zone
2 Zone in transition
3 Residential : low class
4 Residential : middle class
5 Residential : commuter zone

(B)

The Sector Model:Hoyt (1939)

Based on a study
of housing in over
140 American cities

(C)

The Multiple-Nuclei Model;
Ullman & Harris (1945)

A modification of
models A & B

Key (to models B and C)

1 Central Business District	2 Light manufacturing area
3 Residential: low class	4 Residential : middle class
5 Residential: high class	6 Heavy manufacturing
7 Outlying business district	8 Residential suburb
9 Industrial suburb	

(D)

Mann's Model (1965)

Based on studies of Huddersfield, Nottingham and Sheffield

Zone
1 Central Business District
2 Transitional zone
3 Residential
   C&D small terraced houses
   B Larger by-law houses
   A Large old houses
4 Post-1918 residential with post-1945 on the periphery
5 Dormitory towns

Sector
A Middle class
B Lower middle class
C Working class (including main local authority estates)
D Industry and lowest working class

prevailing wind

Fig 10.6 Urban models

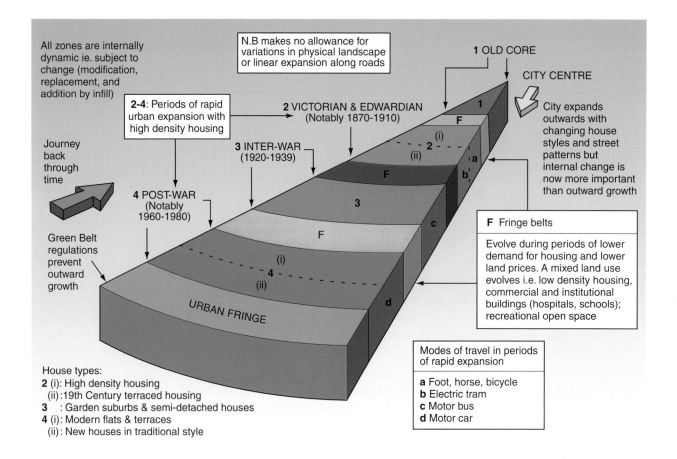

All zones are internally dynamic ie. subject to change (modification, replacement, and addition by infill)

N.B makes no allowance for variations in physical landscape or linear expansion along roads

1 OLD CORE

CITY CENTRE

2-4: Periods of rapid urban expansion with high density housing

2 VICTORIAN & EDWARDIAN (Notably 1870-1910)

City expands outwards with changing house styles and street patterns but internal change is now more important than outward growth

Journey back through time

3 INTER-WAR (1920-1939)

4 POST-WAR (Notably 1960-1980)

Green Belt regulations prevent outward growth

F Fringe belts

Evolve during periods of lower demand for housing and lower land prices. A mixed land use evolves i.e. low density housing, commercial and institutional buildings (hospitals, schools); recreational open space

URBAN FRINGE

House types:
2 (i): High density housing
   (ii): 19th Century terraced housing
3   : Garden suburbs & semi-detached houses
4 (i): Modern flats & terraces
   (ii): New houses in traditional style

Modes of travel in periods of rapid expansion

a Foot, horse, bicycle
b Electric tram
c Motor bus
d Motor car

Fig 10.7 Urban growth: Whitehand's changing townscape model

although this model is based on historical evolution, it is not a static model since

◆ all the zones are subject to internal change – old buildings are adapted for new uses, or are demolished and replaced, while empty spaces, for example, large private gardens, playing fields and brownfield sites, are used for new buildings

◆ internal change, that is, within each zone and therefore the city as a whole, was much more important than outward expansion because of **Green Belt** legislation in the last decades of the twentieth century.

The Models Applied to Edinburgh

The model which is perhaps least applicable to the growth of Edinburgh is the **concentric model**. This model was based on a study of Chicago where the gentle relief allowed the city to grow in concentric semi-circles from the shores of Lake Michigan. In the case of Edinburgh, the much more varied relief with

steeply sloping igneous hills and deeply incised rivers made concentric growth unlikely. In reality, aspects of all the models are likely to be found in most cities. The **multi-nuclei model** does apply to Edinburgh in a sense, since there were two large urban coastal areas, Portobello and Leith, which eventually became part of the city. Each had its own recognisable Central Business District, and each had zones of contrasting urban land use, age and townscape. They were already expanding along the coast and inland when they were absorbed by extensions to the city's boundary. There were also very minor nuclei in the villages of varying size which eventually became part of the city (see Figures 10.5 and 10.8).

The **sectoral model** may also be identified in the expansion of Edinburgh since the city did grow outwards in radial fashion. This was not however in broad segments, but along the main roads which radiated from the **Central Business District**, and to a lesser extent, along the main railway lines. Much of this linear development took place between 1919 and 1939, when villas and bungalows were built in

Fig 10.8 Duddlington, a former weaving village with a twelfth century kirk, on a dry-point site at the foot of Arthur's Seat

narrow strips along the A1, the A8, and the A90, with large areas of open countryside left between them (see Figure 10.9). Such areas, fringe wedges rather than fringe belts, were subsequently built up, often in very different styles from the earlier villas and bungalows, in the second half of the twentieth century.

The **Mann Model** makes much of a west–east

Fig 10.9 Ribbon development of villas along Lasswade Road in south Edinburgh with infilling behind by 1960s tower blocks

division, with wealthier people and better housing being found in the western sectors, but this is only partially true when applied to Edinburgh. There are areas of very expensive housing in the north west in Barnton and Cramond (low density) and the south west in Morningside (mostly high density), but there are also areas of high-density, low-cost housing in between at Sighthill and Wester Hailes, which are among the poorest areas in the city (see Figure 10.10).

The **Whitehand Model** has some relevance to Edinburgh's growth, but more perhaps in the general principles than in the actual details since the house types are particular to England. However, in transects along the main roads out of the CBD, it is possible to identify some of the stages in the evolution of Edinburgh in the contrasting house types (see Figure 10.11). There are also many examples of new private housing estates being built on former playing fields and market gardens, that is, in Fringe Belts, and perhaps even more examples of redundant buildings being converted into housing: even the oldest areas of the city are internally dynamic.

Although its emphasis on the townscape is welcome, this model is of limited use since it concentrates on buildings, especially housing, and makes no reference to the location of industry or commerce within the city. Edinburgh, like all cities, has clearly differentiated urban zones, apart from the outer residential areas

◆ the **Central Business District (CBD)**

◆ the **Inner City**

◆ the industrial zones.

10.8 Edinburgh's Functions

The location and nature of these zones is linked with the changes in Edinburgh's functions (see Figure 10.12). Edinburgh like most large cities has many different functions but the relative emphasis of these has changed through time. Despite the movement to London, first of the royal court in 1603, and then of the parliament in 1707, its function as capital has remained. Its role as a national and international financial centre began with the establishment of the

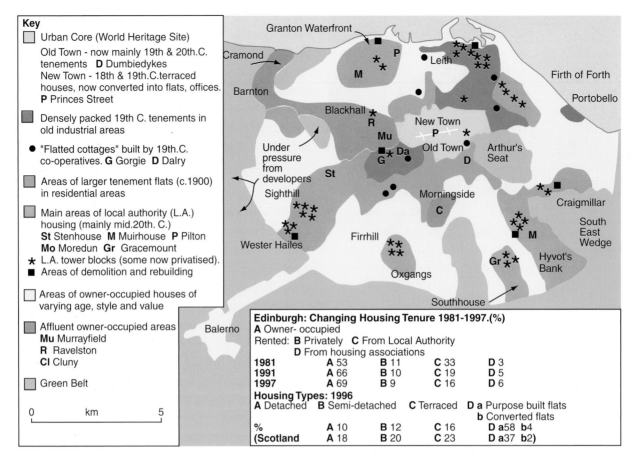

Key

Urban Core (World Heritage Site)

Old Town - now mainly 19th & 20th.C. tenements **D** Dumbiedykes
New Town - 18th & 19th.C.terraced houses, now converted into flats, offices. **P** Princes Street

Densely packed 19th C. tenements in old industrial areas

● "Flatted cottages" built by 19th.C. co-operatives. **G** Gorgie **D** Dalry

Areas of larger tenement flats (c.1900) in residential areas

Main areas of local authority (L.A.) housing (mainly mid.20th. C.)
St Stenhouse **M** Muirhouse **P** Pilton
Mo Moredun **Gr** Gracemount
★ L.A. tower blocks (some now privatised).
■ Areas of demolition and rebuilding

Areas of owner-occupied houses of varying age, style and value

Affluent owner-occupied areas
Mu Murrayfield
R Ravelston
Cl Cluny

Green Belt

0   km   5

Edinburgh: Changing Housing Tenure 1981-1997.(%)
A Owner- occupied
Rented: **B** Privately **C** From Local Authority
        D From housing associations

	A	B	C	D
1981	A 53	B 11	C 33	D 3
1991	A 66	B 10	C 19	D 5
1997	A 69	B 9	C 16	D 6

Housing Types: 1996
A Detached **B** Semi-detached **C** Terraced **D a** Purpose built flats
                                                       b Converted flats

	A	B	C	D
%	A 10	B 12	C 16	D a58 b4
(Scotland	A 18	B 20	C 23	D a37 b2)

Fig 10.10 Generalised map of housing types in Edinburgh

Bank of Scotland in 1694, but has grown spectacularly, especially in the twentieth century. It has been renowned centre of education and culture since the foundation of Edinburgh University in 1585, and the period of the late eighteenth to early nineteenth century during which the New Town was built, further enhanced its reputation. It had a trading function, both locally and internationally (through the Port of Leith), which has now diminished. Industry expanded with the port and especially with the building of the railway network, but it has also declined, as well as changing, both in type and location. Edinburgh has also a minor function both as a religious and a military centre. As other functions have changed or declined, Edinburgh's function as a centre of tourism has boomed in the last 50 years. The recognition of the townscape of the Old and New Towns as a World Heritage Site could enhance this function.

10.9 The Central Business District (see Figure 10.13)

The Central Business District is found in the most accessible location within a city. As a city expands, the appearance and the extent of the zone may change. Within the CBD, there may be a variety of functions:

◆ commercial and retail, with high-order shops such as department stores and fashionable boutiques

◆ business in different forms (e.g. banks, building societies, estate agents, law offices, insurance offices)

◆ an administrative function with local and/or national government offices

◆ a cultural function with museums, galleries, theatres and concert halls in addition to university buildings.

In some cities, the CBD may also incorporate a wider recreation and entertainment function with large cinemas, hotels, restaurants, cafes, clubs and bars, although this function might be found in a separate, but adjoining zone. Throughout the CBD there is usually a shortage of suitable office premises, and because of this demand, rents are very high.

In the case of Edinburgh, the first CBD, in a very simple form, was located in the Old Town. It was then extended north and south along the high bridges built in the late eighteenth century over the deep valleys on either side of the crag and tail. The New Town was originally planned as a mainly residential area, but the movement of the wealthier people from the Old Town was followed by

businesses. By the end of the nineteenth century the commercialisation of the First New Town, from Princes Street to Queen Street, was complete. Banks, insurance offices, the stock exchange, publishers' offices, department stores, hotels and restaurants were established. In some cases, especially on Princes Street, ornate replacement buildings were constructed, but in other cases the original Georgian three-storey houses and their gardens were converted for commercial use. The first national government office in the New Town was Register House at the end of the eighteenth century, but the local government offices remained in the High Street of the Old Town. At the same time as the evolution of

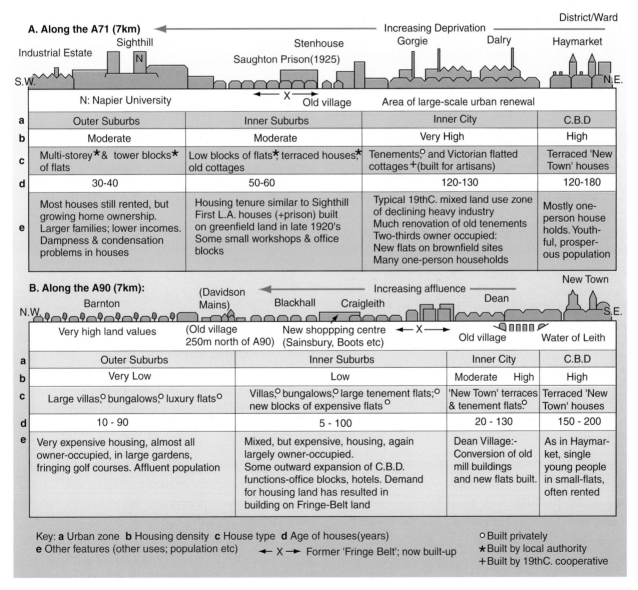

A. Along the A71 (7km) — District/Ward — Increasing Deprivation

Industrial Estate · Sighthill · Saughton Prison(1925) · Stenhouse · Gorgie · Dalry · Haymarket

N: Napier University — X Old village — Area of large-scale urban renewal

	Outer Suburbs	Inner Suburbs	Inner City	C.B.D
a				
b	Moderate	Moderate	Very High	High
c	Multi-storey* & tower blocks* of flats	Low blocks of flats* terraced houses* old cottages	Tenements° and Victorian flatted cottages + (built for artisans)	Terraced 'New Town' houses
d	30-40	50-60	120-130	120-180
e	Most houses still rented, but growing home ownership. Larger families; lower incomes. Dampness & condensation problems in houses	Housing tenure similar to Sighthill First L.A. houses (+prison) built on greenfield land in late 1920's Some small workshops & office blocks	Typical 19thC. mixed land use zone of declining heavy industry Much renovation of old tenements Two-thirds owner occupied: New flats on brownfield sites Many one-person households	Mostly one-person households. Youthful, prosperous population

B. Along the A90 (7km): — New Town — Increasing affluence

Barnton · (Davidson Mains) · Blackhall · Craigleith · Dean

Very high land values — (Old village 250m north of A90) · New shopping centre (Sainsbury, Boots etc) — X — Old village · Water of Leith

	Outer Suburbs	Inner Suburbs	Inner City		C.B.D
a					
b	Very Low	Low	Moderate	High	High
c	Large villas,° bungalows,° luxury flats°	Villas,° bungalows,° large tenement flats;° new blocks of expensive flats °	'New Town' terraces & tenement flats.°		Terraced 'New Town' houses
d	10 - 90	5 - 100	20 - 130		150 - 200
e	Very expensive housing, almost all owner-occupied, in large gardens, fringing golf courses. Affluent population	Mixed, but expensive, housing, again largely owner-occupied. Some outward expansion of C.B.D. functions-office blocks, hotels. Demand for housing land has resulted in building on Fringe-Belt land	Dean Village:- Conversion of old mill buildings and new flats built.		As in Haymarket, single young people in small-flats, often rented

Key: **a** Urban zone **b** Housing density **c** House type **d** Age of houses(years) **e** Other features (other uses; population etc) — X — Former 'Fringe Belt'; now built-up

° Built privately
* Built by local authority
+ Built by 19thC. cooperative

Fig 10.11 Contrasting townscapes in Edinburgh

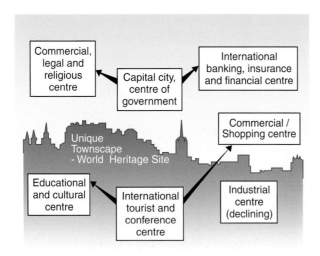

Fig 10.12 Edinburgh's functions

Edinburgh's CBD, similar zones developed, on a smaller scale, in both Leith and Portobello. As Edinburgh expanded with the building of tenements along the main roads (to Leith, for example), small shop units were constructed at ground floor level. This resulted in linear development of low-order shopping services radiating outwards from the CBD.

In the twentieth century the changes in Edinburgh's CBD continued and accelerated (see Figure 10.13). Large government offices were established on the site of the former Calton Jail in the 1930s, and following belated slum clearance in the 1960s, the Scottish Office had new headquarters in the St James Centre, east of St Andrew's Square. Large new government offices were also built on the site of railway workshops at Meadowbank. The St James Centre also houses a large hotel, a major department store, branches of chain stores, as well as many smaller shop units, and its bulk considerably modified the townscape. Another new shopping centre was built in the 1980s below street level at the Waverley Station, the only part of Princes Street with shops on the southern side.

The 1950s and 1960s saw many changes in Princes Street in the nature of the shops, with the closure of long-established businesses, the demolition of their premises, and the appearance of chain stores in the new concrete blocks which matched the existing skyline, if not the architecture. To the north, expanding insurance companies had new blocks built in St Andrew's Square and George Street, but then

were forced to extend their activities into new or converted premises in the Old Town, the South Side, Canonmills and even Leith. While the permanent population of the CBD had dwindled as more houses north of Queen Street were converted for use as offices, the number of jobs within the CBD continued to increase. The disadvantages of high rents were compounded by traffic and parking problems, and therefore in the last decade of the twentieth century, banks, insurance companies and legal firms moved a large number of jobs to:

♦ Saltire Court in Castle Terrace off Lothian Road

♦ the new financial centre created on **brownfield sites** at The Exchange where the Western Approach Road joins Lothian Road and at Port Hamilton, one block further south (see Figure 10.15)

♦ the **greenfield site** at Edinburgh Park near the point where the Edinburgh Bypass meets the M8 (see Figures 10.16 and 10.17).

Most companies concerned still retain a presence in the CBD, although the major banks have sold some of their city centre properties which have been converted into fashionable bars and clubs, and even pharmacies (see Figure 10.30). One small government office block has been converted into a budget hotel and the St James Centre office has been closed, because the Scottish Office has moved entirely to a new office block on Commercial Quay in Leith Docks. This is the best evidence of the outward migration of some of the functions of the CBD (see Figure 10.19). The retailing importance of the CBD remains, despite the development of large suburban shopping centres such as Cameron Toll and the Gyle (both on greenfield sites), and Kinnaird Park and Edinburgh Fort (brownfield sites). They have easy access from the bypass, easy parking, and rents are only one-third of those in Princes Street. These, and the other smaller centres developed closer to the city centre, for example, at Craigleith (on the site of the former quarry) and Meadowbank (on the site of an iron foundry and brewery) have had a more damaging effect on the linear shopping services mentioned earlier (e.g. Newington Road, St John's Road), than on the CBD. The Gyle and other such centres complement rather than compete with the

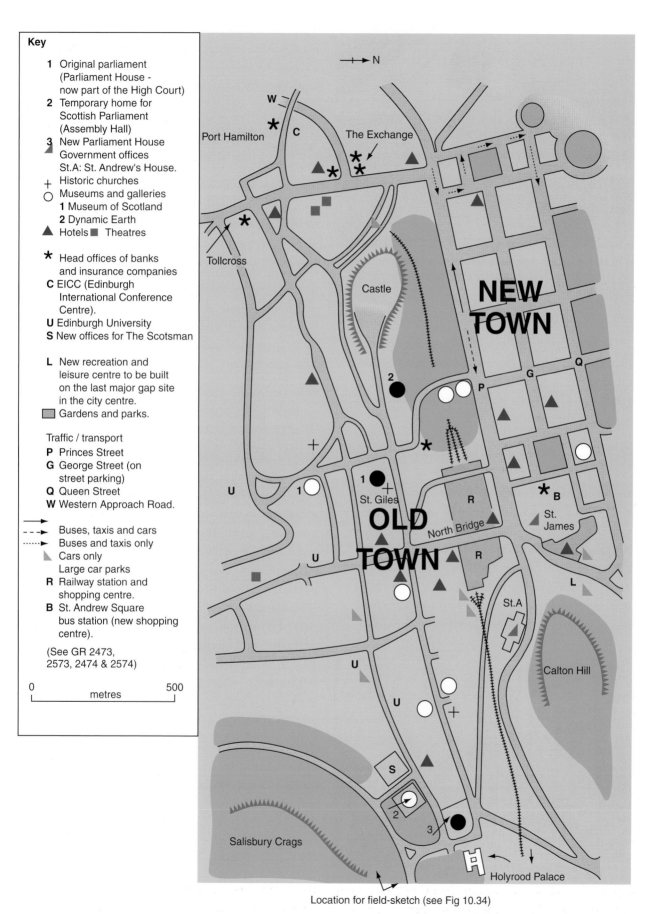

Key

1 Original parliament
(Parliament House -
now part of the High Court)

2 Temporary home for
Scottish Parliament
(Assembly Hall)

3 New Parliament House
Government offices
St.A: St. Andrew's House.

+ Historic churches

O Museums and galleries
1 Museum of Scotland
2 Dynamic Earth

▲ Hotels ■ Theatres

* Head offices of banks
and insurance companies

C EICC (Edinburgh
International Conference
Centre).

U Edinburgh University

S New offices for The Scotsman

L New recreation and
leisure centre to be built
on the last major gap site
in the city centre.

▢ Gardens and parks.

Traffic / transport

P Princes Street

G George Street (on
street parking)

Q Queen Street

W Western Approach Road.

→ Buses, taxis and cars

----► Buses and taxis only

◣ Cars only
Large car parks

R Railway station and
shopping centre.

B St. Andrew Square
bus station (new shopping
centre).

(See GR 2473,
2573, 2474 & 2574)

0 ————— metres ————— 500

N

Port Hamilton

W

C

The Exchange

Tollcross

Castle

NEW TOWN

Q

G

P

2

1
St. Giles

OLD TOWN

U

U

U

U

R

R

North Bridge

St. James

* B

St.A

L

Calton Hill

S

2

3

Salisbury Crags

Holyrood Palace

Location for field-sketch (see Fig 10.34)

Fig 10.13 Selected aspects of Edinburgh's CBD (compare this with Figure 10.3B)

Fig 10.14 The opening of the new Scottish Executive offices on Victoria Quay at Leith Docks are a good example of the outward movement of the CBD

CBD. The nature of shopping in the city centre has changed dramatically, with many fast food outlets, and its many shops selling knitwear, leisure and sportswear. Despite its perceived lack of designer stores, it is still a major attraction to people from the city, the Lothians and Fife as well as to the many tourists from all over the world. More than 145 000 people work in the city centre, 8000 of them in shops; more than 26 000 people walk along Princes Street each day, and it is estimated that more than 11 million day trips are made each year, mainly for shopping. The proposals to extend and up-grade shopping facilities in the CBD by creating a store for Harvey Nichols in the redevelopment of the St Andrew's Square bus station, and for the Princes Street Galleries underground shopping centre, are

Fig 10.15 As rents and traffic have increased in the CBD, a new financial centre has been created on a brownfield site on Lothian road

further evidence of the dynamic, ever-changing nature of Edinburgh's CBD. Because of planning restrictions, developers have not been able to build high-rise office blocks as in other cities.

10.10 Edinburgh's Industries

Edinburgh has had an industrial function which is now in decline in terms of numbers of factories and employees, but which has perhaps increased in relative value of its output. The first industrial areas were waterfront or water-based (see Figure 10.20) located either:

◆ along the rivers, making use of water power – the grain mills at Colinton and Dean, and the paper mills at Balerno along the Water of Leith, and the iron mills in the Almond Gorge at Cramond, or

◆ beside wells which provided water for the brewing industry on the edge of Holyrood Park at Holyrood and Craigmillar.

The raised beaches at Leith and Granton became the location for other waterfront and port-based industries (e.g. flour milling), and then the building of railways in the nineteenth century resulted in industrial zones being established in Gorgie, Dalry, Fountainbridge, and Leith. Centrally located, these were essentially areas of heavy, coal-based industry, with breweries, distilleries, bakeries, chemical, engineering, iron, rubber and printing works surrounded by densely packed tenements built for their workers, typical examples of old industrial landscapes (see Figure 10.21). Other industrial areas were established, such as at Portobello where glass, pottery and bricks were made, while some of the outlying villages were associated with weaving and coal mining.

The twentieth century saw first of all the emergence of new industrial areas, for instance, the electrical engineering works on or near Ferry Road, and then the creation of new estates for light industries on what was then the edge of the city at Sighthill. They used electricity rather than steam power, and depended on road rather than rail transport, and workers may have travelled long distances across the city by bus to what were then suburban locations.

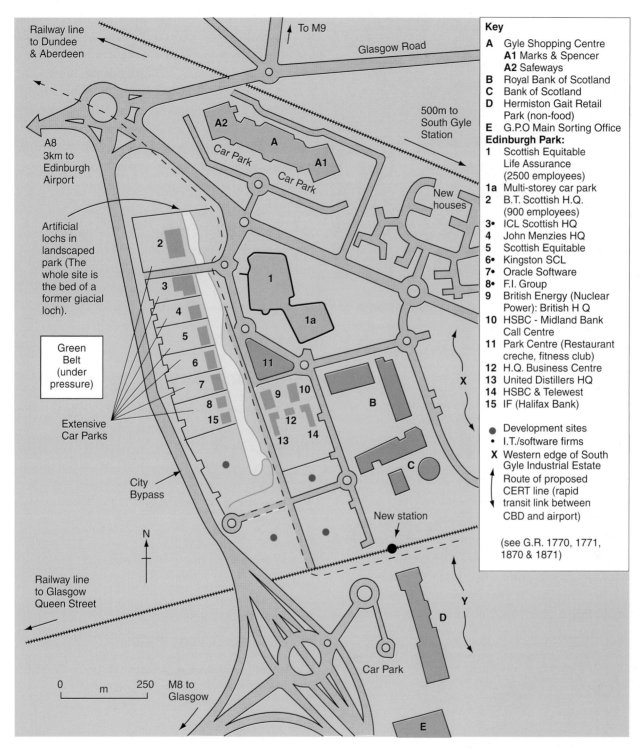

Key

A Gyle Shopping Centre
A1 Marks & Spencer
A2 Safeways
B Royal Bank of Scotland
C Bank of Scotland
D Hermiston Gait Retail Park (non-food)
E G.P.O Main Sorting Office
Edinburgh Park:
1 Scottish Equitable Life Assurance (2500 employees)
1a Multi-storey car park
2 B.T. Scottish H.Q. (900 employees)
3• ICL Scottish HQ
4 John Menzies HQ
5 Scottish Equitable
6• Kingston SCL
7• Oracle Software
8• F.I. Group
9 British Energy (Nuclear Power): British H Q
10 HSBC - Midland Bank Call Centre
11 Park Centre (Restaurant creche, fitness club)
12 H.Q. Business Centre
13 United Distillers HQ
14 HSBC & Telewest
15 IF (Halifax Bank)

● Development sites
• I.T./software firms
X Western edge of South Gyle Industrial Estate
↕ Route of proposed CERT line (rapid transit link between CBD and airport)

(see G.R. 1770, 1771, 1870 & 1871)

Fig 10.16 Edinburgh Park and Gyle Shopping Centre – part of the West Edinburgh 'Growthpole' (see GR 1770, 1771, 1870 and 1871)

Later still, changing economic circumstances saw the virtual disappearance of almost all the riverside, port and railway-based heavy industries. Most closed down; some moved out to new premises beyond what was then the city boundary, for example, tyre manufacture from Fountainbridge to Newbridge, and Edinburgh Crystal from Easter Road to Penicuik.

Edinburgh suffered from the lack of development area status, and also from the inducements offered to industry by the new town of Livingston. Even the once flourishing electronic industry has been affected, but has survived in the presence of BAe (defence avionics) on Ferry Road, Granton and South Gyle and Hewlett-Packard (personal

Fig 10.17 The carefully landscaped setting provides a good working environment for the thousands of office workers and researchers at Edinburgh Park

computers) and Motorola at South Queensferry. With relatively easy access to

- the airport
- the motorways
- the bypass and the Forth Road Bridge
- the universities
- other electronics firms,

these are examples of route-oriented industries which are also located in the suburbs. The creation of the two university '**technopoles**' in the Green Belt is the latest stage in the changing industrial location in Edinburgh. Figure 10.22 shows the location of the

Fig 10.18 Familiar high street stores at a new shopping centre, Edinburgh Fort, on a brownfield site in south-east Edinburgh (see Figure 10.19)

surviving patches of manufacturing industry, notably the '**growth pole**' in west Edinburgh. Most of the industrial areas are now in the suburbs and are also route-oriented. The old industries of the Inner City, that is, centrally located within both Edinburgh and Leith, have largely disappeared, and the areas have now been transformed either by demolition and rebuilding, or by conversion of premises. Waterfront industries have mostly gone, certainly along the rivers, but an important residue is found at Leith and Granton.

As a result of all these changes, Edinburgh is now a city where fewer than one in ten persons works in manufacturing industry, but eight out of every 10 people, both residents and commuters, work in tertiary industry (see Figure 10.23).

10.11 Inner City

Large urban areas such as Edinburgh usually have an area or areas, which have distinctive inner city features i.e.

- mixed land use
- a high density of buildings
- limited open space
- a large proportion of old houses
- buildings of many ages and styles
- many derelict buildings
- much waste land (e.g. gap sites)
- old, declining, industries
- low-order shopping
- many small workshops
- limited recreational facilities
- a declining population
- many one-person households
- a significant ethnic minority in the population
- a higher than average unemployment rate
- many people with a low disposable income.

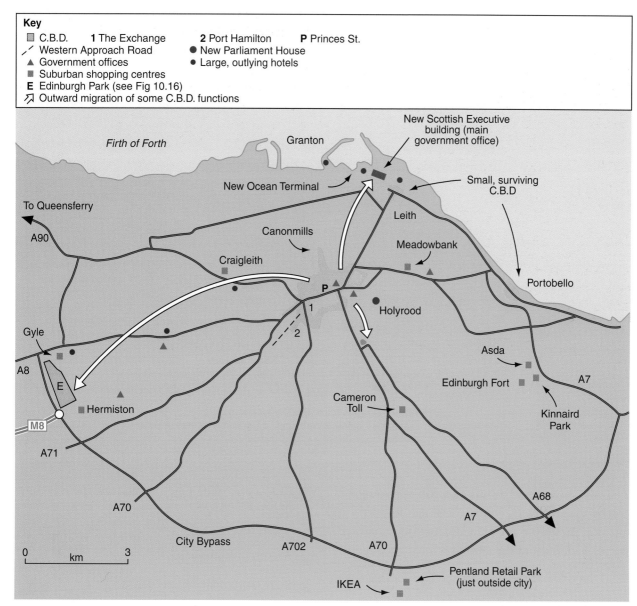

Key
- ▨ C.B.D.   **1** The Exchange   **2** Port Hamilton   **P** Princes St.
- ╱ Western Approach Road   ● New Parliament House
- ▲ Government offices   ● Large, outlying hotels
- ▪ Suburban shopping centres
- **E** Edinburgh Park (see Fig 10.16)
- ⤢ Outward migration of some C.B.D. functions

Fig 10.19 Edinburgh's migrating CBD

These are characteristics which are common to inner city areas in other urban concentrations in the developed world, but again it should be stressed that each city is unique. In any urban model, the inner city would be considered as being a more-or-less continuous area around the CBD. In the case of Edinburgh, several areas are recognised as having Inner City characteristics requiring action – Dalry-Gorgie, the Southside, parts of Portobello and Central Leith (see Figure 10.10). The map shows them to be widely separated, but they are all areas of largely tenement housing, built more than 80 years ago (see Figure 10.24).

Edinburgh's Old Town had many of the characteristics of the Inner City, although it had some of the functions of the CBD and many historic buildings, but the massive redevelopment which took place in the second half of the twentieth century has totally transformed it. The depopulation of the mid-twentieth century has long ended and the Old Town is now an area of in-migration, mostly of younger people. Most inner city areas have urban regeneration programmes in which buildings are modernised, converted, demolished and replaced, and gap sites filled. One of the areas which is undergoing such a transformation is Central Leith (see Figure 10.25).

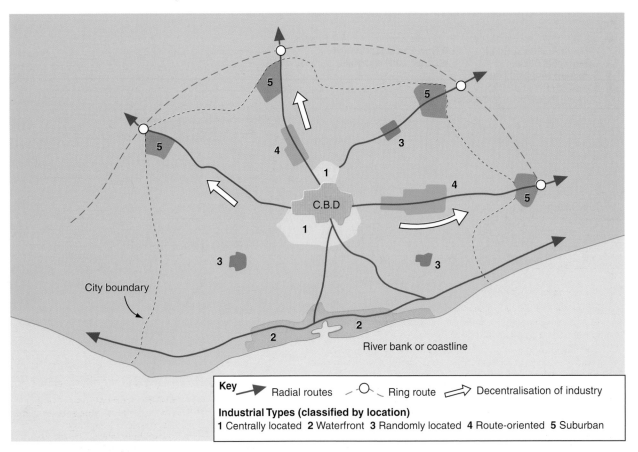

Fig 10.20 A model of intra-urban industrial location in a Western city

The area shown is what was essentially the walled port in the sixteenth century. Although its port function makes this a slightly different inner city, it has, or had, most of the attributes listed above:

◆ as the trade of the port declined, so too have its industries. Some have gone completely – shipbuilding, fertilisers, distilling

Fig 10.21 Old industrial landscape in Gorgie in the 1970s with tightly packed tenements around the factory buildings, at that time modernised for use by Ferranti Electronics, but now largely re-developed

◆ large areas of the port area and the town became derelict

◆ there were many sub-standard houses in the mid-twentieth century

◆ there were many empty buildings

◆ it had limited shopping facilities

◆ there were many people living alone

◆ the population declined and was ageing.

At least two features made it different from other inner cities. It included Leith's original small CBD and, just to the east, there was an extensive area of public open space in Leith Links. It also lay between the port industrial area, and a second industrial area to the south, along the river and the railway lines.

The problems of housing and shopping were identified in the 1960s, and resulted in the building of high-rise blocks of flats and a new shopping centre, then later, the renovation of some of the old tenement blocks. By the end of the twentieth

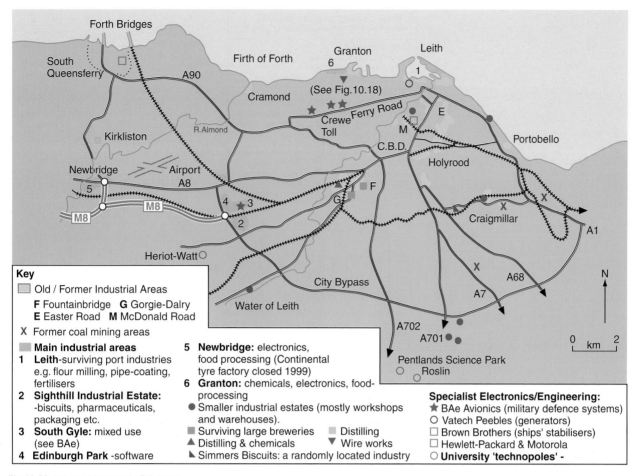

Fig 10.22 Industrial location in Edinburgh

Key diagram text:

Key

☐ Old / Former Industrial Areas

F Fountainbridge **G** Gorgie-Dalry
E Easter Road **M** McDonald Road

X Former coal mining areas

▨ **Main industrial areas**
1 **Leith**-surviving port industries
e.g. flour milling, pipe-coating,
fertilisers
2 **Sighthill Industrial Estate:**
-biscuits, pharmaceuticals,
packaging etc.
3 **South Gyle:** mixed use
(see BAe)
4 **Edinburgh Park** -software

5 **Newbridge:** electronics,
food processing (Continental
tyre factory closed 1999)
6 **Granton:** chemicals, electronics, food-
processing
● Smaller industrial estates (mostly workshops
and warehouses).
■ Surviving large breweries ▨ Distilling
▲ Distilling & chemicals ▼ Wire works
◣ Simmers Biscuits: a randomly located industry

Specialist Electronics/Engineering:
★ BAe Avionics (military defence systems)
○ Vatech Peebles (generators)
☐ Brown Brothers (ships' stabilisers)
☐ Hewlett-Packard & Motorola
○ **University 'technopoles' -**

century, there were still **gap sites** to be filled and old houses to be renovated, notably along the river. Some of the improvements, such as the opening of expensive restaurants along the harbour, and the relocation of businesses from the CBD, were the result of individual initiatives. Others, especially the relocation of the Scottish Executive to new premises on Victoria Quay, were the result of official planning. The area has further benefited from the development of the passenger shipping trade; the choice of Leith as the final anchorage for the Royal Yacht Britannia and the decision to build an ocean terminal for visiting liners. The potential of the old, stone-built whisky bonds was soon realised, and they have been converted into open-plan offices and luxury flats above boutiques and restaurants. There are three new large hotels for the expected tourist trade while the riverside and Inner Basin have been renovated (see Figure 10.26). Leith Walk, the main street leading to Edinburgh's CBD, has been improved, even to the extent of tree-planting. At the same time, there are still many low-order or empty shops in Great Junction Street, although there are plans to up-grade the inadequate shopping centre.

The general effect of all these changes has been to

◆ increase the daytime population significantly as civil servants travel into work in the Scottish Office from other areas of the city and beyond

◆ attract people from the rest of Edinburgh to spend their leisure time in Leith

◆ attract younger, wealthier people to live in Leith

◆ create an affluent dock and riverside fringe around a core which, though improved, has still many of the attributes of an inner city.

One of these, not yet discussed, is the problem of atmospheric pollution in the narrow, tenement-lined Great Junction Street, as a result of traffic congestion. Traffic is a major issue in all large urban areas, and Edinburgh is no exception.

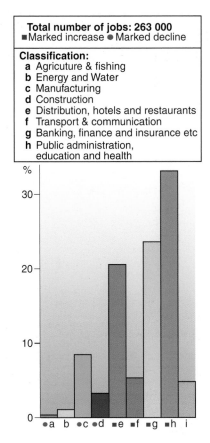

Total number of jobs: 263 000
■ Marked increase ● Marked decline

Classification:
 a Agriculture & fishing
 b Energy and Water
 c Manufacturing
 d Construction
 e Distribution, hotels and restaurants
 f Transport & communication
 g Banking, finance and insurance etc
 h Public administration,
   education and health

Fig 10.23 Employment in Edinburgh

Traffic

In the case of Edinburgh (Figure 10.27), the traffic problem is the result of:

◆ growing car ownership, related to high disposable income

◆ the great increase in the number of service buses in the 1990s, especially in the CBD (Edinburgh has also many tourist buses, which cause particular problems in the Royal Mile)

◆ the increase in number and size of lorries as rail freight has declined in importance

◆ the old road pattern, road width and road surfaces

◆ the presence of conservation areas which prevented the building of new roads or widening existing ones

◆ the many commuters travelling in cars to the city from dormitory towns and villages in Fife, the Lothians and the Borders

◆ the creation of suburban shopping centres such as

The Gyle, where the 80 000 vehicles each week greatly increase local traffic flows.

The increase in traffic has caused traffic jams on all the commuter routes (see Figure 10.28), especially the A90 to and from the Forth Road Bridge, and a great shortage of parking spaces in the CBD, as well as serious pollution problems where traffic is heaviest and slowest moving (see Table 1).

Solutions which have been attempted to solve the problem of parking in the CBD include (i) the construction of multi-storey car parks; (ii) the use of gap sites, disused railway land and an old bus garage as car parks; (iii) the introduction of traffic meters and traffic wardens to control kerbside parking. Private parking spaces in the CBD cost more than houses. Table 2 shows the types of parking spaces available in the city centre.

The problem of traffic flows has been a concern for many decades. There were plans in the 1970s to drive motorway-style roads close to the city centre, but these were never implemented. A short section of express-way was built along the line of a disused

Table 1: Edinburgh's polluted streets (mg per cubic metre of NO$_2$)				
		European Guidelines		
West Maitland Street (Haymarket–A8)	305			
St John's Road (A8)	290			
		Upper Limit		*200*
Gorgie Road (A71)	190			
Leith Street	175			
Leith Walk	160			
		Unhealthy		*135*

railway, and provides a fast elevated route from Gorgie to Lothian Road where the new financial district has now been established. The A90 to and from the forth Road Bridge was improved by building a dual carriageway within the Green Belt, but congestion continues and increases on the narrower stretch of the road within the built-up area. A major step forward was to build the Edinburgh Bypass along the foot of the Pentlands linking up with the M8, the M9 and the Forth Road Bridge, and taking the heaviest traffic out of the city centre. Unfortunately, the volume of traffic has increased so much (perhaps because of the development of Edinburgh Park and The Gyle) that there are frequent delays on the bypass itself. Further east, at Straiton, the expansion of the Pentland Retail Park, the building of many new houses, and the decision of IKEA, the Swedish furniture company, to site its first Scottish store near there, have all greatly increased the volume of traffic. Less than 10% of the bypass traffic consists of goods vehicles.

Within the city, in the late 1990s, **Greenways** (see Figure 10.27) were designated along five of the main bus routes in to the city centre (4000 buses travel along Princes Street each day). These are bus-only lanes which operate all day. They have produced much faster journey times into the CBD, and have significantly increased the number of passengers carried but they have also

◆ increased car traffic in the outer lane

Table 2: Types and numbers of parking spaces available in the central area controlled zone	
Type	**Number**
On-street	5504
Residents	5563
Off-street car parks	4361

Fig 10.24 Gorgie townscape in 2000 with renovated tenements and more expensive replacement housing built on the cleared site shown in Figure 10.21

◆ adversely affected the trade of small shops such as those at Haymarket, because of new parking restrictions.

There have also been major changes on Princes Street itself with only buses and taxis being allowed

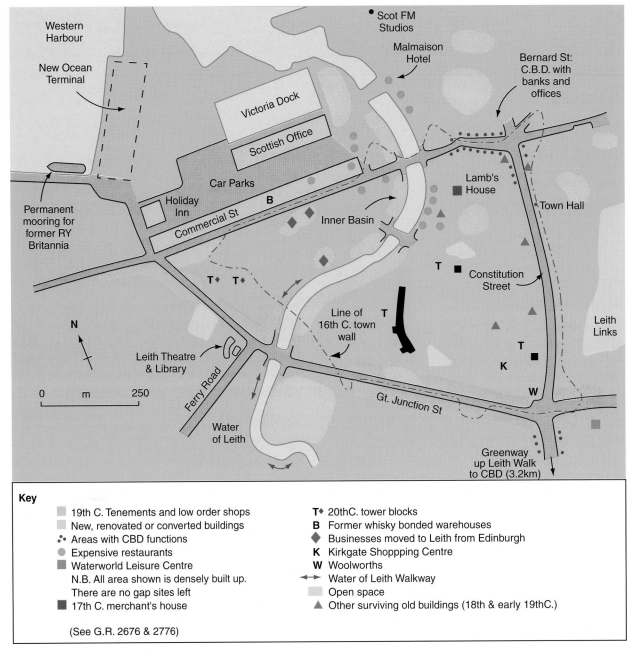

Fig 10.25 Leith's inner city (generalised)

to travel in both directions, while cars can travel westwards but not eastwards. Cars going east are diverted around Charlotte Square and along Queen Street (see Figure 10.13), but have also been using some of the other streets in the New Town. Other attempts to reduce the use of cars include:

◆ the building of a car-free housing development at Tollcross

◆ the possibility of charging tolls on cars travelling into the city centre

◆ improving the rail services into Edinburgh from Lothian and Fife

◆ the possible re-opening the suburban railway line, and building new stations where necessary

◆ providing park and ride facilities on the edge of the city, especially linked to the **CERT (City of Edinburgh Rapid Transport)** system. This would link Edinburgh Airport with the CBD (a 15 minute journey) via Edinburgh Park, and then go on, it is hoped, to the rejuvenated Leith Waterfront.

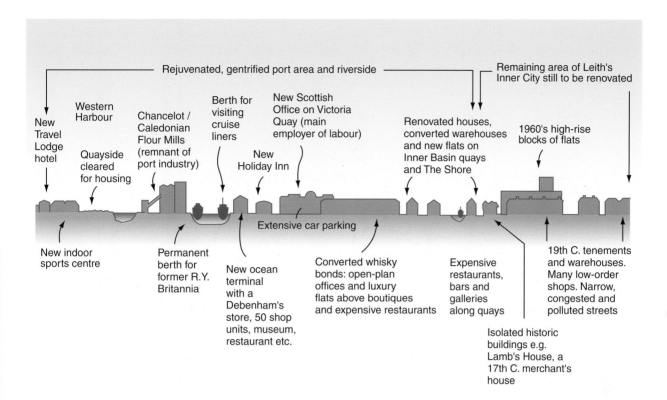

New Travel Lodge hotel

Western Harbour

Quayside cleared for housing

Chancelot / Caledonian Flour Mills (remnant of port industry)

Berth for visiting cruise liners

New Scottish Office on Victoria Quay (main employer of labour)

New Holiday Inn

Rejuvenated, gentrified port area and riverside

Remaining area of Leith's Inner City still to be renovated

Renovated houses, converted warehouses and new flats on Inner Basin quays and The Shore

1960's high-rise blocks of flats

Extensive car parking

New indoor sports centre

Permanent berth for former R.Y. Britannia

New ocean terminal with a Debenham's store, 50 shop units, museum, restaurant etc.

Converted whisky bonds: open-plan offices and luxury flats above boutiques and expensive restaurants

Expensive restaurants, bars and galleries along quays

Isolated historic buildings e.g. Lamb's House, a 17th C. merchant's house

19th C. tenements and warehouses. Many low-order shops. Narrow, congested and polluted streets

Fig 10.26 Selected elements in the townscape of Leith

Future Developments

There have been many recent changes in Edinburgh's townscape. A city which was essentially stone-built is now being modified by widespread use of brick for new buildings; redundant schools, warehouses, breweries, distilleries and hostels have been converted into housing; old houses have been renovated; old industrial areas have been cleared for new housing which is also found on old gap-sites and playing fields (see Figures 10.29 and 10.30). This infilling is evidence of internal change without the building of 'streets in the sky' (skyscraper blocks) but there is evidence also of changes which will take place in the Green Belt (see Figure 10.31A). As around most cities in Britain, there are already inappropriate land uses in Edinburgh's Green Belt:

significant areas have already been lost to motorway and bypass construction, and there are plans for a major transformation of the South-East Wedge. These will involve not only the building of the new infirmary, but also the creation of housing for 12 000 people, mostly in '**urban villages**'. The existing farmland will be covered by housing, industrial estates and new recreational areas (see Figure 10.31B). The other major development which is expected to take place is the renovation of Edinburgh's '**blue belt**' (the coastline from Cramond to Portobello). Figure 10.32 shows part of this area at Granton where a very run-down waterfront industrial area, which evolved in a very haphazard way, is to be transformed into an area of water-based recreation, business, varied industry and new housing.

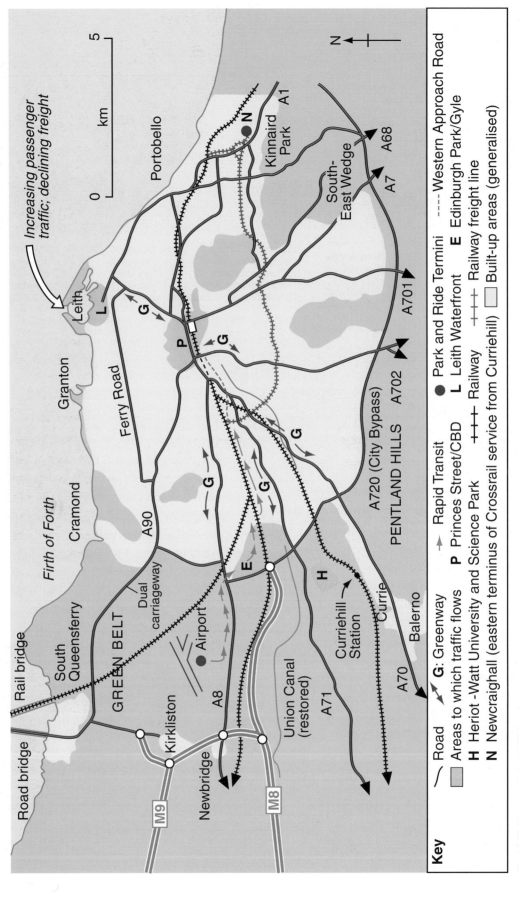

Fig 10.27 Transport links in Edinburgh

Key

Road — G: Greenway — Rapid Transit ● Park and Ride Termini — Western Approach Road

Areas to which traffic flows ↗ Princes Street/CBD L Leith Waterfront E Edinburgh Park/Gyle

Built-up areas (generalised)

H Heriot-Watt University and Science Park +++ Railway +++ Railway freight line

N Newcraighall (eastern terminus of Crossrail service from Curriehill)

Fig 10.28 Evening rush hour traffic congestion at the Princes Street–Lothian Road junction produces high levels of noise and atmospheric pollution

Fig 10.29 Brick-built houses are now becoming a common feature of Edinburgh's townscape

Fig 10.30 As some of the functions of the CBD migrate outwards, former office premises in George Street have been turned into fashionable bars

Tenements line traffic-congested Morningside Road

Ribbon development: bunglows below the Braid Hills 'Urban Lung'

Fringe Belt conversion: A convent and chapel are converted into luxury housing

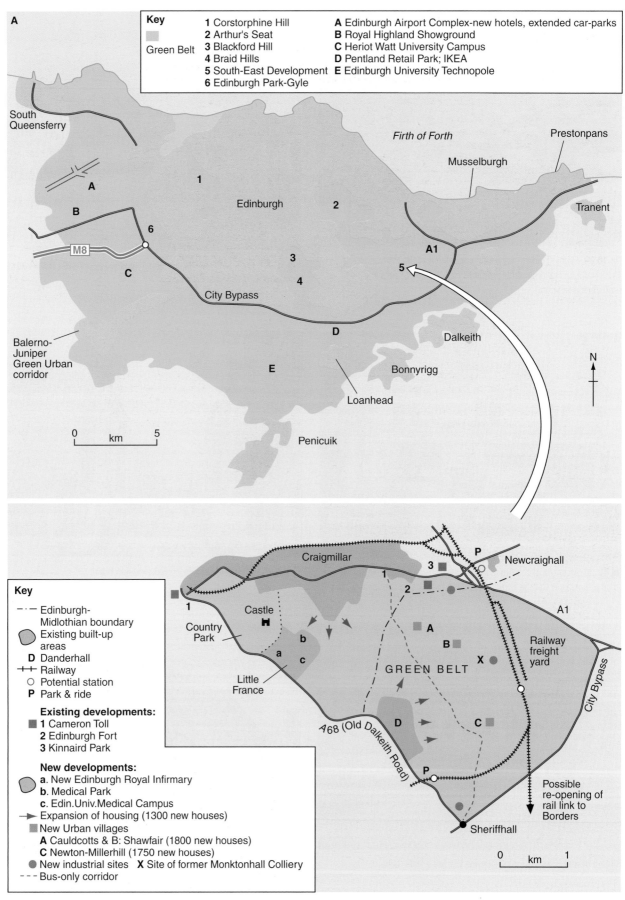

Fig 10.31 (A) Pressures on Edinburgh's Greenbelt (B) South-East Wedge

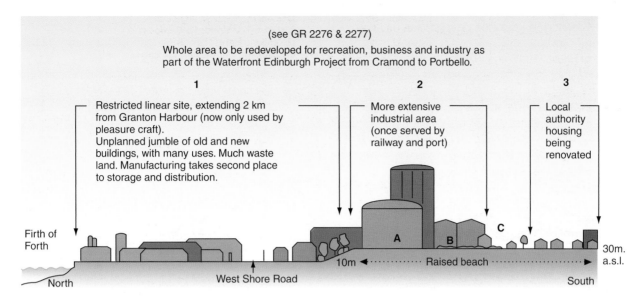

(see GR 2276 & 2277)
Whole area to be redeveloped for recreation, business and industry as part of the Waterfront Edinburgh Project from Cramond to Portbello.

1
Restricted linear site, extending 2 km from Granton Harbour (now only used by pleasure craft).
Unplanned jumble of old and new buildings, with many uses. Much waste land. Manufacturing takes second place to storage and distribution.

2
More extensive industrial area (once served by railway and port)

3
Local authority housing being renovated

Firth of Forth

North   West Shore Road

10m ◄·········· Raised beach ··········► 30m. a.s.l.

South

Manufacturing: GEC Marconi electro-optics, chemicals, food processing
Processing: Waste and waste water
Storage and distribution: pipes, cement, bottled gas, building materials, books.
Motor vehicles: car-hire depots, tyre fitters vehicle testing
New trading estate (engineering workshops) and new business centre

Remnants of old industrial landscape
A: Gas Board/British Gas (now Transco depot)
B: Wire works
Once also an iron works, a chemicl works and an oil depot therefore much contaminated land, now being cleared
C: Caroline Park House - an important 17 C. mansion lost in this industrial wasteland

Fig 10.32 Waterfront Granton – a declining industrial area undergoing renewal (see GR 2276 and 2277)

ASSIGNMENTS

Analysis of (land use) maps and transects, **annotation & analysis** of photographs and field sketches of urban landscapes and **analysis** of survey data.

1 Study Figures 10.1A and 10.1B
a) Name the urban settlement (which later became of major importance) which does not appear on Figure 1A. **Name** the settlements which are shown on 1A but not on 1B. **Suggest reasons** for these discrepancies.
b) Describe the distribution of the urban settlements shown by the two maps, identifying similarities and differences.

2 Study Figures 10.3A and 10.3B and the text.
Draw a large summary diagram (a spider diagram) on an A4 page
a) to **explain** the difference between **site** and **situation**
b) to **show** the essential features of Edinburgh's site and situation.

c) to **identify** the physical features which originally might have been
(i) an advantage **(ii)** a disadvantage.

3 a) Study Figure 10.7 and **describe** how an urban townscape changes (according to the Whitehand model) from the **urban fringe** to the **CBD**.
b) Study Figure 10.11 and for **either** the transect along the A71 or the transect along the A90, **identify** similarities to, and differences from, the model.

4 Study Figures 10.12 and 10.13
a) Describe and explain the differences in street pattern within the Central Business District (CBD).
b) In **tabular form**, **match** the **functions** of Edinburgh with the **uses** of the Central Business District.

5 Study Figures 10.16 and 10.19:
a) Suggest at least **two** reasons why some of the functions of the CBD might have migrated to outlying locations.

b) Explain, with examples, how the 'West Edinburgh Growthpole' might be described as multi-functional.

6 Study Figures 10.20 and 10.22. The model suggests different locations for industries within a large urban area.
   a) In **tabular form**, **match** the types shown in the model with the industries shown on the map.
   b) What form of transport, not shown the model, might have been a major factor in the location of the old industrial areas in Edinburgh?

7 Study Figure 10.23, which shows, in graph form, employment in Edinburgh at the end of the twentieth century.
   a) Explain why there were still people within Edinburgh employed in agriculture.
   b) Suggest two reasons why this small percentage was declining.
   c) What proportion of the workforce was employed in manufacturing? Would this proportion have been higher or lower 50 years earlier?
   d) What evidence does this graph show of Edinburgh's importance as a centre of tertiary employment and of its varied functions?

8 Study the population and employment figures for Edinburgh given in the table below.

Population structure 1997		
Up to 14 yrs	15–64 yrs	65 and over
73 162	306 426	69 262

Number of people in employment (1997)
263 000

   a) Calculate the **dependency ratio** for Edinburgh.
   b) Suggest at least **four** reasons why all the available jobs in Edinburgh would not have been filled by local people.

c) If the number of jobs exceeded the number of local people available, how was this deficit overcome? What was one consequence of this?

9 Study Figures 10.25 and 10.26
The townscape of the 'inner city' of Leith has been totally transformed in recent years. Make a large **summary diagram (spider diagram)** in note form which lists
   a) the surviving features typical of an inner city area
   b) examples of old buildings converted to new uses
   c) examples of major new developments
   d) new functions appearing in what was once a port industrial area.
At the centre of this diagram, name the two developments which you consider have had most to do with the regeneration and gentrification of this area.

10 Study Figure 10.27, the text and the information in the tables below.

Edinburgh: Travel to work (%)				
Train	3.2	Car driver	42.1	Bicycle/motor bike 2.0
Bus	31.1	Car passenger	6.9	
Walk	11.8	Work at home	2.9	

Vehicles entering Edinburgh's CBD between 8.00–9.00am (i.e. the rush hour):			
Car	84.1%	Bicycles/motor-bikes	4.0%
Bus	2.7%	Vans/lorries	9.2%

   a) Draw graphs to illustrate the above figures
   b) Describe and explain the traffic and associated problems (shown in the two tables) faced by Edinburgh, and **comment** on the attempted solutions.

Fig 10.33 Field sketch inner city townscape (GR 239732)

11 Study the information given in the table below:

Distance travelled from home to work in Edinburgh (% of workforce)							
km	<2	2–4	5–9	10–19	20–29	30–39	40+
%	20.8	27.8	22.0	16.8	6.1	2.4	4.1

a) Look at Figure 10.27 again. How far would people have to travel to work from **(i)** Portobello to Edinburgh Park, **(ii)** Currie to Leith Waterfront **(iii)** Cramond to the CBD?
From the table above, are these typical of the distances people have to travel to work?

b) With the aid of a road map or an atlas, find out into which distance category in the table above the following places would fit:
Penicuik; Haddington; Dalgety Bay; Livingston; Peebles; Bathgate; Musselburgh; Dunfermline; Linlithgow; Kirkcaldy; Broxburn; Dalkeith.

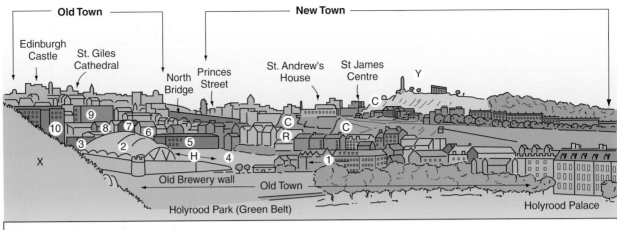

Key
1 Site of new Scottish Parliament building (former brewery offices being demolished).
2 Dynamic Earth exhibition
3 Barclay House: new offices for 'The Scotsman', on site of former gas works.
4 Site cleared for luxury flats
5 New four-star hotel
6 Housing association flats
7 New luxury flats
8 New student residences

9 Edinburgh University (Moray House)
10 Local authority tower blocks
   (to be renovated).
H Holyrood Road
R Royal Mile
C Crags

Fig 10.34 Field sketch: changing townscape, Central Edinburgh, Spring 1999 (looking WSW from GR 272738)

From which of these places would it be possible to travel to Edinburgh by rail?

c) From a study of the atlas/road map, **suggest** which settlements more than 40 km from Edinburgh might supply some of the long-distance commuters.

d) These travel-to-work distances give some idea of Edinburgh's **sphere of influence**. In what other ways might this be measured? (Figure 10.12 might give some ideas). Which function gives Edinburgh a world-wide sphere of influence?

12 Study Figure 10.31 and the text.
a) **Explain** what is meant by the 'Green Belt'.
b) Less than 1% of the land within Edinburgh available for development is either vacant or derelict. **Suggest reasons** why developers are very interested in obtaining land on Corstorphine Hill, including the site of Edinburgh Zoo (see also Figure 10.10 : Housing Types in Edinburgh).
c) **List** the evidence from Figure 10.31 which shows that Edinburgh's Green Belt is under pressure. **Comment** on the suitability of the proposals for the development of the South East Wedge.

13 Study the data about housing in Edinburgh on Figure 10.10 and the map itself.
a) **Explain** the relationship between population density and work (in the past) with the distribution of tenements and flatted cottages.
b) **Draw** graphs to show
(i) the **changes** in housing tenure in Edinburgh between 1981 and 1997

(ii) the **differences** in house type between Edinburgh and Scotland as a whole.

c) **Comment** on the changing nature of housing in Edinburgh.

14 Study the field sketch of an Inner City townscape (GR 239732, Figure 10.33). This shows an area near Haymarket, just outside the CBD, and alongside the main railway line to Glasgow. There are three contrasting buildings.
a) **Copy** this sketch and **annotate** it to
(i) **identify** the three buildings labelled a–c, choosing from the following descriptions:
20th Century cash-and-carry warehouse;
mid-19th Century distillery (derelict);
18th Century house converted into offices
(ii) **indicate** where you might expect to find: red brick walls; slate roof; white-harled stone walls; bare stone walls.
b) **Suggest** reasons why
(i) the old house has been converted into modern offices
(ii) the distillery site was being cleared in 1999 to build expensive houses.
c) Is this a planned or an unplanned townscape? Give reasons for your answer.

15 Study the field sketch of the changing townscape of Central Edinburgh (Figure 10.34). If you have the 1:50 000 map (or quadrant), find the location of the viewpoint at GR 272738.
a) **Compare** this sketch with Figure 10.13. **Locate** on

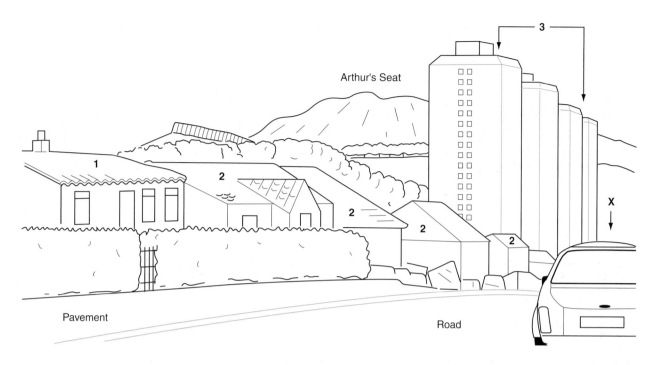

Arthur's Seat

Pavement

Road

X

Fig 10.35 An area of mixed residential types in South-East Edinburgh (GR 2969)

the map the features named on the field sketch. **Name** the two hill areas labelled X and Y on the sketch.

b) With the aid of Figure 10.3, **state** two pieces of evidence of the nature of the site which are shown in the sketch.

c) From the evidence in the sketch, **list** at least **two** ways in which the townscape of the Old Town differs from that of the New Town.

d) **List** the Inner City characteristics which are suggested in the sketch. What is the evidence that '**gentrification**' is taking place in this area?

e) **State** the major change which will take place in this townscape in the next two years.

16 Study the field sketch of an area in south east Edinburgh (GR2969), Figure 10.35, which was first developed after 1945. X marks the site of the new Edinburgh Royal Infirmary under construction in 1999.

a) **Copy** the sketch and **annotate** it to

(i) **show** the urban lung and area of green belt (including woodland) in the background.

(ii) **describe** the main feature of the **site** on which all the buildings are located.

(iii) **identify** the different types of housing 1–3, choosing from the following list: **renovated 1960s L.A. tower blocks; renovated 1940s prefabricated houses ('prefab'); villas under construction on 'brownfield' site (i.e. where prefabs. have been demolished).**

b) **Comment** on the differences in population density within this small area.

17 Study the maps (Figure 10.36) which show changes over a 30-year period in a small area of mixed residential/industrial use between the New Town and Leith.

a) **List** the uses which have remained the same

b) **List** the changes which have taken place

c) **List** the ways in which you think this area is typical of the Inner City.

d) **Suggest** two reasons why the up-market furniture business should have moved here from George Street in the CBD.

e) What evidence is there that this area is still changing in character?

18 Most townscapes (unless they have become fossilised) are constantly changing, yet retain elements from their historic past.

a) **Draw** a summary (spider) diagram, with named examples, to show that Edinburgh's townscape consists of both changing and historic elements.

b) **Discuss** the value of both of these elements to the city.

c) For any other large urban area known to you, make a tabular summary of how the townscape consists, with named examples, of both changing and unchanged elements.

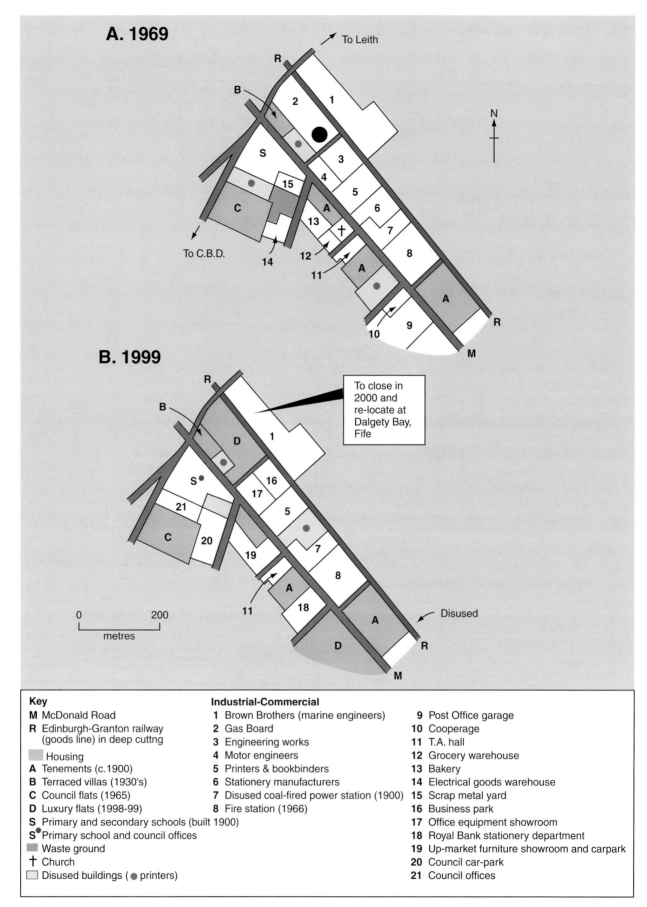

A. 1969

To Leith

R

B

2 1

N

S

3

4

15

5

C

6

A

13

7

To C.B.D.

14

12

†

8

11

A

A

10 9

R

M

B. 1999

R

B

To close in
2000 and
re-locate at
Dalgety Bay,
Fife

D

1

S

16

17

21

5

C

20

7

19

8

A

11

18

Disused

A

D

R

M

0   200

metres

Key

M McDonald Road
R Edinburgh-Granton railway
(goods line) in deep cuttng

   Housing
A Tenements (c.1900)
B Terraced villas (1930's)
C Council flats (1965)
D Luxury flats (1998-99)
S Primary and secondary schools (built 1900)
S Primary school and council offices
   Waste ground
† Church
   Disused buildings (● printers)

Industrial-Commercial

1 Brown Brothers (marine engineers)
2 Gas Board
3 Engineering works
4 Motor engineers
5 Printers & bookbinders
6 Stationery manufacturers
7 Disued coal-fired power station (1900)
8 Fire station (1966)

9 Post Office garage
10 Cooperage
11 T.A. hall
12 Grocery warehouse
13 Bakery
14 Electrical goods warehouse
15 Scrap metal yard
16 Business park
17 Office equipment showroom
18 Royal Bank stationery department
19 Up-market furniture showroom and carpark
20 Council car-park
21 Council offices

Fig 10.36 Changing land use: Mixed residential-industrial area in North Edinburgh (generalised) (see GR 2575 and 2675)

Key terms and concepts

- 'bluebelt' (p. 330)
- 'brownfield' site (p. 319)
- burgh (p. 307)
- Central Business District (CBD) (p. 317)
- CERT (p. 329)
- conurbation (p. 309)
- core (p. 307)
- fringe belt (p. 315)
- functions (p. 308)
- gap site (p. 326)
- gentrification (p. 330)
- Green Belt (p. 315)
- 'greenfield' site (p. 319)
- 'Greenways' (p. 328)
- 'growthpole' (p. 323)
- inherited urban fabric (p. 306)
- Inner City (p. 323)
- Mann urban model (p. 316)
- morphology (p. 308)
- New Towns (p. 310)
- (the) New Town (p. 312)
- pre-urban nucleus (p. 307)
- site (p. 306)
- situation (p. 306)
- 'technopole' (p. 323)
- tenement (p. 308)
- tofts (p. 308)
- townscape (p. 308)
- urban fringe (p. 315)
- urban hierarchy (p. 308)
- urban 'lung' (p. 312)
- urban models (p. 313)
- urban renewal (p. 310)
- urban villages (p. 330)
- Whitehand urban model (p. 315)

Suggested reading

Broadley, E & Cunningham, R (1991) *Core Themes in Geography – Human* Oliver & Boyd

Clark, D (2000) *World Urban Development: Processes and Patterns at the end of the 20th Century* **Geography**: Vol. 85, Pt. 1

Guinness, P & Nagle, G (1999) *Advanced Geography: Concepts and Cases* Hodder & Stoughton

Maclean, K & Thomson, N (1988) *Landscapes and Peoples of Western Europe* OUP

Matthews, H (1991) *British Inner Cities* OUP

Nagle, G (1998) *Changing Settlements* Nelson

Nagle, G & Spencer, K (1996) *A Geography of the European Union: A Regional and Economic Perspective* OUP

Waugh, D (1994) *Geography: An Integrated Approach* Nelson

Waugh, D (1997) *The U.K. and Europe* Nelson

Waugh, D (1998) *The New Wider World* Nelson

Witherick, M (1995) *Environment and People* Stanley Thorne

Internet sources

Edinburgh Facts & Figures: www.edinburgh.gov.uk.

INDEX